Springer
Proceedings in Physics   20

W0044011

Managing Editor: H. K. V. Lotsch

Springer Proceedings in Physics is a new series dedicated to the publication of conference proceedings. Each volume is produced on the basis of camera-ready manuscripts prepared by conference contributors. In this way, publication can be achieved very soon after the conference and costs are kept low; the quality of visual presentation is, nevertheless, very high. We believe that such a series is preferable to the method of publishing conference proceedings in journals, where the typesetting requires time and considerable expense, and results in a longer publication period. Springer Proceedings in Physics can be considered as a journal in every other way: it should be cited in publications of research papers as *Springer Proc. Phys.*, followed by the respective volume number, page number and year.

Primary Processes in Photobiology

Proceedings of the 12th Taniguchi Symposium,
Fujiyoshida, Yamanashi Prefecture, Japan,
December 7–12, 1986

Editor: T. Kobayashi

With 153 Figures

Springer-Verlag Berlin Heidelberg GmbH

Professor Takayoshi Kobayashi

Department of Physics, Faculty of Science, University of Tokyo, Bunkyo-ku, Tokyo 113, Japan

ISBN 978-3-642-72837-2   ISBN 978-3-642-72835-8 (eBook)
DOI 10.1007/978-3-642-72835-8

© Springer-Verlag Berlin Heidelberg 1987
Originally published by Springer-Verlag Berlin Heidelberg New York 1987
Softcover reprint of the hardcover 1st edition 1987

Offset printing: Weihert-Druck GmbH, D-6100 Darmstadt

2153/3150-543210

Preface

The 12th Taniguchi International Symposium, Biophysics Division, was held in the Jinzai-Kaihatsu Center at Fujiyoshida in the Yamanashi Prefecture, 7–12 December, 1986. The title of the symposium was "Primary Processes in Photobiology".

Life utilizes solar energy in two ways: in the transmission of information and in the conversion of light energy to chemical energy. In order to elucidate the molecular mechanisms of highly sensitive visual responses and other photosensitive responses of biological systems, and of highly efficient photoenergy transduction to chemical energy in photosynthesis, it is important to observe molecular processes in biological systems. Biological chromophores should have large absorption cross sections in the visible region, where solar energy transmitted to the earth is most intense, in order to utilize light energy in that region. Examples of light-absorbing substances are rhodopsin in the visual pigments of many animals, chlorophyll-protein complexes in photosynthetic pigments of green plants, and bacteriorhodopsin in proton-pumping pigments of halobacteria. The chromophores in rhodopsin and bacteriorhodopsin are protonated Schiff bases of retinal isomers (11-cis or all-trans), and those in chlorophyll-protein complexes are chlorophylls. The chromophores have the electronic structure of an extended π-electronic system, as in various organic dye molecules. The chromophores and organic dye molecules have an intense transition between the ground state and the lowest excited singlet state in the visible region. Because of this large absorption cross section in the 400–700 nm region, the radiative lifetime of the excited singlet state in the light absorbing pigments is between 1 and 10 ns. This is a consequence of quantum mechanical requirements and Einstein's radiation theory.

In order that biological systems efficiently utilize the light energy absorbed by the pigments without losing the energy by photoemission processes, the initial photobiological reaction must be completed in a time shorter than the radiative lifetimes. This is why a time-resolved method is invaluable for the study of the primary processes of vision, photosynthesis, and other photobiological processes. Using highly developed laser spectroscopy techniques, great progress has recently been achieved in the area of various primary processes in photobiology. It was therefore an excellent time to hold a symposium on the Primary Processes in Photobiology. I have the impression that the symposium greatly facilitated contact between participants from various fields of research.

On behalf of the participants, I would like to express grateful thanks to Mr. Toyosaburo Taniguchi and the Taniguchi Foundation for financial support of the symposium. We are very much indebted to Professors M. Kotani (Chairperson), S. Ebashi, F. Oosawa, and A. Wada, members of the planning committee, for giving us the exceptional opportunity, and Professor A. Tasaki, Secretary General of the planning committee, for his encouragement and advice throughout the symposium.

Tokyo, Japan
February 1987

Takayoshi Kobayashi

1.	M. Stockburger	16.	T. Hattori
2.	T. Kitagawa	17.	T. Hayakawa
3.	M. Ottolenghi	18.	K. Ichimura
4.	M.A. El-Sayed	19.	T. Takahashi
5.	T. Kobayashi	20.	A. Maeda
6.	T. Kakitani	21.	M. Mimuro
7.	F. Tokunaga	22.	N. Kamo
8.	R.A. Mathies	23.	W.W. Parson
9.	A. Ikegami	24.	A. Migus
10.	G. Atkinson	25.	K. Minoshima
11.	Y. Inoue	26.	K. Ogasawara
12.	I. Yamazaki	27.	T. Kouyama
13.	S.G. Boxer	28.	S. Koshihara
14.	J.L. Martin	29.	A. Terasaki
15.	H. Ohtani		

Contents

Part I

New Experimental Techniques

Vibrational and Electronic Dephasing Time Measurement with the Use of Temporally Incoherent Light

T. Hattori, A. Terasaki, and T. Kobayashi

Department of Physics, Faculty of Science, University of Tokyo,
7-3-1 Hongo, Bunkyo-ku, Tokyo 113, Japan

1. Introduction

The dynamical properties of matter have been studied by increasing number of scientists, and information with higher time resolution is being obtained by the development of picosecond and femtosecond spectroscopies. Since picosecond light pulses were first emitted from passively mode-locked ruby laser in 1965 [1], continuous efforts to get shorter pulses have been made, and recently optical pulses as short as 8 fs were obtained [2] by the method of pulse compression of the output from a group-velocity-dispersion-compensated colliding-pulse mode-locked laser. Time-resolved coherent and conventional spectroscopies have been applied to several systems using ultrashort light pulses with pulse width of a few tens to a hundred femtoseconds. However, there are several difficulties in the study of the ultrafast phenomena using such short pulses: (i) Laser systems for the generation of ultrashort pulses are necessarily very expensive and complicated. (ii) The wavelengths of femtosecond laser pulses are limited in the region around 615-625 nm because of the lack of appropriate combination of saturable absorber and gain medium, and the tunability of each laser is generally poor. (iii) It is difficult to maintain a short pulse width in actual optical systems, since shorter pulse has broader power spectrum and suffers from dispersion broadening when it passes through ordinary dispersive or nonlinear materials.

Recently a new spectroscopic technique with incoherent light utilizing coherent transient optical effects has been presented and verified experimentally, [3-7]. Since ordinary electronic devices do not have subpicosecond or femtosecond time resolution, optical experiments using that time region are usually performed using nonlinear optical phenomena. In these methods, the signal light generated or modulated by nonlinear optical effects is detected for the measurement of response of matter, using the correlation between excitation and probe light beams. In typical experimental systems, an optical pulse is split into two beams, and they meet again in a sample after passing through variable and fixed optical delay lines, and the intensity of signal or probe light is measured as a function of the delay time. Generally the signal intensity is expressed as a function of the field amplitude (or intensity) and the response function of the matter.

In the studies of the dynamics taking place in matter, therefore, the time resolution is expected to be determined not by the pulse duration of the light but by the correlation time. According to this principle, extremely high time resolution may be easily obtained by using light having a short enough correlation time, or a broad enough spectral width, for the time region to be measured. The availability of this principle for short-time measurement has been verified for the dephasing time measurement by degenerate four-wave mixing (DFWM) spectroscopy [3-7].

This technique was also utilized for the study in the field of biophysics. MEECH et al. [8] studied the decay of the primary donor in the reaction center of a photosynthetic bacterium after resonant excitation. The decay lifetime at 1.5 K was instrument-limited in their experiment, implying a population relaxation time much shorter than 100 fs.

In the first half of this article, we describe the study of dephasing in a polydiacetylene film measured by DFWM [9]. Dephasing times in a polydiacetylene (poly-3BCMU) film were resolved for the first time at two wavelengths by DFWM using incoherent light. The measured dephasing times, 30 fs at 648 nm and 130 fs at 582 nm, correspond to excitons in chains of the polymer with different conjugation lengths.

Though transient DFWM spectroscopy, both using coherent short pulses [10] and incoherent cw light sources as mentioned above, is powerful for the study of dynamic properties of matters, it is not applicable to optically forbidden transition, and the range of the available wavelength is limited. Dephasing of Raman active vibrational modes in molecules can be investigated by so-called transient CARS or, more generally, transient coherent Raman spectroscopy [11-13], where a pair of picosecond pulses excites a vibrational system coherently and a second pulse of the higher frequency probes the coherence of the system after a certain delay time. The information about the dephasing dynamics of the system can be obtained by the dependence of the CARS intensity on the delay time.

We studied theoretically a possible application of the principle that the correlation time determines the resolution time, to transient coherent Raman spectroscopy. Theoretical derivation of the delay-time dependence of the coherent Raman intensity, and the experimental demonstration of the measurement of the dephasing of the 2915-cm^{-1} mode in dimethylsulfoxide are presented in the second half [14].

2. Electronic Dephasing in Polydiacetylene Measured by Degenerate Four-Wave Mixing

The dynamical properties of the excited states of polydiacetylenes (PDAs) have recently gained much interest of many scientists. They have been studied experimentally by time-resolved absorption, reflection and emission spectroscopy [15-20], and excited state lifetimes obtained were 9 ± 3 ps in aqueous solution [15] and 2 ps in crystalline phase [16]. Knowledge of the dephasing dynamics of PDAs is of great importance not only for elucidating the properties of excited states and the origins of their large optical nonlinearity but also for various applications such as optical switching and optical signal processing. DFWM was applied to the dephasing time measurement [21-23], but dephasing times have not been resolved so far. DENNIS et al. [21] observed DFWM from two PDAs (2d and 2j) solutions with 180 ps pulses, but they found that the response times were faster than their resolution time. CARTER et al. [22] observed DFWM from a PDA (PTS) crystal with 6 ps pulses, and they concluded that the response time was faster than 6 ps. RAO et al. [23] performed similar measurements on a PDA (poly-4BCMU) film with 500 fs pulses, but they could not resolve the dephasing time either.

In this study, we applied DFWM with incoherent light to the measurement of the dephasing times in a film of a PDA, poly(4,6-decadiyne-1,10-diol bis((n-butoxycarbonyl)-methyl) urethane), which is abbreviated as poly-3BCMU. By detecting signals diffracted in two directions simultaneously, we could resolve a dephasing time as short as 30 fs. We measured dephasing times of the sample at two wavelengths, 648 nm and 582 nm, and found that the dephasing of the exciton in a polymer chain with a longer conjugation length (at 648 nm) is four times faster than that in a chain with a shorter conjugation length (at 582 nm). This result may be related to the lower fluorescence efficiencies of the rod-like form of poly-3BCMU with longer conjugation length than coil-like form.

2.1. Experimental

The experimental apparatus used for the dephasing time measurement is shown in Fig. 1. The incoherent light source was a broad-band dye laser pumped by a N_2 laser. The oscillator cavity of the broad-band dye laser with a spectral width (FWHM) of 8 nm was constructed with an aluminum mirror and a glass plate, and no tuning element was placed in the cavity. Dye laser light was linearly polarized by a Glan-Thompson prism. It was then divided into two beams, n_1 and n_2, by a beam splitter, and n_2 was delayed with respect to n_1 by a variable delay line. The polarization planes of the two beams could be rotated independently by the use of coupled Fresnel rhombs for the retardation of one of the two beams by half a wave. Degenerate four-wave mixing signals diffracted in two directions, $2n_2-n_1$ and $2n_1-n_2$, were detected simultaneously by photodiodes to obtain resolution times shorter than the correlation time of the incoherent light [7]. The sample was a film of poly-3BCMU cast on a glass plate from chloroform solution. The absorption spectrum of the sample film is shown in Fig. 2.

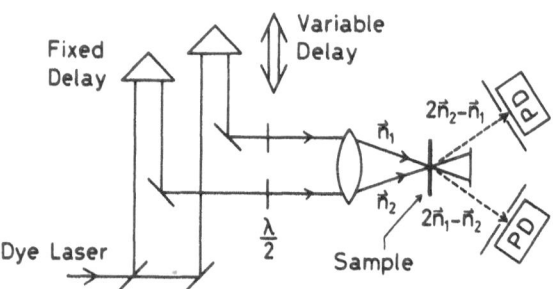

Fig. 1. Experimental setup for degenerate four-wave mixing measurement. PD and $\lambda/2$ stand for a photodiode and a half-wave plate (Fresnel rhomb), respectively. The vectors, n_1 and n_2, are the unit direction vectors of the two excitation beams

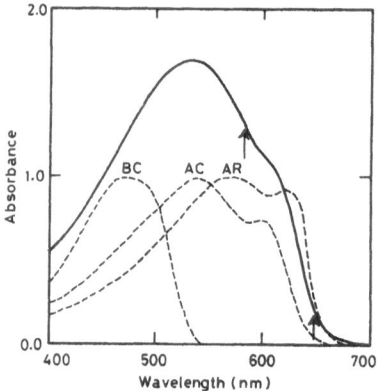

Fig. 2. Absorption spectrum of the sample film of poly-3BCMU (solid line) used in the present study. The absorption spectra of three forms in solution are reproduced from Ref. [24] (broken lines). AR, AC, and BC denote acetylenic type in rod-like conformation, acetylenic type in coil-like conformation, and butatrienic type in coil-like conformation, respectively. Two excition wavelengths, 648 nm and 582 nm, are indicated by arrows

2.2. Results and Discussion

Figure 3 shows the data which were obtained with the two excitation beams under parallel polarization conditions. In this signal curve, no detectable peak shift or asymmetric tail was observed. The peak intensity of the signal was about thirty times higher than those obtained with perpendicular polarizations. Hence the signal obtained with parallel polarizations can be attributed almost exclusively to diffraction from a thermal grating, which is generated only when

4

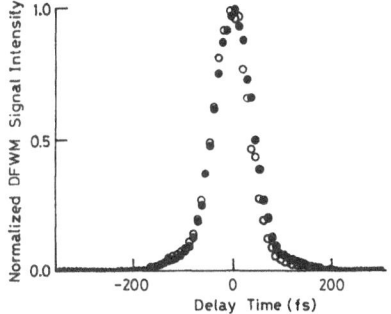

Fig. 3. DFWM signals obtained with the two excitation beams of polarizations parallel to each other. The wavelength of the excitation light was 648 nm. Open and closed circles show the delay-time dependent intensities of the signal diffracted in the directions $2\mathbf{n}_1-\mathbf{n}_2$ and $2\mathbf{n}_2-\mathbf{n}_1$, respectively

mutual coherence exists between the two excitation beams, and has much longer lifetime than the grating formed by electronic nonlinear optical processes [10].

The contribution of the thermal grating to the DFWM signals is eliminated when the polarizations of the two excitation beams are perpendicular to each other, and we can obtain DFWM signals which are only due to electronic nonlinear susceptibilities by the perpendicular polarization [7]. The data obtained with perpendicular polarizations are shown in Fig. 4. There are shifts of delay times between the peaks of the signal intensities of the two directions. The peak shifts were 30 and 90 fs at 648 and 582 nm, respectively. There are no pronounced asymmetric tails, which are expected for dephasing times much longer than the correlation time. The tails seen in the data at 648 nm are the same as those seen in thermal grating signals, and they reflect the shape of the laser spectrum.

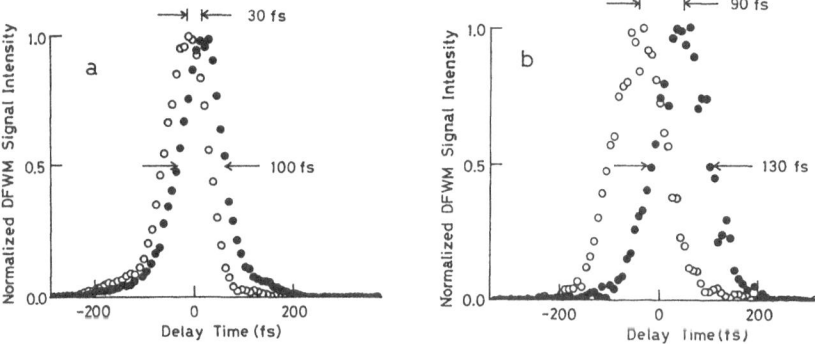

Fig. 4. DFWM signals with the two excitation beams of polarizations perpendicular to each other. Open and closed circles show the intensity of the signal diffracted in the directions $2\mathbf{n}_1-\mathbf{n}_2$ and $2\mathbf{n}_2-\mathbf{n}_1$, respectively. The wavelengths of the excitation light are a) 648 nm and b) 582 nm

Under resonance excitation condition, the delay-time dependence of the signal intensities is expressed by the following equation [6]:

$$I(t_d) \propto \int_0^\infty dt \int_0^\infty dt' G(t'-t)G(t-t_d)G^*(t'-t_d)\exp[-2(t+t')/T_2] . \qquad (1)$$

Here T_2 is the dephasing time, t_d is the delay time, and $G(t)$ is the autocorrelation function of the incoherent light field. When the dephasing time is much longer than the correlation time of the light field, the signal decays exponentially at the rate of $4/T_2$. On the other hand, when the dephasing time is comparable with the correlation time, which is the case in the present study, the dephasing time of the matter can also be obtained from the peak shift of the two signals diffracted in two directions even though the signal shapes have no prominent tails [7,10]. Using (1) and the observed peak shifts, the dephasing times are calculated to be 30 fs at 648 nm and 130 fs at 582 nm by assuming the autocorrelation functions to be Gaussian. The effect of spectral diffusion on these dephasing times can be neglected, since spectral diffusion in polymeric systems such as PDA may take place in subpicoseconds or slower.

This result, that the dephasing time at a longer wavelength is shorter than that at a shorter wavelength, is contrary to those of previous studies [7,10]. The peak shift was reported to be larger at a longer wavelength than at a shorter wavelength with cresyl fast violet in cellulose, and the result was explained in terms of the difference between the rates of intramolecular relaxation processes at the two wavelengths [7]. Dephasing time measurements of three dyes, cresyl violet, Nile blue, and oxazine 720, in polymethylmethacrylate (PMMA) at 15 K at 620 nm were also reported [10]. The peak shift for cresyl violet was found to be 60 fs, whereas for the other two dyes shifts were shorter than 20 fs. The difference in the dephasing times was attributed to that in the excess energy of the exciting photon from the absorption edge of each sample.

The present sample has its own characteristics, different from ordinary dyes [24,25]. It is known that poly-3BCMU in solution has two conformations, rod-like and coil-like, both of which can have two isomeric bond structures, the acetylenic and butatrienic types [24]. These four forms are realized by changing the temperature and the composition of the solvent. The shoulder of the absorption spectrum of the sample at 620 nm (2.0 eV) (Fig. 1) corresponds to the AR (acetylenic type in rod-like conformation) band in Fig. 3 of Ref. [24], and the peak at 530 nm (2.3 eV) corresponds to the AC (acetylenic type in coil-like conformation) band. They are attributed to the π-π exciton transition in each type of the polymer chain. Therefore, our sample is thought to be a mixture of a coil-like conformation (with shorter conjugation lengths of π-electron) and a rod-like conformation (with longer conjugation lengths) in the acetylenic bond structure, or a mixture of the polymer chains with continuously distributed conjugation lengths between these two extreme forms realized in solution.

In the present experiment, the excitation wavelength of 648 nm is on the absorption edge of the rod-like form exciton, while that of 582 nm is in resonance with the exciton in polymer chains with shorter conjugation lengths. Therefore, the present experimental results indicate that the exciton dephasing is about four times faster in the rod-like form than in a chain with a shorter conjugation length.

The difference in the dephasing time between the two forms can be explained in the following two ways. One explanation is that phase change occurs more frequently because of larger mobility of excitons in longer conjugated chains than shorter ones, where excitons do not move over long distances. The other explanation is that exciton levels lie more closely in longer conjugated chains than in shorter conjugated chains, and therefore, dephasing due to multilevel excitation is faster [10]. We cannot determine which of the two is the case from the present data only. An extended study with other excitation wavelengths and temperatures is in progress.

It has been reported that the fluorescence intensities are suppressed when the solution of poly-3BCMU is converted from a yellow (butatrienic, coil-like) form to a blue (acetylenic, rod-like) form, and when the solution of PDA (poly-4BCMU) is converted from the coil form to the rod form [24]. It is also reported that only partially polymerized crystal of PDA (PTS) emits fluorescence [26]. Our results support the explanation [24,26] by which the changes in the fluorescence quantum efficiencies were attributed to the increase in the nonradiative decay rates with exciton delocalization.

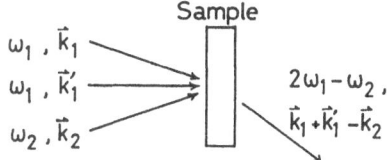

Sample

$\omega_1, \vec{k}_1$

$\omega_1, \vec{k}_1'$

$\omega_2, \vec{k}_2$

$2\omega_1 - \omega_2,$

$\vec{k}_1 + \vec{k}_1' - \vec{k}_2$

Fig. 5. Schematic of CSRS experiment for vibrational dephasing measurement

3. Vibrational Dephasing Measurement by CSRS

A new method for the observation of picosecond or subpicosecond vibrational dephasing dynamics using incoherent or broad-band light is presented theoretically and experimentally. It is based on a transient coherent Raman process with three beams (see Fig. 5). In our experimental scheme, two of them are incoherent light from a single broad-band laser, and a delay time between the two is variable. A third beam has a higher frequency than the incoherent light by a vibrational energy in a molecule of interest, and is coherent in the delay-time range of the measurement. The delay-time dependence of coherent Stokes Raman scattering (CSRS) intensity offers the information about the coherence dynamics of the vibrational transition with a resolution time limited by the correlation time of the incoherent light. For the theoretical calculation, a three-level model of molecular system with homogeneous broadening is used, and the delay-time dependence of CSRS signal intensity was calculated. Dephasing dynamics of the 2915-cm^{-1} mode in dimethylsulfoxide was observed experimentally by the new method using nanosecond laser pulses, and the result was found to agree well with that obtained with picosecond pulses.

3.1. Description of a Model and Time Dependence of Signal Intensity

Until now all reported studies of transient spectroscopy with incoherent light, including photon echo and pump-probe spectroscopy, were concerned with broad-band (about 100 cm^{-1}) with only one center frequency which is resonant with a two-level system. However, there exist various coherent transient phenomena where light beams of two or more different frequencies are concerned, and the time resolution of the transient coherent spectroscopies using these phenomena is also expected to be determined not by the duration of light pulses used but by the correlation time of the radiation field.

In this section, theoretical expectation values of the coherent Raman signal intensity using incoherent light will be presented with a simple model. Calculation of the time-dependent intensity of coherent Stokes Raman scattering will be presented for the correspondence with the experimental study described in the following section, although the time dependence of the intensity of coherent anti-Stokes Raman scattering is substantially the same.

A simple model of a three-level system (see Fig. 6) is usually taken for theoretical considerations of coherent Raman phenomena [27,28]. Two vibrational levels |1> and |2> belong to the ground electronic state, whereas level |3> belongs to an electronically excited state. Levels |1> and |3> and levels |2> and |3> are connected with each other by electronic transition dipoles. The energy difference between |3> and |1> is $\hbar\Omega_3$, and that between |2> and |1> is $\hbar\Omega_2$. In ordinary coherent Raman experiments, this system is placed in a radiation field which consists of light beams of two frequencies, the difference between which is resonant with the transition between |1> and |2>. The higher frequency is denoted by ω_{AS} and the lower by ω_L.

CSRS is a third-order effect, and the intensity is proportional to the light intensity of frequency ω_{AS} and to the squared light intensity of frequency ω_L. In usual CSRS (or CARS) experiments two beams are used, but for the purpose of time-resolved measurement, a triple-beam (BOXCARS) configuration was applied [29], where two beams of frequency ω_L are used, and three waves are mixed to generate a wave of frequency $2\omega_L - \omega_{AS}$.

7

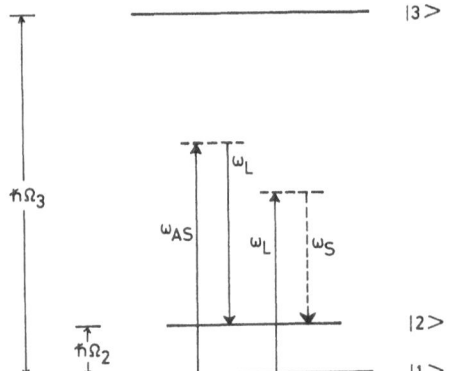

Fig. 6. Energy diagram of the model system

The electric field is given as

$$E(\mathbf{r},t) = E_{AS}(t)\exp[i(\mathbf{k}_{AS}\mathbf{r} - \omega_{AS}t)] + E_{L1}(t)\exp[i(\mathbf{k}_{L1}\mathbf{r} - \omega_L t)]$$

$$+ E_{L2}(t)\exp[i(\mathbf{k}_{L2}\mathbf{r} - \omega_L t)] + c.c. \quad , \tag{2}$$

where c.c. stands for complex conjugates of the preceding terms and $E_{AS}(t)$, $E_{L1}(t)$, and $E_{L2}(t)$ are functions of t slowly varying compared to the optical frequencies. In BOXCARS experiments, light of frequency $\omega_S = 2\omega_L - \omega_{AS}$ and with wave vector $\mathbf{k}_S = \mathbf{k}_{L1} + \mathbf{k}_{L2} - \mathbf{k}_{AS}$ is detected. This can be easily performed by separating spatially the signal from the other light beams because of the directionality of the signal and the laser beams.

The polarization with frequency ω_S and wave vector $\mathbf{k}_S$ is derived by a perturbational method. The following conditions are assumed; (i) Light frequencies are tuned exactly to the vibrational energy. (ii) The light of frequencies ω_{AS} and ω_L are off-resonance with electronic transitions. (iii) The broadening of the relevant energy level of the molecular system is homogeneous. Under these conditions, a third-order polarization $P^{(3)}(\mathbf{k}_S, \omega_S)$ is given by

$$P^{(3)}(\mathbf{k}_S, \omega_S) = C\exp[i(\mathbf{k}_S\mathbf{r} - \omega_S t)] \int_{-\infty}^{t} dt' \exp[-(t-t')/T_2]$$

$$\times [E_{L1}(t)E_{L2}(t') + E_{L2}(t)E_{L1}(t')]E_{AS}^*(t') \quad , \tag{3}$$

where T_2 is the dephasing time of the vibrational transition, and C is a time-independent proportionality coefficient. This expression has two terms, in which E_{L1} and E_{L2} are exchanged with each other.

In the present CSRS experiments using incoherent light, two incoherent light beams of central frequency ω_L are obtained by splitting a beam from a broad-band dye laser. One of the two beams is delayed to the other with a delay time t_d, and the light of frequency ω_{AS} is assumed to be coherent in the time scale of observation.

When one assumes that the correlation time of the incoherent light is much shorter than the dephasing time, and that the stochastic property of the incoherent light is expressed in terms of a Gaussian random process, one can derive a simple expression for the signal intensity as

$$I(t_d) = 1 + G(t_d) + (2\tau_c/T_2)\exp(-2|t_d|/T_2) \quad . \tag{4}$$

Here G(t) is the autocorrelation function of the incoherent light field amplitude normalized to unity at its peak, and the correlation time τ_c is defined by

$$\tau_c = \int_0^\infty G(t)dt \quad . \tag{5}$$

In the above expression, the fraction of the third term in the signal intensity is approximately proportional to the ratio of the correlation time to the dephasing time. Therefore, for the dephasing time to be determined clearly, incoherent light with an adequate correlation time must be used, since small values of the intensity cannot be distinguished from the background intensity due to inevitable noises from various sources.

3.2. Experiment and Results

CSRS signal by the symmetric CH-stretching vibration of dimethylsulfoxide (DMSO) with a wavenumber 2915 cm^{-1} was measured by the apparatus shown in Fig. 7. Coherent light source was the second harmonic (532 nm) of a Q-switched Nd:YAG laser operated at 8 Hz. Since the correlation time of this light source is estimated to be about 30 ps, it can be regarded safely as coherent for the time period shorter than 10 ps. The main part (about 90% intensity) of this beam was split off and used to pump a broad-band dye laser. By changing the laser dye concentration, the oscillation wavelength was tuned to 630 nm which is resonant with the Raman mode of DMSO . The spectral width (FWHM) of the laser light was 7 nm, which corresponds to a correlation time of about 100 fs.

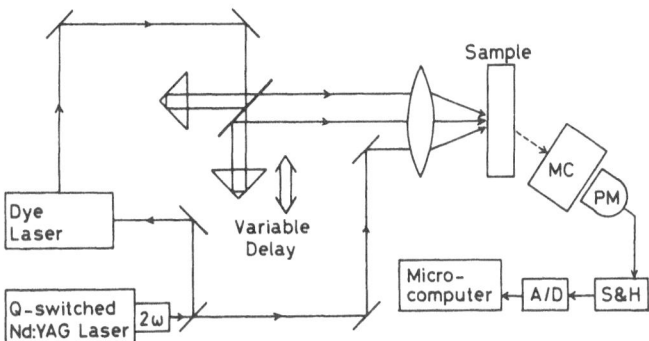

Fig. 7. Experimental setup for CSRS measurement using incoherent and coherent light. MC is the monochromator; PM, the photomultiplier

The dye laser beam was split into two beams with nearly equal intensity using a beam splitter. Both of them were made parallel again by the same beam splitter after passing through delay lines. The length of one of the delay lines was varied by a stepping motor. The three beams were focused in DMSO in a 10-mm sample cell by a lens of 20-cm focal length. To eliminate the scattered laser light, an iris with 5-mm diameter, a 10-cm monochromator, and color filters were placed between the sample and the photomultiplier.

In Fig. 8, the CSRS signal intensity is plotted as a function of the delay time in a semilogarithmic scale after the background is subtracted. From the slope of the trailing part, the vibrational dephasing time T_2 is obtained to be 1.4 ps, which is in agreement with the value obtained from the experiment using picosecond pulses [12]. The intensity ratio between the peak and the tail extrapolated to $t_d=0$ is about seven, which is also in good agreement with the theoretical ratio with the values of T_2 (1.4 ps) and the correlation time (100 fs) obtained above.

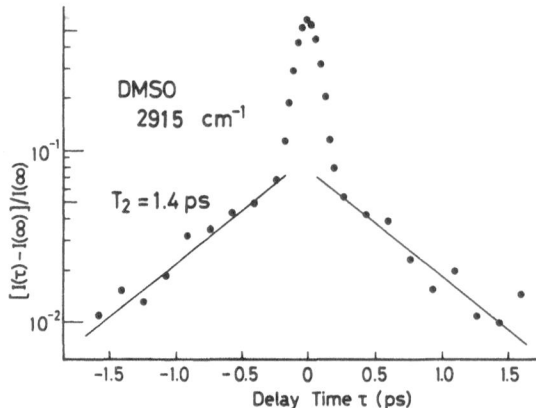

Fig. 8. The delay-time dependence of the CSRS intensity normalized by the background intensity. The background intensity is subtracted

Since CSRS or CARS does not include a rephasing process, which takes place in photon echoes, the obtained value of the dephasing time of three-level systems has some ambiguities when the system is inhomogeneously broadened [30,31]. Raman echo [27] can provide the information about the dephasing dynamics without the ambiguities [28]. However, it is a difficult experiment since it is a higher-order process. Raman echo experiments have been carried out in solids [32], gases [33,34], and liquid nitrogen [35], but not yet in liquids at room temperature, and the first experiment may possibly be performed with the use of incoherent light applying a similar method to the present study.

Acknowledgements

We wish to thank Professor T. Kotaka for providing us with poly-3BCMU, and Professor E. Hanamura for helpful discussion. This work is partially supported by a Grant-in-Aid for Special Distinguished Research (56222005) from the Ministry of Education, Science and Culture of Japan.

References

1. H.W. Mocker and R.J. Collins, Appl. Phys. Lett. 7, 270 (1965)
2. W.H. Knox, R.L. Fork, M.C. Downer, R.H. Stolen, and C.V. Shank, Appl. Phys. Lett. 46, 1120 (1985)
3. N. Morita and T. Yajima, Phys. Rev. A 30, 2525 (1984)
4. S. Asaka, H. Nakatsuka, N. Fujiwara, and M. Matsuoka, Phys. Rev. A 29, 2286 (1984)
5. R. Beach and S.R. Hartmann, Phys. Rev. Lett. 53, 663 (1984)
6. H. Nakatsuka, M. Tomita, M. Fujiwara, and S. Asaka, Opt. Commun. 52, 150 (1984)
7. M. Fujiwara, R. Kuroda and H. Nakatsuka, J. Opt. Soc. Am. B 2, 1634 (1985)
8. S.R. Meech, A.J. Hoff, and D.A. Wiersma, Chem. Phys. Lett. 121, 287 (1985)
9. T. Hattori and T. Kobayashi, Chem. Phys. Lett. 133, 230 (1987)
10. A.M. Weiner, S.De Silvestri, and E.P. Ippen, J. Opt. Soc. Am. B 2, 654 (1985)
11. A. Laubereau and W. Kaiser, Rev. Mod. Phys. 50, 607 (1978)
12. S.M. George, H. Auwester, and C.B. Harris, J. Chem. Phys. 73, 5573 (1980)
13. S.M. George, A.L. Harris, M. Berg, and C.B. Harris, J. Chem. Phys. 80, 83 (1984)
14. T. Hattori, A. Terasaki, and T. Kobayashi, Phys. Rev. A 35, 715 (1987)

15. S. Koshihara, T. Kobayashi, H. Uchiki, T. Kotaka, and H. Ohnuma, Chem. Phys. Lett. 114, 446 (1985)
16. G.M. Carter, J.V. Hryniewicz, M.K. Thakur, Y.J. Chen, and S.E. Meyler, Appl. Phys. Lett. 49, 998 (1986)
17. T. Kobayashi, J. Iwai, and M. Yoshizawa, Chem. Phys. Lett. 112, 360 (1984)
18. T. Kobayashi, H. Ikeda, and S. Tsuneyuki, Chem. Phys. Lett. 116, 515 (1985)
19. J. Orenstein, S. Etemad, and G.L. Baker, J. Phys. C 17, L297 (1984)
20. L. Robins, J. Orenstein, and R. Superfine, Phys. Rev. Lett. 56, 1850 (1986)
21. W.M. Dennis, W. Blau, and D.J. Bradley, Appl. Phys. Lett. 47, 200 (1985)
22. G.M. Carter, M.K. Thakur, Y.J. Chen, and J.V. Hryniewicz, Appl. Phys. Lett. 47, 457 (1985)
23. D.N. Rao, P. Chopra, S.K. Ghoshal, J. Swiatkiewicz, and P.N. Prasad, J. Chem. Phys. 84, 7049 (1986)
24. T. Kanetake, Y. Tokura, T. Koda, T. Kotaka, and H. Ohnuma, J. Phys. Soc. Japan 54, 4014 (1985)
25. R.R. Chance, G.N. Patel, and J.D. Witt, J. Chem. Phys. 71, 206 (1979)
26. H. Sixl and R. Warta, Chem. Phys. Lett. 116, 307 (1985)
27. S.R. Hartmann, IEEE. J. Quant. Electron. QE-4, 802 (1968)
28. R.F. Loring and S. Mukamel, J. Chem. Phys. 83, 2116 (1985)
29. A.C. Eckbreth, Appl. Phys. Lett. 32, 421 (1978)
30. W. Zinth, H.-J. Polland, A. Laubereau, and W. Kaiser, Appl. Phys. B 26, 77 (1981)
31. S.M. George and C.B. Harris, Phys. Rev. A 28, 863 (1983)
32. P. Hu, S. Geschwind, and T.M. Jedju, Phys. Rev. Lett. 37, 1357 (1976)
33. K.P. Leung, T.W. Mossberg, and S.R. Hartmann, Phys. Rev. A 25, 3097 (1982)
34. V. Brückner, E.A.J.M. Bente, J. Langelaar, and D. Bebelaar, Opt. Commun. 51, 49 (1984)
35. J.D.W. van Voorst, D. Brandt, and B.L. van Hensbergen, in Technical Digest of Topical Meeting on Ultrafast Phenomena, (1986)

Part II

Photosynthesis and
Phototransformation

Energy Gap Law in Electron Transfer Reaction

T. Kakitani

Department of Physics, Nagoya University,
Furo-cho, Chikusa-ku, Nagoya 464, Japan

1. Introduction

Electron transfer is one of the most fundamental processes in chemical and biological reactions. Photoinduced electron transfer and related chemical reactions in solution have been the subject of lively investigation. In this article, we first survey the theoretical and experimental problems related to this topic. In the second, we propose a new aspect of the role of the polar solvent mode played in the electron transfer reaction and try to resolve the above problems.

2. Mechanism of Electron Transfer in Solution

The mechanism of the electron transfer in polar solution is schematically shown in Fig.1. Acceptor and donor molecules diffuse until they collide with each other to form an encounter complex A···B. At the special orientation of solvent molecules, the energy of A···B (state (c)) becomes equal to that of A^-···B^+ (state (d)). It corresponds to the crossing point of potential curves of the initial and final states, called the transition state. The electron transfer reaction proceeds through this transition state.

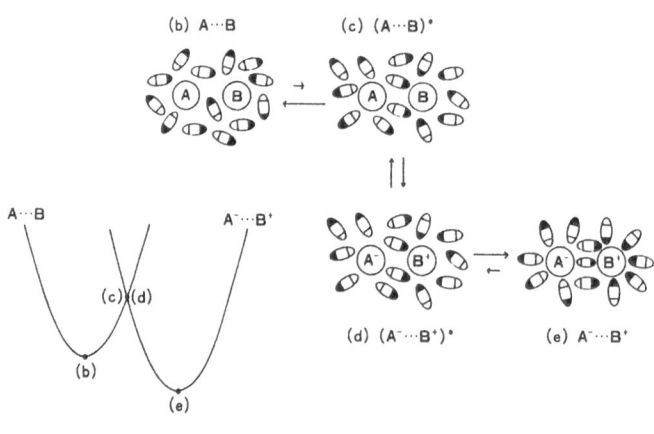

Fig.1 Schematic diagram of the electron transfer process in solution. A and B are reactants. Spheroids are polar solvent molecules.

3. Energy Gap Law by Current Theory

The theory of the electron transfer rate in this mechanism was first formulated by Marcus in 1956 [1]. He based on the dielectric continuum model of solvents and formulated the activation free energy for the non-equilibrium polarization. His theory was later rewritten by Dogonadze and Levich [2,3] using the polaron model. A possible important role of the intramolecular quantum mode of the donor and acceptor molecules was suggested by Kestner, Logan and Jortner in 1974 [4].

All the above theories predicted the energy gap law shown in the lower part of Fig.2. The abscissa is the energy gap ΔE and the ordinate is the logarithm of the electron transfer rate W. It is nearly bell-shaped with the maximum at $\Delta E = E_r$, E_r being the reorientation energy. This energy gap law is qualitatively explained by the following. For $\Delta E < E_r$, the relation of the energy surfaces of the initial and final states is like (a) in Fig.2. So that, it needs the activation energy to reach the transition state. For $\Delta E = E_r$ in the case of (b), no activation energy is necessary and W becomes the maximum. In the case $\Delta E > E_r$ as shown in (c), again the activation energy is necessary. The energy gap region $\Delta E > E_r$ where W decreases with ΔE is called the inverted region.

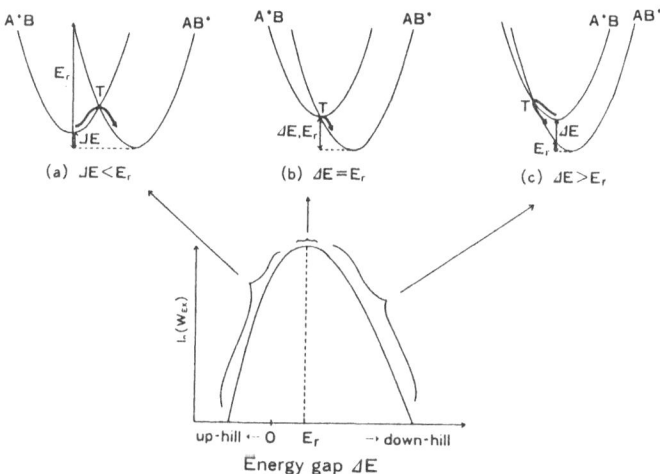

Fig.2 (Top) Relative position of potential surfaces of the initial (A*B) and final (AB*) states depending on the magnitude of ΔE and E_r. (Bottom) Typical energy gap law obtained by the current theory.

4. Rehm and Weller's Experiment

A systematic experimental study of the photoinduced electron transfer was first done by Rehm and Weller in 1970 [5]. Its reaction involves the diffusion process to form an encounter complex and the electron transfer process from the excited state of the fluorescer to the quenching molecule, as shown in the following scheme:

15

$$A^* + B \underset{k_{21}}{\overset{k_{12}}{\rightleftharpoons}} A^* \cdots B \underset{k_{32}}{\overset{k_{23}}{\rightleftharpoons}} A^{\pm} \cdots B^{\mp} \xrightarrow{k_{30}} \quad . \tag{1}$$

The quenching rate k_q is expressed by

$$K_q = \frac{k_{12}}{1+(k_{21}/k_{23})(1+k_{32}/k_{30})} , \tag{2}$$

where k_{23} is the electron transfer rate. The experimental data for the various combination of fluorescers and quenchers are shown in Fig.3. ΔG_{23} is the free energy change due to the electron transfer. Its negative value corresponds to a positive energy gap. The black circles represent the experimental points. This graph clearly shows that from a little positive value of ΔG_{23} to a little negative value of ΔG_{23} the quenching rate drastically increases and it becomes a constant value for a large negative value of ΔG_{23}, i.e., for a large energy gap region. The constant part of the quenching rate might reflect the diffusion-control mechanism. So, the diffusion process may be thought to have masked the intrinsic electron transfer process. However, if we apply the above current theory to k_{23}, we obtain such a theoretical curve as solid curve for $\beta=\infty$ in Fig.3. That is, the intrinsic electron transfer process should become visible at the large energy gap, corresponding to the inverted region of the theoretical curve. Therefore, the non-existence of the inverted region in the Rehm and Weller's experiment has been one of the unsolved problems.

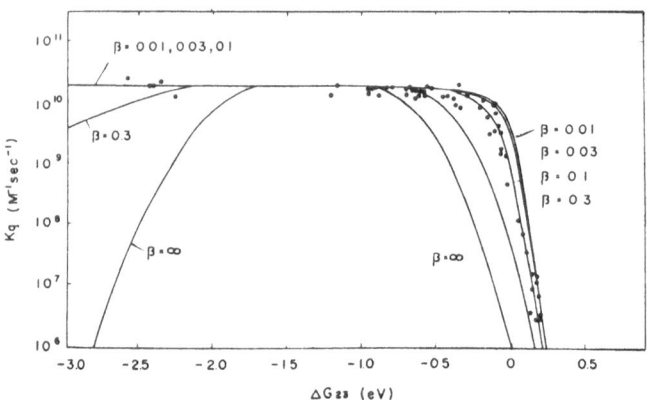

Fig.3 Quenching rate as a function of energy gap. Black circles are the experimental data obtained by Rehm and Weller [5]. Solid curves are the theoretical calculations. The definition of β is given in eq.(8). $\beta=\infty$ corresponds to the current treatment.

5. Linking of Donor and Acceptor Molecules

Very recently, the other types of experiments were done. These were aimed to eliminate the diffusion process, by fixing the donor and acceptor molecules to both ends of the spacer. In the Wasielewski et al.'s experiment [6], the tetraphenylporphyrin is the donor and some

kinds of quinones are acceptors. They observed the photoinduced charge
separation (CS) rate and the charge recombination (CR) rate. According
to their result, the inverted region is clearly seen in the CR reaction.
In the experiment by Miller et al. [7], the donor is the biphenyl anion
radical and the acceptor is some aromatic molecules and quinones.
Spacer is a rigid steroidal molecule. They observed the electron
transfer from anion to a neutral acceptor. This electron transfer
process is called the charge shift (CSH) reaction, hereafter. Their
result also showed a moderate inverted region.

The experimental results by Wasielewski et al. and Miller et al.
apparently show that the inverted region becomes visible, since donor and
acceptor molecules are linked by a spacer molecule. However, this is not
the case. Mataga et al. [8-11] have recently measured the CR rate of
separated, charged donor and acceptor molecules which were produced by
photoinduced electron transfer in solution. The result showed that very
sharp inverted region appeared for this reaction. Therefore, the linking
of donor and acceptor molecules has nothing to do with existence of the
inverted region. Instead, the problem of the inverted region seems to be
correlated with the type of electron transfer reaction.

6. New Aspect of the Solvent Mode

Here, we would like to point out a new aspect of the solvent mode
[12]: The orientational motion of solvent molecules surrounding the
charged reactant should be more restricted than that surrounding the
neutral reactant. In other words, the potential curvature k_c as a
function of the solvent coordinate surrounding the charged reactant
should be much larger than the potential curvature k_n as a function of
the solvent coordinate surrounding the neutral reactant.

For the photoinduced charge separation reaction, the potential
curvature of the initial state (A*···B) is much smaller than that of the
final state (A$^{\mp}$···B$^{\pm}$) as shown in Fig.4(a). So that, the crossing
between the initial and final potential curves is obtained as long as the
energy gap is positive. Once this condition is fulfilled, the crossing

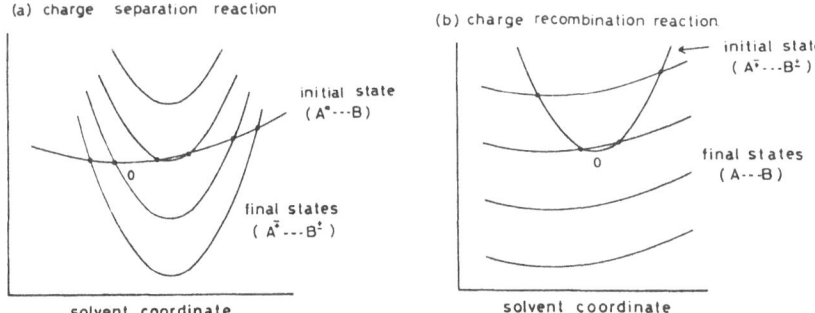

Fig.4 Proposed schematic energy surfaces of the solvent mode in (a)
photoinduced charge separation and (b) charge recombination reactions.
The curvature of the neutral state of reactants is taken to be much
smaller than that of the charged state of reactants.

occurs at a small energy height from the minimum of the initial state
even if the energy gap is very large. This property contributes to the
non-existence of the inverted region in the charge separation reaction.

In the case of the charge recombination reaction, on the other hand,
the potential curvature of the initial state is much larger than that of
the final state.as shown in Fig.4(b). The crossing between potential
curves of the initial and final states occurs so far as the energy gap
is negative. Therefore, the crossing energy level is very high unless
the energy gap is small in negative value. This fact indicates that the
solvent mode is nearly silent for the energy gap law and the inverted
region should be prominently observed in the charge recombination
reaction.

7. Theoretical Formula

For the quantitative calculations, we formulate the electron
transfer rate using the Fermi   Golden rule as follows:

$$W(\Delta E) = AZ^{-1}\exp[-E_u/kT]\sum_v|<u|v>|^2\delta(E_u-E_v+\Delta E) , \qquad (3)$$

where A is the electron-tunneling matrix element, E_u and E_v vibrational
energies of the initial and final states, Z the partition function, and
$|<u|v>|$ the Franck-Condon factor. Decoupling the vibrational modes into
the intramolecular quantum mode and the solvent mode, we can write $W(\Delta E)$
by a convolution form as

$$W(\Delta E) = \int_{-\infty}^{\infty} S(\epsilon) W_q(\Delta E-\epsilon)d\epsilon , \qquad (4)$$

where W_q is the electron transfer rate when only the quantum mode is
considered. $S(\epsilon)$ is the density function due to the Franck-Condon
factor of the solvent mode. The vibrational frequency of the quantum
mode is almost the same between the initial and final states. Then, we
can evaluate W_q quantum-mechanically as follows [13]:

$$W_q(\Delta E-\epsilon) = \bar{A} \exp[-S(2\bar{v}+1)]\cdot I_{|p|}(2S\sqrt{\bar{v}(\bar{v}+1)})\times[(\bar{v}+1)/\bar{v}]^{p/2} , \qquad (5)$$

where

$$\bar{v} = [\exp(\hbar<\omega>/kT)-1]^{-1} ,$$

$$p = (\Delta E-\epsilon)/\hbar<\omega> ,$$

$$S = \delta^2/2 , \qquad \bar{A} = A/\hbar<\omega> . \qquad (6)$$

In the above, $<\omega>$ and δ are the average angular frequency of the quantum
mode and the displacement of the normal coordinate between the initial
and final states, respectively.

18

Since we assume that the vibrational frequency of the solvent mode of the charged state is much larger than that of the neutral state, we cannot treat this case quantum-mechanically in general. We treat the solvent mode contribution classically, following the Hopfield method as follows:

$$S(\varepsilon) = \int_{-\infty}^{\infty} D_a'(E)D_b(E)dE , \tag{7}$$

where D_a' and D_b are the electron insertion spectrum of A*(or A^+) and the electron removal spectrum of B (or B^-), respectively, corresponding to the reactions

$$\begin{cases} A* + B & \to & A^- + B^+ & \text{(CS reaction)} \\ A^+ + B^- & \to & A\ + B & \text{(CR reaction)} \\ A^+ + B & \to & A\ + B^+ & \text{(CSH reaction)} . \end{cases}$$

Here, we define an important parameter β as

$$\beta = k_n/(k_c-k_n) . \tag{8}$$

Then, if $k_c \gg k_n$ holds, $\beta \ll 1$. If $k_c=k_n$ holds as assumed in the current theory, $\beta=\infty$. We also define another parameter Δ for the reorientation energy of the solvent accompanying the electron transfer reaction.

Finally we obtain formulas which are valid for $\beta \ll 1$ as follows: In the CS reaction

$$S(\varepsilon) = \frac{\beta}{4kT} \exp[-\beta(\sqrt{\varepsilon+\beta\Delta} - \sqrt{(1+\beta)\Delta})^2/kT]$$

$$+ \exp[-\beta(\sqrt{\varepsilon+\beta\Delta} + \sqrt{(1+\beta)\Delta})^2/kT]$$

$$+ 2\exp[-\beta\{\varepsilon+(1+2\beta)\Delta\}/kT]\cdot I_0(\beta\sqrt{(1+\beta)\Delta(\varepsilon+\beta\Delta)} /kT)\}$$

$$\text{for} \quad \varepsilon > -\beta\Delta ,$$

$$S(\varepsilon) = 0 \qquad\qquad\qquad \text{for} \quad \varepsilon \le -\beta\Delta . \tag{9}$$

In the CR reaction

$$S(\varepsilon) = \frac{1+\beta}{4kT} \{\exp[-(\sqrt{(1+\beta)(\beta\Delta-\varepsilon)} - \beta\Delta)^2/kT]$$

$$+ \exp[-(\sqrt{(1+\beta)(\beta\Delta-\varepsilon)} + \beta\sqrt{\Delta})^2/kT]$$

$$+ 2\exp[-\{\beta(1+2\beta)\Delta-(1+\beta)\varepsilon\}/kT]\cdot I_0(\beta\sqrt{(1+\beta)\Delta(\beta\Delta-\varepsilon)} /kT)\}$$

$$\text{for} \quad \varepsilon < \beta\Delta ,$$

$$S(\varepsilon) = 0 \qquad\qquad\qquad \text{for} \quad \varepsilon \ge \beta\Delta . \tag{10}$$

In the CSH reaction

$$S(\varepsilon) = \frac{\sqrt{\beta(1+\beta)}}{\pi kT} \exp[\{\varepsilon - \beta(1+2\beta)\Delta\}/2kT]$$

$$\times \cosh(\beta\sqrt{2(1+\beta)\Delta|\varepsilon|}/kT) \cdot K_{\beta\sqrt{2(1+\beta)\Delta|\varepsilon|}/2kT}((1+2\beta)|\varepsilon|/2kT) \ .$$

(11)

8. Numerical Results

In the numerical calculations, we have chosen typical values of parameters as follows: $\bar{A}=2\times10^{12}s^{-1}$, $\hbar\langle\omega\rangle=0.10eV$, $kT=0.025eV$, $S=3$, $\Delta=1.0eV$. The numerical result of $S(\varepsilon)$ for the CS reaction is shown in Fig.5A for some values of β. The dotted curve is for $\beta=\infty$ which corresponds to $k_c=k_n$ as in the current theory. According as β becomes small, $S(\varepsilon)$ becomes step-wise function. The total rate W obtained by the convolution of W_q and $S(\varepsilon)$ is shown in Fig.5B. When β is small, rapid increase of W is seen in the small energy gap region and no appreciable inverted region is obtained, as we expected. By taking into account the diffusion process, the Rehm and Weller's experimental data can be well reproduced by our theoretical curve with $\beta\approx0.1$ as shown in Fig.1 [12].

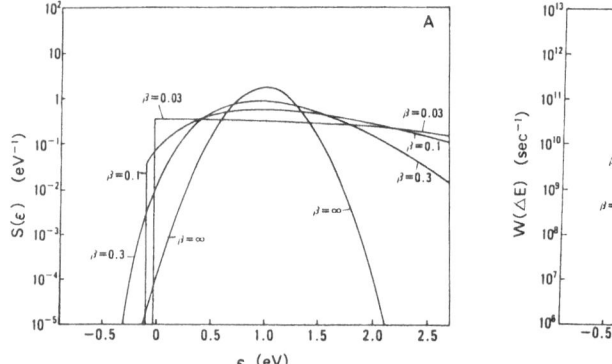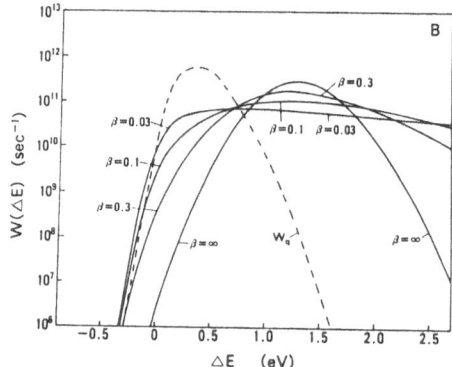

Fig.5 Calculated values of (A) $S(\varepsilon)$ and (B) $W(\Delta E)$ in the case of charge separation reaction. W_q is shown by the dotted curve.

The numerical result of $S(\varepsilon)$ in the case of CR reaction is plotted for some values of β in Fig.6A. It is clearly seen that $S(\varepsilon)$ becomes a very sharp triangle or delta-like function when β is small. By taking the convolution with W_q, we obtain W in Fig.6B. When β is small, W is close to W_q, giving rise to a remarkable inverted region. This result is consistent with the experimental data of Wasielewski et al. [6] and Mataga et al. [8-11].

In the case of the CSH reaction, the calculated result of $S(\varepsilon)$ has some anomaly at $\varepsilon=0$, but for the other region of ε, it is rather intermediate of the cases of CS and CR reactions, as shown in Fig.7A.

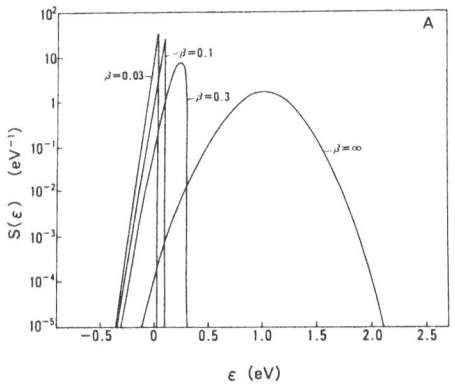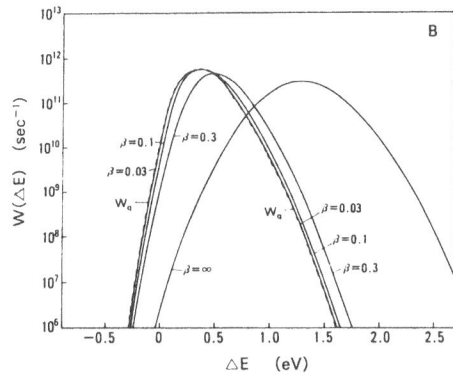

Fig.6 Calculated values of (A) $S(\varepsilon)$ and (B) $W(\Delta E)$ in the case of charge recombination reaction. W_q is shown by the dotted curve.

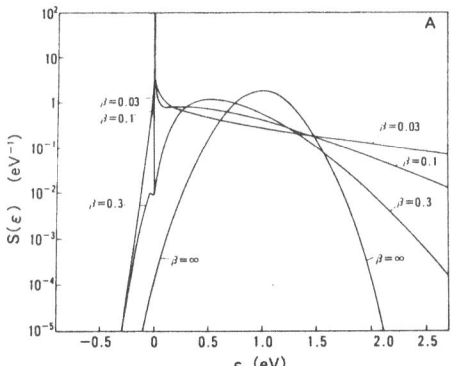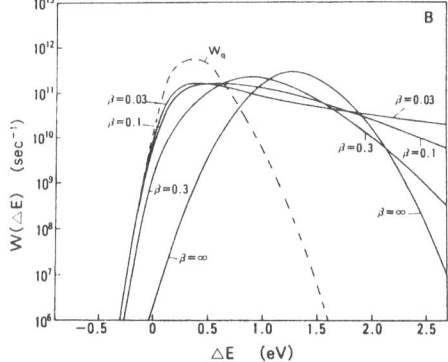

Fig.7 Calculated values of (A) $S(\varepsilon)$ and (B) $W(\Delta E)$ in the case of charge shift reaction. W_q is shown by the dotted curve.

When one takes convolution of $S(\varepsilon)$ and W_q, the anomaly disappears and W has a weak inverted region depending on the value of β, as shown in Fig.7B. This result is consistent with the experimental data of Miller et al. [7].

9. Discussion

We can summarize the above results as follows: Taking into account the great difference between the force constant of the solvent mode surrounding a neutral solute and that surrounding a charged solute, we have derived three different energy gap dependences of the electron transfer rate, corresponding to the CS, CR and CSH reactions. The experimental data by Rehm and Weller [5], Wasielewski et al. [6], Miller et al. [7] and Mataga et al. [8-11] could be consistently explained by our theory: The non-existence of the inverted region in the CS reaction is due to the step-wise function of $S(\varepsilon)$, while the existence of a remarkable inverted region in the CR reaction is due to the delta-like function of $S(\varepsilon)$.

Based on the distinctive energy gap dependences of the CS and CR reactions, we can argue as follows: The charge separation can easily occur but the charge recombination rarely occurs unless the matching condition $\Delta E = E_r$ is nearly satisfied. We are now investigating whether a remarkable regulation mechanism of the electron transfer, as seen in the reaction center of photosynthetic bacteria, can be explained by our energy gap laws. For this purpose, it is essential to make clear whether the dynamical behavior of dielectrics of the reactions center is similar to polar solvent or not. It is our future problem.

Here, it will be worthwhile to point out that the great difference of the force constant of the solvent mode between the charged and neutral states of reactant is related to the problem of dielectric saturation of the solvent. Unless there is dielectric saturation, the force constants are equal to each other [14]. Only when a considerable amount of dielectric saturation exists, do the force constants differ greatly. We have already noticed the dielectric saturation in the polarization of the dipole moment around a charged reactant, by a theoretical study based on the dielectric continuum model and the Lorentz field method [15]. More detailed investigation by the Monte Carlo simulation is in progress.

References

1. R.A. Marcus: J. Chem. Phys. 24, 966 (1956)
2. R.R. Dogonadze: Doklad. Chem. 124, 9 (1959)
3. V.G. Levich: Advan. Electrochem. Electrochem. Eng. 4, 249 (1966)
4. N.R. Kestner, J. Logan, J. Jortner: J. Phys. Chem. 78, 21 (1974)
5. D. Rehm, A. Weller: Israel J. Chem. 8, 259 (1970)
6. M.R. Wasielewski, M.P. Niemczyk, W.A. Svec, E.B. Pewitt: J. Am. Chem. Soc. 107, 1080 (1985)
7. J.R. Miller, L.T. Calcaterra, G.L. Closs: J. Am. Chem. Soc. 106, 3047 (1984)
8. N. Mataga: Pure Appl. Chem. 56, 1255 (1984)
9. N. Mataga, A. Karen, T. Okada, S. Nishitani, N. Kurata, S. Misumi: J. Am. Chem. Soc. 106, 2442 (1984)
10. N. Mataga, A. Karen, T. Okada, S. Nishitani, Y. Sakata, S. Misumi: J. Phys. Chem. 88, 4650 (1984)
11. T. Kakitani, N. Mataga: J. Phys. Chem. 90, 993 (1986)
12. T. Kakitani, N. Mataga: Chem. Phys. 93, 381 (1985)
13. J. Jortner: J. Chem. Phys. 64, 4860 (1976)
14. R.A. Marcus: J. Chem. Phys. 24, 979 (1956)
15. T. Kakitani, N. Mataga: Chem. Phys. Letters 124, 437 (1986)

Analysis of the Excitation Energy Transfer in Spinach Chloroplasts at Room Temperature

Identification of the Component Bands by the Time-Resolved Fluorescence Spectrum and by Convolution of the Decay Kinetics

M. Mimuro[1], I. Yamazaki[2], N. Tamai[2], T. Yamazaki[2], and Y. Fujita[1]

[1]National Institute for Basic Biology, Myodaiji, Okazaki, Aichi 444, Japan
[2]Institute for Molecular Science, Myodaiji, Okazaki, Aichi 444, Japan

1. Introduction

Photosynthesis is the chain of reactions which converts solar energy into chemical energy. This process is triggered by the absorption of light by photosynthetic pigments and subsequent energy transfer to reaction centers where photochemical reactions take place. In higher plants and algae which show oxygenic photosynthesis, two kinds of reaction centers are known, reaction center I (RC I) and reaction center II (RC II), and these RCs are connected by an electron transfer chain. In both RCs, the primary electron donor consists of two molecules of chlorophyll a (chl a), called special pair, and the primary electron acceptor, phaeophytin [1]. Three-dimensional structure of RC polypeptide and the arrangement of pigments are now clearly known by the X-ray analysis in 3 A resolution in the case of photosynthetic bacterium [2, 3]. The portion of pigments associated with RC is less than 1 %. Almost all the pigments act in antennae which transfer the energy to RC. These pigments are not free in vivo, but are bound to proteins giving rise to pigment proteins, called light-harvesting pigment protein complex. These complexes are organized to form higher order structures called photosystems I and II (PS I and II) for the energy transfer to the corresponding RCs. The mechanism of the energy transfer within the pigment proteins is the main subject of this article.

In the analysis of excitation energy transfer, fluorescence has been used as the essential index. Every pigment, except for carotenoids, fluoresce in the course of energy transfer. Development of picosecond laser and a time-correlated single photon counting method enables the direct measurement of time behaviour of fluorescence which precisely reflects relaxation processes of the excited molecules. This allows for the analysis of the energy transfer mechanism. Precise measurement of the decay kinetics will yield rate constants for individual relaxation processes, when experimental system is simple. However when plural components are present, measurement of the decay kinetics is not necessarily enough to clarify the time behaviour of individual components. Other indices are required for an exact description of the transfer process. Therefore, we tried to measure the time-resolved fluorescence spectrum in picosecond (ps) time range. This enables the clarification of the presence of component bands, and changes in the fluorescence spectrum more directly indicate the occurrence of energy transfer.

On the phycobilin-chl a system, we have measured the time-resolved fluorescence spectrum and described the rise and decay curve of individual components by deconvolution of the spectra [4, 5]. This treatment was successful because of a good separation of fluorescence for each component. In contrast, fluorescence bands are expected to overlap each other in the case of chl a/b system including spinach. Separation to component bands, that is, identification of component bands is the primary step for the analysis of the energy transfer. For this purpose, we tried two kinds of measurements. One is the time-resolved fluorescence spectra and the other is identification by convolution of the decay kinetics, as was reported by many authors [6 - 10]. Comparison of results obtained by two independent methods will give much clearer information on the presence of components and their contribution to the total spectrum, which is expected to be identical irrespective of the measuring methods.

2. Materials and method

Spinach chloroplasts were isolated basically following the procedure reported by Jensen and Bassham [11]. The isolation medium contained 50 mM tricine-NaOH (pH 7.5), 0.33 M sucrose, 10 mM NaCl and 2 mM $MgCl_2$. Subparticles were isolated by the method of Sane et al. [12] (see also ref. 13) with a combination of mechanical disruption and differential centrifugation.

Absorption and fluorescence spectra were measured with a Hitachi 557 dual wavelength spectrophotometer and a Hitachi 850 spectrofluorometer, respectively. When necessary, the base line correction and correction of the spectral sensitivity of the apparatus were done by a microcomputer. For the fluorescence polarization spectrum, glass plate polarizers were used. The correction factors of the apparatus for the polarized light [14] were estimated by separate measurements using concentrated rhodamine B.

Time-resolved fluorescence spectra were measured at 22°C by the apparatus reported earlier [4, 5]. Briefly, the light source was Ar-ion laser and the repetition rate was 82 MHz. This light is fed to a synchronously pumped dye laser. The repetition rate was reduced to 800 KHz by a cavity-dumper. Output pulses have a duration of 6 ps (FWHM). The light intensity was in a range of 10^8 to 10^9 photons/cm^2 per pulse. This is low enough to avoid singlet-singlet annihilation, which shortens the fluorescence lifetime. Fluorescence was detected by a time-correlated single photon counting method after being dispersed by a monochromator. The detector was a microchannel plate photomultiplier (R1294U, Hamamatsu Photonics), which has a better time resolution than usual photomultipliers. In most cases, the excitation pulse was detected as a pulse of 50 to 60 ps (FWHM). This value insures a time resolution better than 10 ps by convolution [15]. For the spectrum, decay kinetics at every 0.625 nm were measured by scanning monochromator and data were stored on floppy disc. After measuring a whole set of decay in a particular wavelength region, the time-resolved fluorescence spectrum was reconstituted [15]. No correction was made for the spectral sensitivity of the apparatus.

3. Results

3.1 Fluorescence properties of chloroplasts

Figure 1 shows absorption and fluorescence spectra of spinach chloroplasts at room temperature. In the absorption spectrum, the maximum due to the lowest excited le-

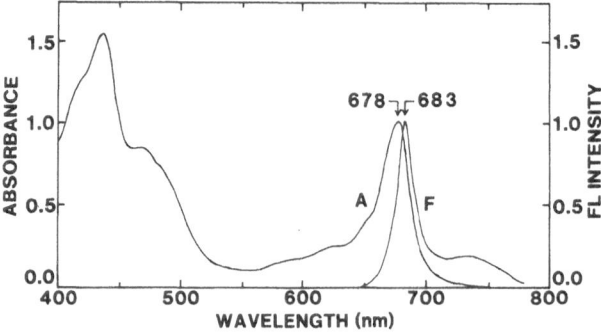

Figure 1. Absorption (A) and fluorescence (F) spectra of spinach chloroplasts at room temperature

vel (Q_y transition) was observed at 678 nm with an asymmetric band shape. This indicates the presence of plural component bands, each of them with a slightly different absorption maximum [16]. Excitation energy transfer occurs among these components. The fluorescence spectrum was a little sharper than the absorption band and its maximum was observed at 683 nm when excited at 435 nm. Vibrational structure was not clear except for the 720 nm band. It is well known that this spectrum is the mixture of fluorescence from both PSs [17] and the fluorescence yield of PS I is very low (about 6 % of that of PS II) [18]. Therefore, the spectrum can be interpreted as if almost all the fluorescence comes from PS II.

The time-resolved fluorescence spectrum of chloroplasts at room temperature is shown in Fig. 2. Numbers in the figure indicate the time after the laser peak in ps. The spectra were drawn after normalization to the maximum intensity in each spectrum. When chloroplasts were excited at 630 nm, the fluorescence maximum was observed at 684 nm. Other bands were also clearly detectable around 705 and 715 nm at 13 ps after the pulse. The maximum shifted to 683 nm within 25 ps and after this time, its location was constant throughout the measuring time upto 1.8 ns. The relative intensity in the wavelength region longer than 700 nm became weaker with time, suggesting shorter lifetimes for these components. The dotted line shows the spectrum at 1.4 ns after the excitation pulse. Compared with this spectrum, all spectra at longer than 200 ps after the pulse were similar. This indicates that the energy transfer completed within 200 ps and only the relaxation process was observed after this time. In this time range, only one fluorescence

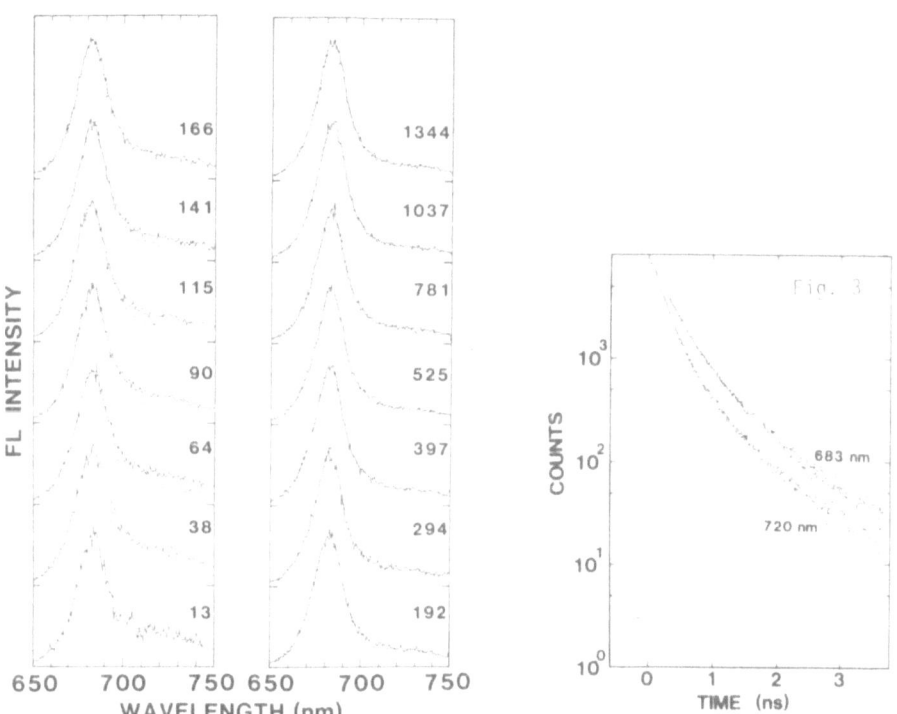

Figure 2 (left) Time-resolved fluorescence spectra of spinach chloroplasts at room temperature. Each spectrum was normalized to the maximum intensity. Numbers indicate the time after the laser pulse in ps.
Figure 3 (right) Decay kinetics of fluorescence of spinach chloroplasts at 683 and 720 nm.

maximum was observed. If plural components with different lifetimes would be present, they should show the same spectrum, suggesting the same origin of those components.

When we measured decay kinetics at fixed wavelengths (Fig. 3), changes in the
kinetics were small irrespective of the wavelengths for observation. It appears
that the overall decay in the wavelength region longer than 700 nm was faster than
that around 683 nm. This agrees well with the clear maximum around 705 and 715 nm
only in the initial time range after excitation.

Analysis of the decay kinetics was done by convolution. We tried to obtain
best fit values in two ways; in the first case, lifetime and amplitude were assumed
to be variables and in the second case, only the amplitude was a variable with the
assumption of fixed lifetimes. When we assumed three components, decay kinetics at
each wavelength were simulated with the chi square in a range of 1.032 to 1.193
(Fig. 4A). The lifetime of the fastest decay component was about 85 ps (Fig. 4B)
and its component spectrum, which was estimated by the amplitude as a function of
the wavelength, showed the maximum around 710 nm (Fig. 4C). This location is close
to that of the component found in the time-resolved spectrum, though two components
were observed in the time-resolved spectra. The lifetimes of second and third
components were about 370 ps and 1.2 ns, respectively and they showed the maxima
around 680 nm (Fig. 4B and 4C).

Two component bands were not resolved in the wavelength region longer than 700
nm even when four components were assumed for the convolution (Fig. 4E). Compared
with the convolution with three components, the chi square values became smaller
(in a range of 0.995 to 1.166). The lifetimes were about 50, 150, 450 and 1400 ps,

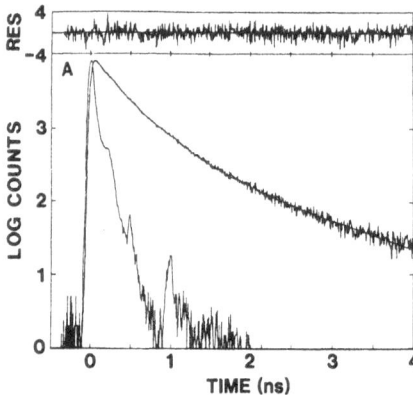

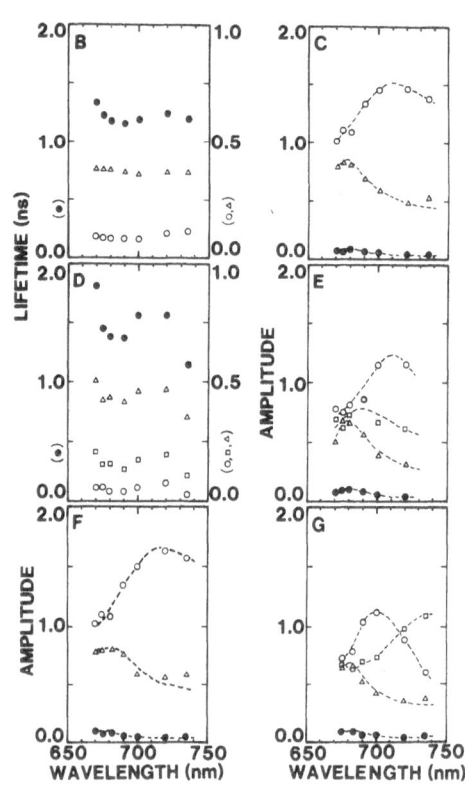

Figure 4 (A) Convolution pattern of
decay kinetics of spinach chloroplasts
at 683 nm. (B) Lifetimes of each component estimated by convolution with
three components and (C) those component spectra. (D and E) the same as
(B and C) but with four-component
convolution. (F and G) Component
spectra estimated with the fixed lifetimes by three and four components,
respectively. Symbols in both figures
correspond to each other.

though small variations were observed depending on the wavelengths (Fig. 4D). The maximum of the shortest lifetime component was located around 710 nm, as in the case of three-component simulation (Fig. 4B). The second component was observed around 690 nm, and the other two, around 680 nm. These indicate that increase in the number of components did not give consistent component bands with those observed by the time-resolved fluorescence spectra.

For the second analysis, we assumed fixed lifetimes of three components and tried to obtain the best fit values of the amplitudes to the observed fluorescence decay. As lifetime estimation, we simply used the average value of those obtained by the previous analysis (87, 372 and 1230 ps). The spectrum thus obtained was shown in Fig. 4F. The component spectra were essentially the same as those obtained by the first analysis. On four-component analysis, the same results were also obtained (Fig. 4G). The lifetimes used for the convolution were 51.4, 157, 436 and 1470 ps, and chi square was in a range of 1.001 to 1.350. The fastest component showed its maximum around 700 nm and the second fastest, at longer than 735 nm. These are not consistent with any of the previous spectra.

In the time-resolved spectrum of chloroplasts, a slight blue shift in the fluorescence maximum was observed (Fig. 2), which did not correlate with the energy transfer. This is partly because chloroplasts contain both PSs. The fluorescence yield of PS I is known to be about 6 % of that of PS II [18, 19], indicating very short lifetimes of PS I fluorescence. The short lifetime components were observed in the wavelength region longer than 700 nm. These components might originate from the PS I chl a. If this is the case, the fractionation into component particles give much clearer results. When detergents are used for the preparation, samples in a higher purity are expected. However, even the lowest concentration of detergent always brings about changes in the spectral properties of the photosynthetic pigments. Therefore, we did not use detergent at all to isolate the subparticles. Because of this, the purity of our samples was not necessarily high; normally 84 % for PS I particles and 74 % for PS II particles (cf. ref. 13).

3.2 Fluorescence properties of each subparticle

Absorption spectra of the separated subparticles were slightly different from that of chloroplasts. The maxima of PS I and II particles were observed at 680 and 677 nm, respectively (data not shown). Fig. 5 shows the steady state fluorescence and fluorescence polarization spectra of the PS I and II particles at room temperature. The maximum intensity per unit amount of chl of the PS I particles is about 25 % of that of the PS II particles, because of the contamination of high fluorescent PS

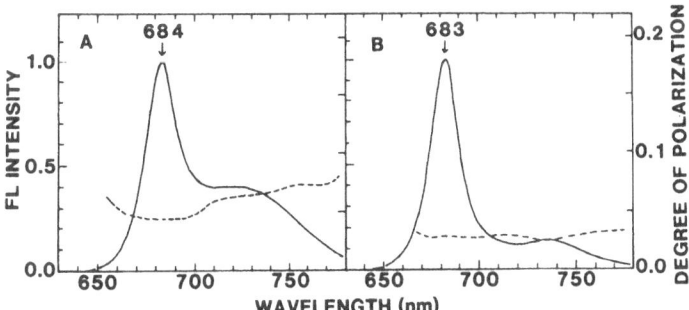

Figure 5. Fluorescence and their polarization spectra of PS I (A) and PS II (B) particles at room temperature. Excitation was 435 nm. The maximum intensity of the PS I particles was one-fourth of that of the PS II particles, when expressed based on the unit amount of chl.

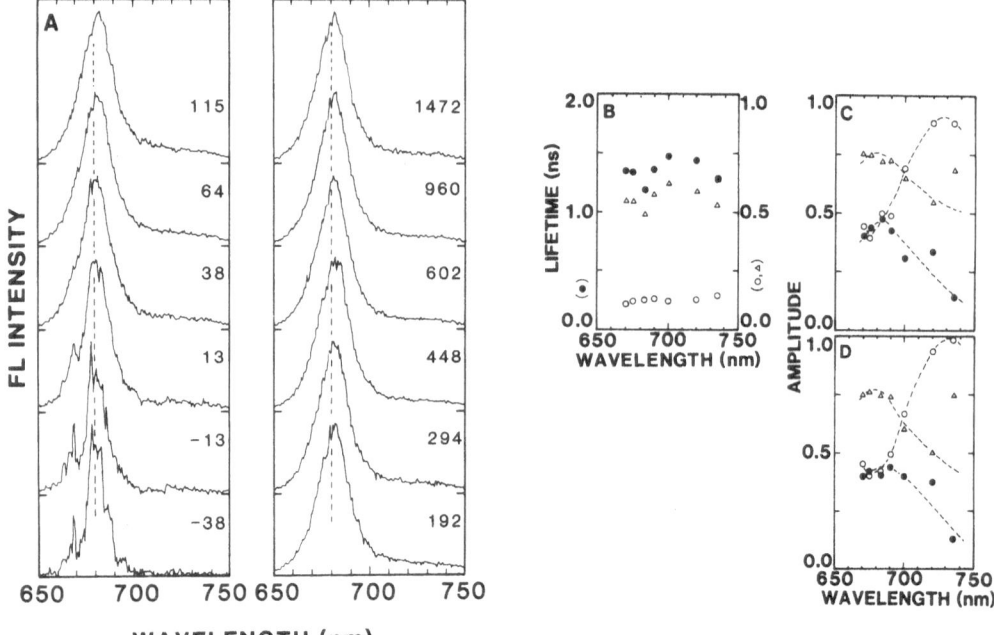

Figure 6. (A) Time-resolved fluorescence spectra of PS II particles at room temperature. (B) Lifetimes and (C) component spectra obtained by convolution with three components, and (D) component spectra simulated with three components of fixed lifetimes.

II. The fluorescence maximum of PS II particles was located at 683 nm, as in the case of chloroplasts, on the other hand, that of the PS I particles, at 684 nm. The intensity in the wavelength region longer than 700 nm was evidently higher than that of PS II particles. The fluorescence polarization spectra indicate that the degree of polarization of PS II particles was low and almost constant in the whole wavelength region of fluorescence. This suggests that the fluorescence originates from one component only. Contrary to this, the degree of polarization of PS I particles was not constant, significantly higher in the longer wavelength region, indicating the presence of additional band(s). This indicates that PS I particles show their own fluorescence properties, even if more than half of the fluorescence of the PS I particles originates from the contaminating PS II.

Time-resolved fluorescence spectrum showed the occurrence of energy transfer in PS II particles (Fig. 6A). When the particles were excited at 630 nm, fluorescence maximum was observed at 680 nm. In the time-resolved spectrum, time zero is set to the maximum of the excitation pulse. The maximum at 680 nm was observed at -38 ps, that is, before the pulse maximum. In most cases, the short lifetime component could be discriminated only in an earlier time range. The 680 nm maximum shifted to 683 nm within 50 ps. A small bump around 670 nm was also detectable in the spectrum after 64 ps, and the spectra after 64 ps were almost the same as that at 1.5 ns, indicating no further changes in the spectrum. The intensity in the wavelength region longer than 700 nm was low. This suggests that the origin of the component band(s) is PS I. The 680 nm fluorescence corresponds to the maximum of the isolated LHC II, that is, the light-harvesting antenna for PS II [20] and the 683 nm fluorescence arises from the RC II complexes [20], as seen in the steady state measurement and time-resolved spectrum of chloroplasts (Fig.

2). Therefore, the shift of the maximum indicates the energy transfer from LHC II to RC II complexes, and hereafter only the relaxation process was observed.

Convolution was done on the decay kinetics at several wavelengths (Fig. 6B). In this case, three components were enough to describe the decay, because four-component analysis did not improve the fit; chi square at 683 nm was 1.007 for three components and 1.008 for four components. The shortest lifetime was about 115 ps, a little longer than that in chloroplasts. The spectrum of the fastest component showed the maximum around 725 nm (Fig. 6C). The second and third components showed lifetimes of about 550 and 1350 ps and the maximum was observed around 680 nm. Convolution by three components with fixed lifetimes (120, 551 and 1340 ps) gave essentially the same spectra (Fig. 6D) as those obtained in the previous analysis (Fig. 6C). The shortest lifetime component shows the maximum around 730 nm, and the other two, around 680 nm. These spectra are again inconsistent with the time-resolved spectrum.

Figure 7A shows the time-resolved fluorescence spectrum of PS I particles at room temperature. Compared with the spectra of chloroplasts (Fig. 2) and PS II particles (Fig. 6A), many bands were found especially in the initial time range after the excitation. At least three components were detected at 688, 698 and 705 nm at -99 ps, and the band at 715 nm was also clear. The intensity of the 688 nm component decreased earlier than the other two components, and the 698 nm component disappeared faster than the 705 nm component, as typically seen in the spectrum at -20 ps. These three components disappeared soon, therefore by 0 ps, the maximum shifted to 683 nm, where the PS II shows its fluorescence maximum. The

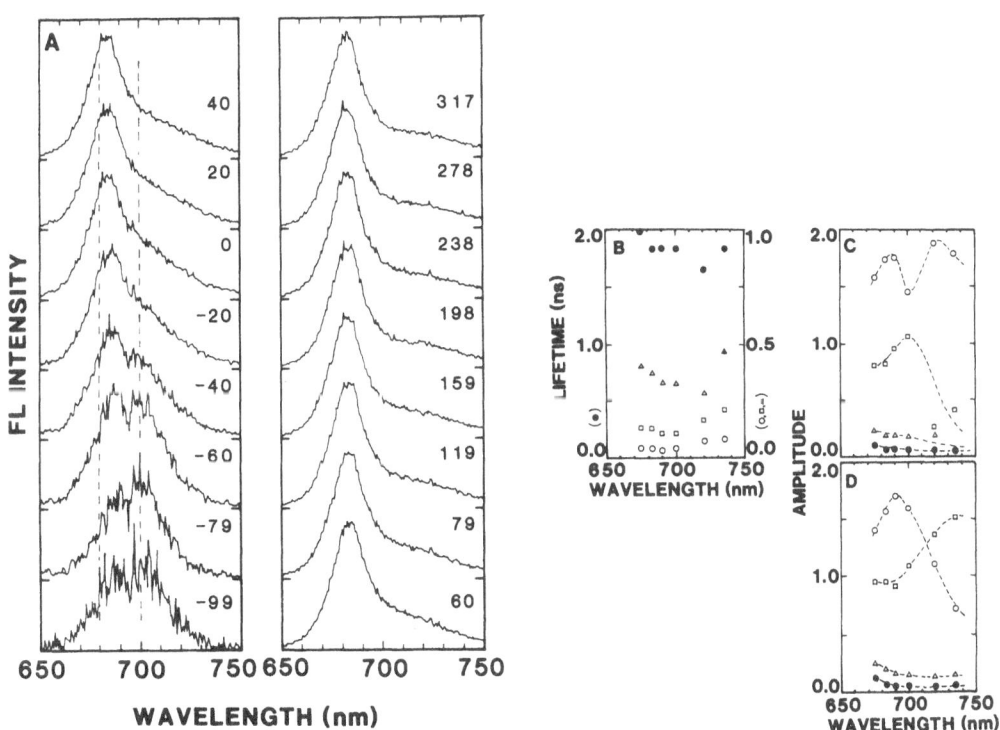

Figure 7. (A) Time-resolved fluorescence spectra of PS I particles at room temperature. (B) Lifetimes and (C) component spectra obtained by convolution with four components, and (D) component spectra simulated with four components of fixed lifetimes.

intensity around 700 nm is still high around 0 ps. In a later time, however, the intensity became lower and consequently the spectrum became similar to the spectrum of PS II particles (cf. Fig. 6A). The longer lifetime component originates most probably from the contaminating PS II. When we see the overall changes, the blue shift of the maximum occurred. However, this is simply due to the slow rise and decay of the PS II fluorescence. We can conclude that in the PS I, energy transfer occurred from the 688 nm component to the 698 nm component, and then to the 705 nm component, and that those fluorescence bands showed very short lifetimes.

On the PS I particles, convolution with four components was applied. Three components did not give good fit in terms of chi square; at 683 nm, chi square was 1.121 for three components and 1.008 for four components. The shortest lifetime was about 35 ps (Fig. 7B) and showed two maxima in the spectrum (Fig. 7C). The presence of this short lifetime component agrees with the low fluorescence yield of PS I chl a. The second component whose lifetime was about 120 ps showed its maximum around 700 nm. The lifetime of this component is very close to the fastest component in PS II particles, however the spectrum was different from that observed in PS II particles. The third and fourth components have lifetimes of about 400 ps and 1.8 ns and both have maxima in the shorter wavelength region. These spectra were again inconsistent with the time-resolved spectrum. The two peaks found in the fastest component indicate the coincidently identical or similar lifetime components, even if the convolution was good enough. We found at least three fast components in PS I, that is, at 688, 698 and 705 nm. They could not be reproduced by the convolution.

Increase in the number of components did not improve the simulation (data not shown). When the decay kinetics were simulated by five components, the variation of the lifetimes at each wavelength became smaller. In this case, simulation gave one negative amplitude, that is, rise term of the fluorescence, as the shortest lifetime component. However, corresponding decay component was not obtained in this simulation. Therefore, we could not exactly elucidate the transfer sequence from these lifetimes.

Inconsistency was not resolved even by the convolution with fixed lifetimes (37.8, 112, 361 and 1860 ps) (Fig. 7D). The shortest lifetime component showed the single maximum around 692 nm, inconsistent with the previous data. The maximum of the second component was longer than 735 nm. The other two have the maxima at shorter than 675 nm. These indicate that the convolution does not necessarily give identical component spectra to those obtained from the time-resolved spectrum.

4. Discussion

Time-resolved fluorescence spectrum clearly shows the presence of component bands; for whole chloroplasts, the 683, 705 and 715 nm components, for PS II, the 680 and 683 nm components, and for PS I, the 688, 698, 705 and 715 nm components. These components were observed by the excitation at 630 nm, where all the chl absorbed light depending on the absorption cross section of respective pigment. A part of the energy was transferred to the other component(s) and a part of it was emitted directly from the pigments which captured energy.

In PS II, the red shift of the maximum was observed from 680 to 683 nm within 50 ps, which reflects energy transfer from LHC II chl a to RC II complex (Fig. 6A). Gillbro et al. [21] reported the transfer time in the LHC II; it takes 6 ps from chl b to chl a and within next 20 ps, redistribution of excitation energy among the chl a molecules takes place. The energy transfer from LHC II chl a to PS II chl a is the next step of the redistribution of the energy. Therefore, our estimation (shorter than 50 ps) was close to the expected time. The time-resolved spectrum showed only one component at longer than 200 ps after the pulse. However, convolution gave two components in this time range irrespective of the number of components; those two showed different lifetimes but the same fluorescence maximum. Heterogeneity of the origin of the fluorescence components is needed to be assumed for the explanation of both phenomena. The additional component around 670 nm was not necessarily clear in the spectra at earlier times. It might be related to uncoupled chl a, artificially formed during the preparation of particles.

In PS I, at least four components were detectable in an initial time range at 688, 698, 705 and 715 nm (Fig. 7A). The former three participate in the energy transfer in this order, judging from the relative intensities of those component bands. The band at 715 nm was sometimes observed around 720 nm [22] and its life-time was relatively longer, compared with the other three. Therefore, its function in the energy transfer was not clearly explainable. The lifetimes of the fast components could not be estimated by the convolution. The 35 and 120 ps components did not correspond to three faster components in PS I. In the spectra, those disappeared within 100 ps.

The component spectra estimated by the convolution were not consistent with those by the time-resolved fluorescence spectra, irrespective whether fixed or variable lifetimes for the components were used. We assumed exponential decay and the number of components for the convolution; three for the case of PS II particles and four or five for PS I particles. Increase in the number of the components brought about the smaller chi square values as an index for a better fit. However, as typically shown in the PS I particles, plural short lifetime components were not reproduced by the convolution analysis. This result also raises the reliability of lifetimes in the case of spinach chloroplasts.

Other groups have measured the fluorescence decay of spinach chloroplasts at room temperature and obtained lifetimes and component spectra [6 - 10] with the same assumption as ours. The lifetimes of the stroma lamellae fraction, comparable to our PS I particles, were very close to our estimation (35, 120, 400 and 1800 ps) except for the longest lifetime component (3300 ps [7]). The component spectra of the shortest lifetime component were always observed at a longer wavelength (690 nm [9] or 700 nm [10]) than those of other components (around 683 nm). These results were also essentially identical to our convolution, except that in our case, a maximum at longer than 730 nm was observed. These ensured that the difference in the component spectrum is not due to the difference in the measuring system used by each group. The main reason for the discrepancy in the component spectrum estimated by two methods is most probably an assumption of the decay function. Therefore, this point should be considered more critically.

Two decay functions have been assumed to describe the time behaviour of fluorescence in the photosynthetic pigment system; one is exponential decay and the other, decay function proportional to the square root of time. Application of the decay function to the observed decay depends on the organization and/or the manner of arrangement of pigments. Recently, Baumann and Fayer [23] reported the theoretical basis of non-exponential decay kinetics in connection with the spatial geometry. At the moment, we do not have a fair basis for why only one decay function is applicable to the in vivo system. It is possible that the observed decay kinetics are a mixture of two decay functions. Inconsistency of the component spectrum obtained by using exponential decay function might be an indication of more complicated decay functions. For the analysis of the energy transfer, we have to describe the rise and decay kinetics of a particular component, not of the decay kinetics at particular wavelength, as precisely as possible. This will lead to a clearer answer.

5. Acknowledgements

The authors express their cordial thanks to the Instrument Center, Institute for Molecular Science for operation of the picosecond laser system. They also thank Dr. Itoh, S., National Institute for Basic Biology for his discussion during this work, and Dr. A. Post, University Amsterdam for his critical reading of the manuscript.

6. Literature

1. Rutherford, A. W. and Heathcote, P. (1985) Photosynthesis Res., 6, 295-316.
2. Deisenhofer, J., Epp, O., Miki, K., Huber, R. and Michel, H. (1984) J. Mol. Biol., 180, 385-398.
3. Michel, H. Epp, O. and Deisenhofer, J. (1986) EMBO J., 5, 2445-2451.
4. Yamazaki, I., Mimuro, M., Murao, T., Yamazaki, T., Yoshihara, K. and Fujita, Y. (1984) Photochem. Photobiol., 39, 233-240.
5. Mimuro, M., Yamazaki, I., Tamai, N., Yamazaki, T. and Fujita, Y. (1985) Photochem. Photobiol., 41, 597-603.
6. Haehnel, W., Nairn, J. A., Reisberg, P. and Sauer, K. (1982) Biochim. Biophys. Acta, 680, 161-173.
7. Holzwarth, A. R., Haehnel, W., Wendler, J., Suter, G. and Ratajczak, R. In Advances in photosynthesis research, (Sybsma, C. Ed.), Vol. 1, p. 73-76. Marinus Nijhoff/Dr, Junk Publishers, The Hague, The Netherlands, 1984.
8. Holzwarth, A. R., Wendler, J. and Haehnel, W. (1985) Biochim. Biophys. Acta, 807, 155-167.
9. Hodges, M. and Moya, I. (1986) Biochim. Biophys. Acta, 849, 193-202.
10. Wittersmann, E., Senger, H. and Holzwarth, A. R. In Proceedings of 7th International Congress on Photosynthesis, In the press.
11. Jensen, R. G. and Bassham, J. A. (1964) Proc. Natl. Acad. Sci. USA., 56, 1095-1101.
12. Sane, P. V., Goodchild, D. J. and Park, R. B. (1970) Biochim. Biophys. Acta, 216, 162-178.
13. Mimuro, M., Tamai, N., Yamazaki, T. and Yamazaki, I. FEBS Letters, in the press.
14. Azumi, T. and McGlynn, S. P. (1962) J. Chem. Phys., 37, 2413-2420.
15. Murao, T., Yamazaki, I. and Yoshihara, K. (1982) Appl. Opt., 21, 2297-2298.
16. French, C. S. (1970) Proc. Natl. Acad. Sci. USA., 68, 2893-2897.
17. Ross, R. T. and Calvin, M. (1967) Biophys. J., 7, 595-614.
18. Kyle, D. J., Baker, N. R. and Arntzen, C. J. (1983) Photobiochem. Photobiophys., 5, 79-85.
19. Beddard, G. S., Fleming, G. R. Porter, G., Searle, G. F. W. and Synowiec, J. A. (1979) Biochim. Biophys. Acta, 545, 165-174.
20. Murata, N. and Satoh, Ki. In Light emission by plants and bacteria, (Govindjee, Amesz, A. and Fork, D. C. Eds.), Academic Press, New York, 1986.
21. Gillbro, T. Sundstrom, V., Sandstrom, A., Spangfort, M. and Anderson, B. (1985) FEBS Letters, 193, 267-270.
22. Mimuro, M., Yamazaki, I., Tamai, N., Yamazaki, T. and Fujita, Y. In Proceedings of 7th International Congress on Photosynthesis, in press.
23. Baumann, J. and Fayer, M. D., (1986) J. Chem. Phys., 85, 4087-4107.

Photonic Energy Transport
in Phycobilin-Chlorophyll System:
A Comparative Study with Artificial Multilayer Films

I. Yamazaki[1], N. Tamai[1], T. Yamazaki[1], M. Mimuro[2], A. Murakami[2], and Y. Fujita[2]

[1]Institute for Molecular Science, Myodaiji, Okazaki, Aichi 444, Japan
[2]National Institute for Basic Biology, Myodaiji, Okazaki, Aichi 444, Japan

1. Introduction

Transport and trapping of electronic excitation-energy have long been the subject of extensive theoretical and experimental works [1,2]. Special attention has recently been paid to the excitation energy in some molecular assemblies of some biological systems as well as artificial organizates. Photosynthetic light-harvesting antenna in plants is characterized by highly efficient absorption and subsequent transport of excitation energy to the reaction center. The antenna pigment systems in red and blue-green algae have accessary pigments, phycobilins, as well as chlorophylls, which are attached on thylakoid membranes [3,4]. A schematic illustration of the structure of phycobilizome is shown in Fig. 1(a). Very little is known about the dipole-dipole resonance (Förster-type) energy transfer in restricted molecular geometries such as biological systems and one- and two-dimensional molecular arrangements. In theoretical approaches, the time-dependent equations for the donor fluorescence decay have been proposed for low dimensional systems by HAUSER et al. [5], ZUMOFEN and BLUMEN [6] and BAUMANN and FAYER [7] The present paper is concerned with the sequential energy transfer in the phycobilin-chlorophyll systems in some algae and also in an artificial Langmuir-Blodgett (LB) multilayers as is shown in Fig. 1(b). A comparative study of the sequential energy transfer has been made by means of a picosecond time-resolved fluorescence spectroscopy.

2. Picosecond Time-Resolved Fluorescence Spectrophotometer

Fluorescence decay curves and time-resolved spectra were measured with a picosecond time-resolved emission spectrophotometric system [8]. The picosecond laser system as

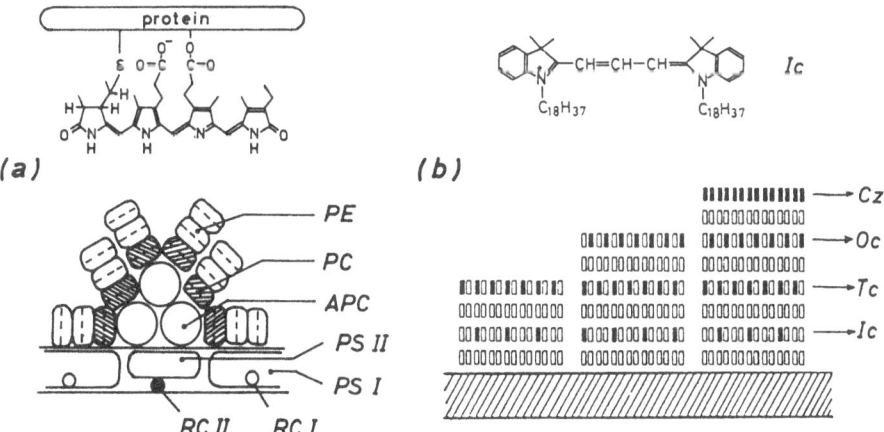

Fig. 1. Schematic representation of multilayered structures of the biological antenna of algae (a) and of the artificial LB films (b). The pigment chromophores are shown above.

an excitation source is composed of a synchronously pumped, cavity-dumped dye laser (Spectra Physics 375 and 344S) and a mode-locked argon-ion laser (Spectra Physics 171-18). The dye laser was operated with a repetition rate of 800 kHz and a single pulse duration of 6 ps (fwhm). The fluorescence decay curves were obtained with a time-to-amplitude converter and a multichannel pulse-height analyzer. A microchannel plate photomultiplier (Hamamatsu R1564-U) was used, which allows us to obtain an instrument response function with 40-ps width (fwhm) for the scattered laser light. The fluorescence lifetime can thus be determined down to 3 ps with an accuracy of ±0.5 ps. The time-resolved spectra were obtained with the minimum time difference of 0.8 ps by plotting the fluorescence intensities at particular delay times as a function of wavelength.

3. Sequential Energy Transfer in Phycobilin-Chlorophyll Systems of Algal Intact Cells

A light-harvesting antenna system of red and blue-green algae consists of phycobilisomes and thylakoid membrane. The phycobilisome is a supramolecular unit involving several kinds of phycobiliproteins having open-chain tetrapyrroles as chromophores (Fig. 1(a)) [3,4]. In the course of the energy transfer from the initially photoexcited phycobiliprotein to the reaction centers (RC) I and II, fluorescence is emitted from almost every type of pigment and can be used as a probe for investigating the mechanism of energy transfer inside the phycobilisome. Recently, a dynamical aspect of this sequential energy transfer has been studied by using a picosecond time-resolved fluorescence spectroscopy [9-15]. In this paper, the time-resolved fluorescence spectra and the fluorescence decay characteristics are reviewed for different kinds of phycobilisomes including species which was obtained by the chromatic adaptation; i.e., a phycoerythrin (PE)-less system grown under red light and a PE-rich system grown under green light [12,16]. Structures of these two phycobilisomes are presented schematically in Fig. 2. The time-behaviors and the rate constants of energy transfer are compared between these two types of phycobilisomes.

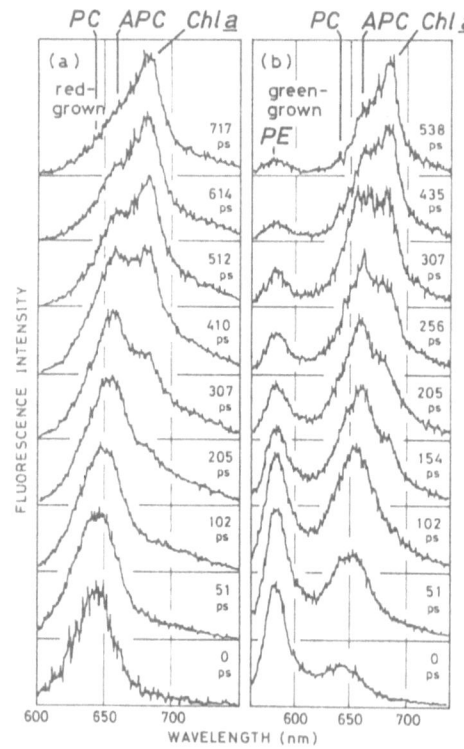

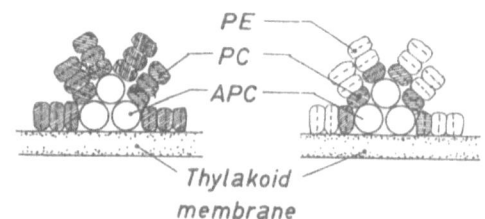

Fig. 2. Time-resolved fluorescence spectra of Flemyella diplosiphon: (a) red-grown (PE-less) and (b) green-grown (PE-rich) intact cells. The schematic representation of the phycobilisome structures are shown above for the red-grown (left) and the green-grown (right) species.

Algal cells were grown autotrophically in the medium of ASP-2 or in the modified Detmer's medium under the continuous illumination of an incandescent light of 2.5 W/cm^2. Cells at the late log-growth phase were used for all measurements. Flemyella diplosiphon (M-100) was grown under red light (2.5 W/cm^2) for establishing the PE-less system and under green light (1.0 W/cm^2) for the PE-rich system. Details of the sample preparation are described elsewhere [11,12,16].

Figure 2 shows the time-resolved fluorescence spectra of two kinds of Flemyella diplosiphon intact cells, i.e., red-grown cells (PE-less) and green-grown cells (PE-rich). The spectrum of PE-less system (Fig. 2(a)) changes with time as follows: (1) Phycocyanin (PC) spectrum appears at 0-100 ps, with its peak being shifted gradually to the red, (2) at 100-400 ps, allophycocyanin (APC) spectrum becomes dominant and then (3) at 500 ps, Chl _a_ spectrum appears clearly and no longer changes after 700 ps. The PE-rich system (Fig. 2b) exhibits PE spectrum in the initial time region (0-200 ps) in addition to sequential appearance of the PC, APC and Chl _a_ spectra.

The time-resolved spectra were analyzed into components to obtain the rise and decay curve of the pigment component itself. Fig. 3 shows the results of the analysis. It can been seen that (1) the fluorescence rises in particular pigments are delayed with its delay time being longer on going from the outer surface to the inner core of the pigment system; (2) in every case the rise time is significantly short compared to the decay time; and (3) fluorescence bands of various pigments appear sequentially in the order of PE-PC-APC-Chl _a_. These sequential time behaviors can be fitted approximately with the decay kinetics of $\exp(-2kt^{1/2})$ type, the exact expressions of which will be given in the next section. The values of the rate constants (k) thus obtained are summarized in Table 1 for different types of phycobilisomes. It is found that the energy transfer at PC level is faster in PE-rich system, whereas those at APC and at Chl _a_ are almost identical. The decay kinetics of $\exp(-2kt^{1/2})$ type can be derived from the Förster kinetics under the condition that the energy transfer occurs with an extremely high efficiency between the donor and acceptor chromophores. This result may suggest that the chromophores are distributed in a special array to form a special channel for the sequential energy transport in the phycobilisomes [12]. Detailed discussion will be given in the next section.

It is seen from Table 1 that there is a regularity in k values among various species: (1) In the case of PE-less systems, the energy transfer of PC → APC is slower than that of APC → Chl _a_, and the former takes almost the same rate constant except for P. cruentum. (2) In the case of PE-rich systems, the process PC → APC is faster than that in PE-less system and is even faster than the process APC → Chl a

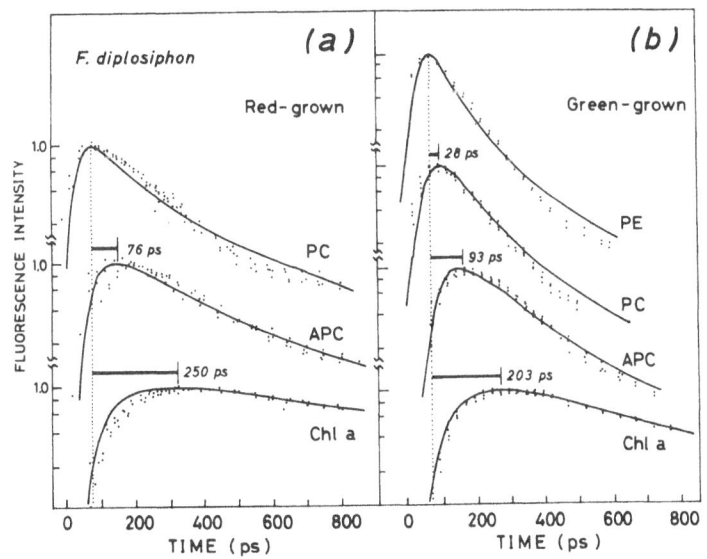

Fig. 3. Fluorescence rise and decay curves for the individual pigments of F. diplosiphon; (a) red-grown (PE-less) and (b) green-grown (PE-rich) species. The dotted lines are obtained experimentally from the spectrum analysis. The solid lines are the best fit curves calculated from (8)-(11).

Table 1. Rate constants of the energy transfer in phycobilin-Chl a system of the algal intact cells

| Pigment system | Algae | Rate constants (k, ps$^{-1/2}$)[a] | | | |
		PE	PC	APC	Chl a
PC-APC-Chl a	Flemyella diplosiphon	–	0.071	0.13	0.042
	Anabaena cylindrica	–	0.065	0.33	0.059
(PE-less)	Anabaena variabilis	–	0.077	0.14	0.057
	Porphyridium aerugineum	–	0.12	0.13	0.091
PE-PC-APC-Chl a	Flemyella diplosiphon	0.11	0.40	0.13	0.056
(PE-rich)	Nostoc sp.	0.17	0.29	0.13	0.063
	Porphyridium cruentum	0.14	0.24	0.27	0.10

a) Rate constants (k) were calculated with the decay kinetics of $\exp(-2k^{1/2})$.

in blue-green systems. It is to be noted that the transfer rate is not necessarily higher at later steps within phycobilisomes. (3) The transfer from APC takes an almost identical rate constant except for two cases. (4) Decays of Chl a fluorescence are similar among blue-green systems, but it is slightly faster in red algal systems. It will depend on the primary photoreaction at the reaction center and the spill-over to photosystem I. Some difference in the phycobilin structure may be suggested between the two algal groups. In conclusion, deviation in kinetic feature is generally small, or negligible, among algal systems of the same type of pigment composition.

The energy transfer in PE-rich systems is characterized by a rapid transfer at PC level in comparison with PE-less systems (Table 1). We should note here that difference in the global structure of phycobilisome little affects the energy transfer kinetics; phycobilisome of P. cruentum is hemispherical, whereas that of Nostoc sp. is hemidiscoidal [17]. Therefore, it is most likely that the presence of PE in the outer surface of phycobilisome is an important factor in the energy transfer process PC ⟶ APC. This is directly proved from the present experiments with the two systems of Flemyella diplosiphon chromatically adapted.

4. Sequential Energy Transfer in LB Multilayer Films

The LB film is a mono- or multi-layered molecular assembly which is prepared by transferring a compressed monolayer spread on a water surface onto a substrate [18]. In the LB film, one can expect to observe new aspects of photochemical and photophysical processes of electronically excited molecules, because a uniform thin monolayer or its multilayers with the two- or three-dimensional order might show configurational and dynamical behaviors quite different from those of homogeneous solution or from the bulk properties of crystals. One can prepare the LB film as multilayered architecture containing chromophoric donor and acceptor molecules with their number densities and interlayer distance being variable in a wide range. In 1967, BÜCHER et al. [18,19] prepared the LB films from successive deposition of monomolecular layers containing cyanine dyes, and examined efficiencies of the interlayer energy transfer on the basis of steady-fluorescence intensity measurement. Recently, LEITNER et al. [20] have examined the same problem with the ps fluorescence decay analyses. In this study, the sequential energy transfer has been studied by means of a ps time-resolved fluorescence spectroscopy for the LB multilayers up to four layers of donor and acceptor. The results were compared with previous studies of the biological antenna.

Three types of the LB films were prepared, consisting of sequences of donor (D) and acceptors (A_1, A_2 and A_3); namely, D-A_1 (hereafter referred to as 2L), D-A_1-A_2 (3L), D-A_1-A_2-A_3 (4L). The structures of these films are illustrated schematically in Fig. 4. Six to eight layers were deposited on a quartz plate in the following

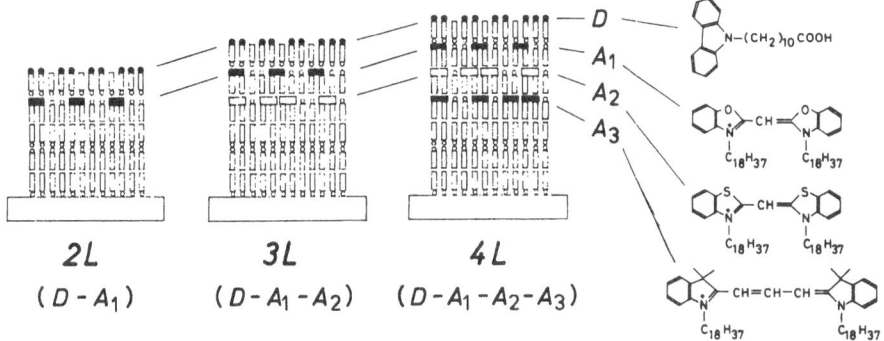

Fig. 4. Schematic illustration of stacking structures of the LB multilayer films used in this study. The concentration of pigment molecules is 5 mol% in each monolayer. The absorption bands corresponding to the lowest excited singlet state are at 350 nm in D, 390 nm in A_1, 440 nm in A_2 and 560 nm in A_3. The distance between donor and acceptor chromophores in adjacent layer is approximately 25 Å.

order: (1) five layers of parmitic acid – cadmium salt, (2) monolayer(s) consisting of parmitic acid and small amounts of dyes (A_1, A_2 and A_3) and (3) monolayer of carbazole (D) and (4) a monolayer of parmitic acid. An outer layer of parmitic acid prevents the multilayered structure from being destroyed. The concentration of pigment molecules was 5 mol% in each layer.

As pigment molecules for D, A_1, A_2 and A_3 layers, we used, respectively, 11-(9-carbazole)undecanoic acid (Molecular Probes, Co., U.S.A.), 1,1'-dioctadecyloxocyanine (Nippon Kankoh Shikiso Kenkyusho, Co., Okayama, Japan), 1,1-dioctadecylthiacyanine (Nippon Kankoh Shikiso), 1,1'-dioctadecyl-3,3,3',3'-tetramethylindocarbocyanine (Molecular Probe). Parmitic acid (Wako Chemical Co. Osaka) was purified five times by recrystallization from ethanol. Mixtures of parmitic acid and small amounts of pigment dissolved in chloroform were spread onto the surface of water subphase containing 3 x 10^{-4} M $CdCl_2$. The subphase conditions were adjusted to a temperature of 17 C and pH 6.3 by adding $NaHCO_3$ buffer solution. Mixed monolayers were deposited on a quartz plate under a constant surface pressure of 2.5 x 10^{-2} Nm^{-1}. The quartz plates used were precoated with five monolayers of parmitic acid-cadmium salt to minimize influence of the substrate surface on energy transfer kinetics.

Figure 5 shows the time-resolved fluorescence spectra of the LB multilayer films. Each spectrum is normalized to the maximum intensity. It is seen that the spectrum changes significantly with time in the picosecond time range. In 2L, following excitation of the D layer at 300 nm, a fluorescence band of D appears at 350 nm, and then A_1 band rises after 100 ps. In 3L, the fluorescence bands of D and A_1 appear in the initial time region, and then A_2 band appears at 470 nm after 10 ps. Similar spectral change can be seen in 4L. In the initial time region, the fluorescence bands due to D, A_1, A_2 appear, and A_3 band rises after 50 ps. It is seen that the fluorescence from the inner layer rises more slowly than those of the outer layers. It is pronounced in the rise of A_1 in 2L, A_2 in 3L and A_3 in 4L. Also noteworthy is that the fluorescence bands of donor and acceptors appear simultaneously just after the excitation; e.g., A_1 band in 2L, 3L and 4L, and A_2 band in 3L and 4L. This is not a consequence of the direct excitation of these acceptors by 300-nm irradiation, but of rapid energy transfer among the multilayer films, because the ratios of the absorbance of the acceptors to that of the D layer are very small (less than 0.05) at this wavelength.

Figures 6(a), (b) and (c) show the fluorescence rise and decay curves for each of the layers in 2L, 3L and 4L, respectively. Similarly to the case of the biological

37

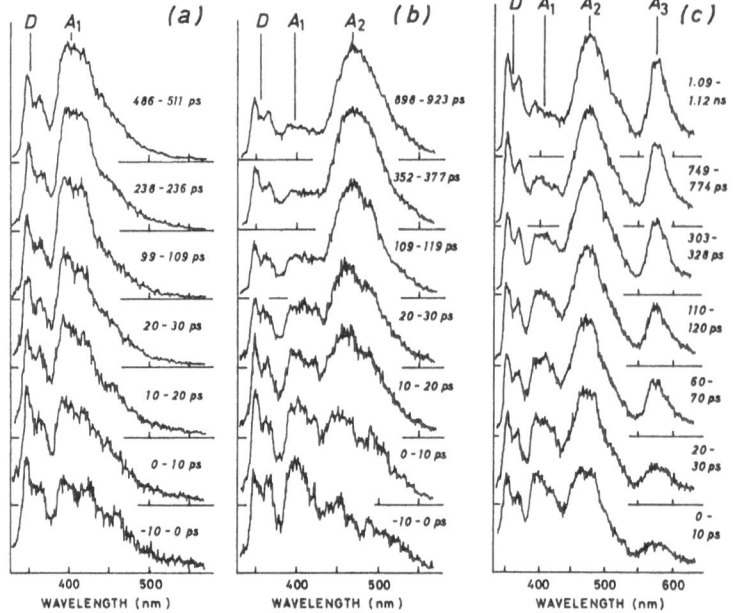

Fig. 5. Time-resolved fluorescence spectra of the LB multilayer films; (a) 2L, D-A$_1$; (b) 3L, D-A$_1$-A$_2$; (c) 4L, D-A$_1$-A$_2$-A$_3$. The excitation wavelength is 300 nm. The structures of these LB films are given in Fig. 4.

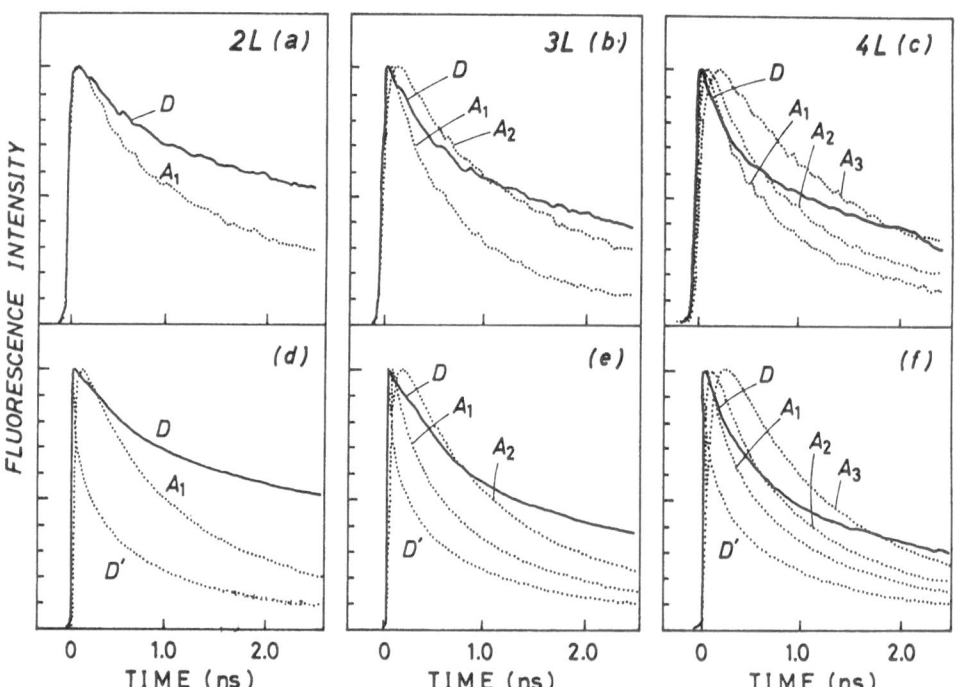

Fig. 6. Fluorescence rise and decay curves (a-c) and theoretical best-fit curves (d-e) of the LB multilayer films of 2L, 3L and 4L. The theoretical curves are obtained from (8)-(11). The decay curves D, A$_1$, A$_2$ and A$_3$ are obtained by monitoring the fluorescence at 350, 420, 470 and 570 nm, respectively.

antenna, all the decay curve profiles are not simple exponential forms and seem to be very complex. In every case, the decay curve of D layer includes a slowly decaying component. It appears that the decay curve of D layer, which is slower in 2L, becomes faster in 3L and 4L. Similarly, the decay of A_1 in 3L and 4L is faster than that in 2L. These facts show that the energy transfer takes place sequentially through D, A_1, A_2 to A_3. According to a curve fitting on the basis of superposition of exponential decays, we can fit only by using four exponentials. The parameter values thus obtained are not systematic throughout the samples concerned here.

Let us consider the decay function of each pigment in the sequential energy transfer. First we should consider the energy transfer in the two-dimensional systems, as is shown schematically in Fig. 7(a). According to HAUSER et al. [5], the donor fluorescence decay function for the two- and three-dimensional systems, in which acceptor molecules are randomly distributed surrounding a donor, are expressed as follows:

$$\rho(t) = \exp[-t/\tau_D - 2g\gamma_A(t/\tau_D)^{1/3}], \quad \text{for two-dimensions} \quad (1)$$

$$\rho(t) = \exp[-t/\tau_D - 2g\gamma_A(t/\tau_D)^{1/2}], \quad \text{for three-dimensions} \quad (2)$$

where

$$\gamma_A = (2/3)\pi n_A R_0, \quad \gamma_A = (2/3)\pi^{3/2} n'_A R_0. \quad (3)$$

τ_D is the lifetime of the donor without acceptor; g is the factor determined by the molecular orientation; n_A and n'_A are the number densities of acceptors in unit area and unit volume, respectively; and R_0 is the critical transfer distance where the rate constant for the energy transfer is equal to that for fluorescence of the donor in the absence of acceptors. Both two equations (1) and (2) are expected to be related to the present case of the stacking structure, as is shown in Fig. 7(b). According to a recent theoretical study by BAUMANN and FAYER [7], the decay curve gradually changes from two- to three-dimensional characters as one increases the number of acceptor layers.

2-D System   Multilayer system

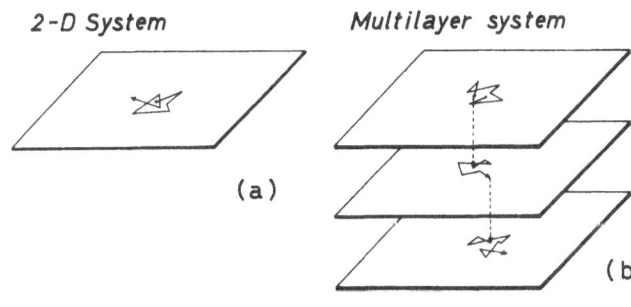

(a)

(b)

Fig. 7. Energy transfer under the restricted geometries of (a) two-dimensional monomolecular layer and (b) stack of monomolecular layers.

In 1978, PORTER et al. [9] examined the sequential fluorescence decay kinetics for intact cells of P. cruentum and proposed that the fluorescence time behaviors of each pigment can be expressed with a decay function of $\exp(-2kt^{1/2})$. This type of equation corresponds to those for the three-dimensional system. If the γ value is large enough to be $2\gamma > (k_D t)^{1/2}$, the exponential term in (2) can be approximated to be $\exp(-2kt^{1/2})$.

Now let us suppose the following differential equations for the sequential energy transfer through the pathway of $D-A_1-A_2-A_3$;

$$dN_1(t)/dt = -2kt^{1/2}N_1(t), \quad (4)$$

$$dN_2(t)/dt = 2kt^{1/2}N_1(t) - 2kt^{1/2}N_2(t), \tag{5}$$

$$dN_3(t)/dt = 2kt^{1/2}N_2(t) - 2kt^{1/2}N_3(t), \tag{6}$$

$$dN_4(t)/dt = 2kt^{1/2}N_3(t) - 2kt^{1/2}N_4(t), \tag{7}$$

where $N(t)$ is the number of the excited molecules at time t for the respective pigments denoted by the suffix; and k_1, k_2, k_3 and k_4 are the energy transfer rate constants in $s^{-1/2}$ units. The solution of the differential equations can be obtained under the initial conditions of $N_1(0) = N_0$, $N_2(0) = N_3(0) = N_4(0) = 0$ as follows:

$$N_1(t) = N_0 \exp(-2k_1 t^{1/2}), \tag{8}$$

$$N_2(t) = \frac{k_1 N_0}{k_2 - k_1} \left\{ \exp(-2k_1 t^{1/2}) - \exp(-2k_2 t^{1/2}) \right\}, \tag{9}$$

$$N_3(t) = \frac{k_1 k_2 N_0}{k_2 - k_1} \left\{ \frac{\exp(-2k_1 t^{1/2}) - \exp(-2k_3 t^{1/2})}{k_3 - k_1} - \frac{\exp(-2k_2 t^{1/2} - \exp(-2k_3 t^{1/2})}{k_3 - k_2} \right\}, \tag{10}$$

$$N_4(t) = \frac{k_1 k_2 k_3 N_0}{k_2 - k_1} \left\{ \frac{\exp(-2k_1 t^{1/2}) - \exp(-2k_4 t^{1/2})}{(k_4 - k_1)(k_3 - k_1)} - \frac{\exp(-2k_4 t^{1/2}) - \exp(-2k_3 t^{1/2})}{(k_3 - k_4)(k_3 - k_1)} \right.$$
$$\left. - \frac{\exp(-2k_4 t^{1/2}) - \exp(-2k_2 t^{1/2})}{(k_2 - k_4)(k_3 - k_2)} + \frac{\exp(-2k_4 t^{1/2}) - \exp(-2k_3 t^{1/2})}{(k_3 - k_4)(k_3 - k_2)} \right\}. \tag{11}$$

In the case of phycobilin-chlorophyll systems, the equations (8)-(11) are found to fit the sequence of the fluorescence rise and decay curves as is shown in Fig. 2. Also in the LB multilayer films, the sequential rise and decay curves can be fitted approximately with equations of (8)-(11). The best-fit curves are shown in Fig. 6 (d), (e) and (f) for 2L, 3L and 4L, respectively. For the curves of D layer, two curves are shown; the one (D') is obtained from the simulating calculation by using (8); the other (D) is from superposition of the calculated curve (D') and the curve of D monolayer without any other acceptor layer. The D monolayer in the absence of acceptor layer shows exponential-like decay with an apparent lifetime of ca. 15 ns [20]. In fact, actual decay curves of D layers in 2L, 3L and 4L all can be fitted reasonably with a superposition of these two curves (Fig. 6). The contribution of the long-decaying component in D can be interpreted by assuming the existence of carbazole chromophores isolated from the energy transfer sequence. The parameter values of the sequential energy transfer are summarized in Table 2. The value of the energy transfer constant, k, falls in between 0.02 and 0.21. By reference to (3), k is determined by the critical transfer distance R_0, the number density of chromophores n_A and the orientation factor g. It is seen that the magnitude of k in the LB films are comparable to those in the biological antenna pigments (Table 1), indicating that the geometrical distribution of chromophores and the critical transfer distance R_0 in the LB films studied are in parallel with those of the phycobilin-chlorophyll system. Interestingly, the k value of D - A_1 is markedly smaller than the values of the other processes. The smaller value in k might be associated with the fact that the molecular orientation of carbazole chromophores (vertical to the substrate surface) in D layer is perpendicular to that of oxacyanine in A_1 layer, on the

Table 2. Rate constants of the energy transfer and fluorescence rise times in the LB multilayer films

Layer	2L		3L		4L	
	k $(ps^{-1/2})$	rise (ps)	k $(ps^{-1/2})$	rise (ps)	k $(ps^{-1/2})$	rise (ps)
D	0.0243	0	0.0243	0	0.0243	0
A_1	0.0904	40	0.211	29	0.211	16
A_2			0.109	99	0.176	57
A_3					0.127	194
α/β [a]	1.2		0.42		0.32	

a) Ratio of the preexponential factors of $\exp(-kt)$ to $\exp(-2kt^{1/2})$ for the curve fitting of fluorescence decays of D layer.

other hand, the orientation of cyanine chromophores in A_1, A_2 and A_3 are parallel to each other and parallel to the substrate plane.

5. Comparison between Biological Antenna and Artificial Multilayer Films

The phenomenological aspects of the sequential energy transfer studied are on the whole parallel between the biological and the artificial systems. However, some significant differences can be seen in the time-resolved spectra, the sequential decay curves and the energy transfer efficiencies. These differences should reflect difference in the spatial distribution of chromophores in these two systems.

Time-resolved spectra   The sequence of appearance of the fluorescence bands, which must be associated with the energy transfer from the outer surface to the inner core, is very clear in the biological system, whereas in the artificial system it is blurred. The initial spectra of the LB films show the acceptor bands; A_1 in 2L, A_1 in 3L and A_1 and A_2 in 4L. It is possible that the irradiation at 300 nm excites simultaneously small amounts of these acceptors which show weak absorption band. However, the absorbance of the acceptor layers at 300 nm is very small in comparison with that of the D layer. Thus the primary reason for this may be the rapid energy transfer among D, A_1 and A_2. In the biological system, by irradiating at 580 nm, only the outer surface phycoerythrin is exclusively excited, although the acceptors have a significant amount of absorbance at this wavelength. This means that the absorption cross section of the outer surface must be large owing to a uniform distribution of chromophores, so that almost all the radiation is absorbed at the surface. Secondly, we should note that, in the LB films, the donor fluorescence bands are still alive at long time region; e.g., D and A_1 bands in 2L and 4L. This may suggest that the distribution of the pigment chromophores in a LB monolayer has an irregular structure so that there exist chromophores which are independent of the energy flow pathway and show slowly decaying components. The irregular distribution of guest molecules in LB monolayer has been demonstrated on the basis of the fluorescence decay curve analyses [20,21]. Existence of such isolated chromophores should reduce the efficiency of the energy transfer.

Rise and decay curves in the energy transfer sequence   Although the energy transfer efficiencies are very high (near unity) in the biological antenna, the fluorescence rise times are found to be rather slow (Fig. 3) relative to those in the LB films (Figs. 6 (a)-(c)). This may be accounted for by the intermolecular distances between chromophores being larger in the biological system than in the artificial one. We should note here that, although the distance between chromophores is fairly large, the energy transfer efficiency is high. It follows from this experimental finding that the chromophores in phycobiliproteins are distributed with a regular array in polypeptide network, and have no energy trap which quenches the excitation

41

before the energy transportation to the next-step acceptor. According to a recent work on the structure of C-phycocyanin by means of a X-ray crystallography with a 0.3 nm resolution [23], an array of phycobilin chromophores should form an energy transfer channel which minimizes the energy dissipation to the medium [24].

On the other hand, the LB monolayer film consisting of two components has an irregular "island" structure in distribution of guest molecules. Even in a low concentration of guest molecule, the island-like distribution makes the effective concentration significantly high and results in forming a number of dimer and/or higher aggregate species on which the excitation energy is trapped [21,22]. In the present case of the LB multilayers, there exist three processes of (1) a rapid energy transfer sequence among nearby donor and acceptor chromophores, (2) an energy trap on dimers and/or higher aggregates, and (3) an intrinsic energy relaxation of isolated chromophores. The processes (2) and (3) reduce the efficiency of the sequential energy transfer in the artificial antenna.

References

1. Th. Förster: Z. Naturf. 4a, 321 (1949)
2. V. M. Agranovich and M. D. Galanin: Electronic Excitation Energy Transfer in Condensed Matter (North-Holland, New York, 1982)
3. E. Gantt: Ann. Rev. Plant. Physiol. 32, 327 (1981)
4. A. N. Glazer: Ann. Rev. Biophys. Biophys. Chem. 14, 47 (1985)
5. M. Hauser, U. K. A. Klein and U. Gösele: Z. Physik. Chem. NF. 101, S255 (1976)
6. G. Zumofen and A. Blumen: J. Chem. Phys. 76, 3713 (1982)
7. J. Baumann and M. D. Fayer: J. Chem. Phys. 85, 4087 (1986)
8. I. Yamazaki, H. Kume, N. Tamai, H. Tsuchiya and K. Oba: Rev. Sci. Instrum. 56, 1187 (1985)
9. G. Porter, C. J. Tredwell, G. F. W. Searle and J. Barber: Biochim. Biophys. Acta 501, 232 (1978)
10. T. Kobayashi, E. O. Degenkokb, R. Bersohn, P. M. Rentzepis, R. MacColl and D. S. Berns: Biochem. 18, 5073 (1979)
11. I. Yamazaki, M. Mimuro, T. Murao, T. Yamazaki, K. Yoshihara and Y. Fujita: Photochem. Photobiol. 39, 233 (1984)
12. M. Mimuro, I. Yamazaki, T. Yamazaki and Y. Fujita: Photochem. Photobiol. 41, 597 (1985)
13. A. R. Holzwarth, J. Wendler and W. Wehrmeyer: Photochem. Photobiol. 36, 479 (1982)
14. J. Wendler, W. John, H. Scheer and A. R. Holzwarth: Photochem. Photobiol. 44, 79 (1986)
15. C. A. Hanzlik, L. E. Hancock, R. S. Knox, D. Guard-Friar and R. MacColl: J. Luminescence, 34, 99 (1985)
16. K. Ohki, M. Watanabe and Y. Fujita: Plant Cell Physiol. 23, 651 (1982)
17. E. Gantt and S. F. Conti: J. Cell Biol. 29, 423 (1966)
18. H. Kuhn, D. Möbius and H. Bücher: In Techniques of Chemistry, ed. by A. Weissberger and B. W. Rossiter, Vol. 1, Part 3B (Wiley, New York, 1972)
19. H. Bücher, K. H. Dreshage, M. Fleck, H. Kuhn, D. Möbius, F. P. Schäfer, J. Sondermann, W. Sperling, P. Tillmann and J. Wiegand: Mol. Crystals, 2, 199 (1967).
20. A. Leitner, M. E. Lippitsch, S. Draxler, M. Riegler and F. R. Aussenegg: Thin Solid Films, 132, 55 (1985)
21. N. Tamai, T. Yamazaki and I. Yamazaki: J. Phys. Chem. 91, (1987) in press.
22. I. Yamazaki, N. Tamai and T. Yamazaki: In Ultrafast Phenomena V, ed. by G. R. Fleming and A. E. Siegman, Springer Ser. Chem. Phys. Vol. 46 (Springer, Berlin, Heidelberg 1986) p.444
23. T. Schirmer, W. Bode, R. Huber, W. Sidler and H. Zuber: J. Mol. Biol. 184, 257 (1985).
24. M. Mimuro, P. Fuglistaller, R. Rumbeli and H. Zuber: Biochim. Biophys. Acta, 848, 155 (1986)

Calculations of Spectroscopic Properties and Electron Transfer Kinetics of Photosynthetic Bacterial Reaction Centers

W.W. Parson[1], S. Creighton[2], and A. Warshel[2]

[1]Department of Biochemistry, University of Washington, Seattle, WA 98195, USA
[2]Department of Chemistry, University of Southern California, Los Angeles, CA 90007, USA

Introduction

The initial electron transfer steps of bacterial photosynthesis occur in pigment-protein complexes called reaction centers (see [1] for a review). The structure of the reaction center from *Rhodopseudomonas viridis* has been solved to 3.0 Å resolution by X-ray diffraction [2-4], and a structure with somewhat lower resolution has been obtained for *Rhodobacter sphaeroides* [5, 6]. In both of these species, and in most of the other species that have been studied, the reaction center contains a cluster of four molecules of bacteriochlorophyll (BChl), two molecules of bacteriopheophytin (BPh), two quinones and one non-heme iron atom, all bound to three polypeptides. The complexes obtained from *Rps. viridis* also have a cytochrome subunit with four *c*-type hemes.

Two of the four BChls in the reaction center (BChl$_{LP}$ and BChl$_{MP}$) are situated very close together. When the pigments are excited with light, this special pair (P) transfers an electron to one of the BPhs (BPh$_L$) with a time constant of 3.0 ± 0.5 ps [7-16]. One of the other two BChls (BChl$_{LA}$) is located approximately in between P and BPh$_L$ and seems likely to play some role in the electron transfer reaction. It has been suggested that the excited complex (P*) first transfers an electron to BChl$_{LA}$, forming an intermediate charge-transfer species (P$^+$BChl$_{LA}^-$) that then passes an electron on to BPh$_L$ [17-20]. However, recent measurements with subpicosecond excitation and probe flashes indicate that the P$^+$BChl$_{LA}^-$ charge-transfer (CT) state is not kinetically resolvable [8-16]. The formation of P$^+$BPh$_L^-$ occurs with essentially the same kinetics as the decay of P*.

The picosecond kinetic measurements would be consistent with a scheme in which P$^+$BChl$_{LA}^-$ does form as an intermediate, but decays too rapidly to be detected. If P* generated P$^+$BChl$_{LA}^-$ with a rate constant (k_1) of about 3×10^{11} s^{-1}, and BChl$_{LA}^-$ transfered an electron to BPh$_L$ with a much higher rate constant (k_2), the concentration of P$^+$BChl$_{LA}^-$ might remain too low for the intermediate state to be resolved:

$$P^* \xrightarrow{\ k_1 \approx 3 \times 10^{11}\ s^{-1}\ } P^+BChl_{LA}^- \xrightarrow{\ k_2 \gg k_1\ } P^+BPh_L^- \qquad (1)$$

Because the electron transfer reaction from P* to P$^+$BPh$_L^-$ occurs at 77K at least as rapidly as it does at 295K [8], this scheme requires that the energy of P$^+$BChl$_{LA}^-$ be similar to, or below, that of P*. As we shall discuss below, it is not clear that the energy of P$^+$BChl$_{LA}^-$ is the proper range.

Another possible view of the initial electron transfer reaction is that the reaction center's lowest excited state itself contains contributions from P$^+$BChl$_{LA}^-$, along with contributions from exciton-type states of the six pigments and CT states of the special pair (BChl$_{MP}^+$BChl$_{LP}^-$ and

43

BChl$_{MP}^-$BChl$_{LP}^+$) [8, 21-24]. In this view, the rate of electron transfer to BPh$_L$ depends largely on the coefficient reflecting the mixing of P$^+$BChl$_{LA}^-$ into P*, and on the orbital overlap of BChl$_{LA}^-$ with BPh$_L$. This mechanism, which is termed "superexchange," would allow the energy of P$^+$BChl$_{LA}^-$ to be above that of P*, although the contribution of P$^+$BChl$_{LA}^-$ to P* will decrease as the CT state increases in energy.

A third view is that, because P* includes contributions from excited states of BChl$_{LA}$ and BPh$_L$ along with the excited states of the special pair, the photochemical reaction could proceed by the initial transfer of an electron from BChl$_{LA}$ to BPh$_L$, generating BChl$_{LA}^+$ BPh$_L^-$ [22]. This step could be followed rapidly by the movement of an electron from BChl$_{LP}$ or BChl$_{MP}$ to BChl$_{LA}$.

Hamiltonian and Absorption Spectrum of the Reaction Center

We have been attempting to develop a theoretical approach that will help to clarify the mechanism of the electron transfer reaction [22, 24, 25]. Our goal is to calculate the rate of electron transfer on the basis of the *Rps. viridis* crystal structure, and to explore how the kinetics depend on the structural features of the reaction center. This is a more difficult task than simply attempting to fit the observed kinetics to a theory with arbitrary parameters. An essential step toward understanding the electron transfer mechanism is to consider how the interactions of the six pigments give rise to the reaction center's complicated spectroscopic properties. To approach this question, we start by writing wavefunctions for the individual BChl and BPh molecules as linear combinations of atomic orbitals. The four main excited states (Q_y, Q_x, B_x and B_y) of the individual molecules are obtained by a configuration-interaction treatment that focuses on the top two filled molecular orbitals and the two lowest empty orbitals. Excited states of the reaction center then are written as linear combinations of the local molecular excited states and intermolecular CT transitions. In the four-orbital model, a pair of pigments *a* and *b* has eight CT transitions (four from molecule *a* to molecule *b* and four from *b* to *a*) in addition to the four local excited states of each molecule; for six pigments, there are 24 local transitions and 120 CT transitions. The coefficients describing the contributions of these different transitions can be obtained by diagonalizing an interaction matrix, **U**.

Expressions for the off-diagonal matrix elements of **U** can be written in terms of semiempirical atomic resonance integrals and electron-electron repulsion integrals [25]. For the exciton-type interactions between local excited states we use a transition monopole treatment, rather than a point-dipole approximation. Once the compositions of the reaction center's excited states have been obtained, absorption, circular dichroism (CD) and linear dichroism spectra of the reaction center can be calculated. Other properties such as the change in permanent dipole moment associated with each absorption band also can be found straightforwardly.

The major parameters that enter into the theory are the transition energies that form the diagonal elements of **U**. For the local Q_y, Q_x, B_x and B_y transitions, we use the energies of the absorption bands of monomeric BChl and BPh in solution, with small adjustments to allow for interactions of the pigments with the protein. (The protein crystal structure has not yet been described in sufficient detail to allow these adjustments to be calculated reliably. For now, we have assumed that the protein affects all four BChls identically, although this is surely an oversimplification.) Lower limits for the energies of the CT transitions can be obtained from the reduction potentials of the pigments in solution and upper limits can be obtained by molecular orbital calculations. Since these limits are broad enough to leave considerable uncertainty, we have varied the CT energies and examined how this affects the calculated spectroscopic properties. Agreement between the calculated and observed properties is obtained by placing the lowest CT transition of the special pair (a transition in which an electron moves from the highest filled orbital of BChl$_{MP}$ to the lowest empty orbital of BChl$_{LP}$, giving BChl$_{MP}^+$BChl$_{LP}^-$) slightly *above* the BChl Q_y transitions in energy. The corresponding transition in the opposite direction (an electron transfer from $BChl_{LP}$ to $BChl_{MP}$, generating $BChl_{MP}^-BChl_{LP}^+$) appears to lie significantly higher in energy [24].

In our analysis, CT transitions of the special pair make major contributions to the reaction center's strong absorption band at long wavelengths. In *Rps. viridis* reaction centers this band occurs at 960 nm, which is strongly red-shifted from the position of the Q_y absorption band of BChl-b in solution (790 nm). We view the red shift as being due largely to the mixing of CT transitions with the exciton transitions of the complex. The coefficients of $BChl_{MP}^+BChl_{LP}^-$ and $BChl_{MP}^-BChl_{LP}^+$ in the lowest excited state are calculated to be on the order of 0.53 and -0.17 respectively. The coefficients obtained for the Q_y transitions of $BChl_{LP}$, $BChlL_{MP}$, $BChl_{LA}$, $BChl_{MA}$, BPh_L and BPh_M are approximately -0.67, 0.46, 0.09, -0.11, -0.03 and 0.03. The other CT transitions and the Q_x, B_x and B_y transitions of the six pigments also make small contributions.

This approach accounts well for the absorption and linear dichroism spectra of *Rps. viridis* reaction centers [24]. The CD spectrum also is obtained reasonably accurately, though the rotational strengths of several bands are overestimated. (The shortcomings of the calculated CD spectrum could be due to our neglect of vibrational effects, or to an oversimplifiied treatment of the mixing of doubly excited states into the ground state; we have not yet explored these points in detail.) Because of the large contributions from CT transitions to the long-wavelength absorption band, the treatment offers an explanation for the special sensitivity of this band to external electric fields. The change in permanent dipole moment associated with the excitation is calculated to be about 8 debye, in good agreement with measurements of the Stark effect in *Rb. sphaeroides* reaction centers [26, 27]. Since CT transitions are expected to be strongly coupled to nuclear motions, our results also are consistent with the broad width of the long-wavelength band, as studied in recent laser hole-burning experiments [28-33].

Calculating electron transfer kinetics from structural information

With a description of the reaction center's excited states in hand, let us return to the electron transfer reaction. In general, the rate constant for an electron transfer transition between two states α and β can be written

$$k_{\alpha\beta} \approx (H_{\alpha\beta}/\hbar)^2 (\pi\hbar^2/k_B T\lambda)^{1/2} \Sigma_{m\alpha,n\beta} \exp(-\Delta g_{m\alpha,n\beta}/k_B T) \qquad (2)$$

where $H_{\alpha\beta}$ is an electronic interaction matrix element, λ is the solvent (protein) reorganization energy, and $\Delta g_{m\alpha,n\beta}$ is the effective activation barrier for a reaction path coupled to vibrational modes m_α in α and n_β in β [34, 35]. The protein reorganization energy cannot yet be calculated for the reactions of interest here because the available structural information is not sufficiently detailed, but the lower limit of λ is likely to be on the order of 200 cm^{-1}. The observation that the initial electron transfer steps in the reaction center are essentially independent of temperature [8] indicates that the overall activation barrier must be much smaller than $k_B T$. In this limit (2) reduces to

$$k_{\alpha\beta} \approx (H_{\alpha\beta}/h)^2 (\pi h^2/k_B T\lambda)^{1/2} \approx (H_{\alpha\beta}/10)^2 (200/\lambda)^{1/2} \times 1.65 \times 10^{11} \text{ s}^{-1} \qquad (3)$$

at 295 K, with $H_{\alpha\beta}$ and λ in units of cm^{-1} [35-37]. Equation 3 gives somewhat lower values for $k_{\alpha\beta}$ than the expressions used recently by MARCUS [37, 38].

It should be possible to calculate λ and the $\Delta g_{m\alpha,n\beta}$ when additional details become available on the crystallographic protein structure [35, 39-41]. For the present, we have focused on the electronic matrix element $H_{\alpha\beta}$ since this depends primarily on the interactions between the pigments.

To evaluate $H_{\alpha\beta}$ one needs to define the initial and final states α and β. This is not entirely straightforward because, as noted above, the excited states of the reaction center include contributions from a great many microscopic transitions. There are several possible approaches to the problem. One possibility is to describe the time-dependent wavefunction of the system as a superposition of *diabatic* states (states that do not diagonalize the interaction matrix **U**). These states then can be used to define the system's absorption spectrum, or any other observable, as a function of time. A time-dependent approach of this nature has been developed recently [35], and its application to the reaction center will be described in detail elsewhere. Here we shall explore a shortcut for finding the optimal set of diabatic states by using the *adiabatic* representation that does diagonalize **U**.

To consider the first step of the mechanism outlined in (1), we wish the diabatic state α to be an excited state corresponding to the 960-nm absorption band (P*), and we want β to be a state that consists predominantly of the lowest-energy CT transition from $BChl_{MP}$ to $BChl_{LA}$ ($BChl_{MP}^{+}BChl_{LA}^{-}$). For consistency with the spectroscopic calculations described above, we assume that this CT transition is below $BChl_{LP}^{+}BChl_{LA}^{-}$ in energy. However, α and β also could include contributions from the other CT transitions, in addition to the locally excited states of the 6 pigments. (One can safely omit some of the CT transitions involving $BChl_{MA}$ and BPh_{M}, which are located at a relatively large distance from $BChl_{LA}$ and BPh_{L}, but this still leaves **U** as an 88x88 matrix. We have retained the 16 CT transitions in which an electron is transferred between P and $BChl_{MA}$, although these also could be omitted without much effect on the results to be described here.)

If α and β are well separated in energy, it generally will be possible to associate them with two of the adiabatic states that are obtained by diagonalizing **U**. This can be done by inspecting the coefficients and energies associated with the adiabatic states. The *solid lines* in Fig. 1A show the calculated energies of two of the reaction center's adiabatic states (υ and η) as a function of the energies that are assumed for the basis transitions in which an electron is transferred from $BChl_{MP}$ or $BChl_{LP}$ to $BChl_{LA}$. Proceeding from left to right in the figure, we decrease these eight CT transitions together in energy, keeping the spacings between them constant; the abscissa gives the energy of $BChl_{MP}^{+}BChl_{LA}^{-}$, the lowest CT transition in the group. (The energies of the corresponding CT transitions in which an electron moves in the opposite direction, from $BChl_{LA}$ to $BChl_{MP}$ or $BChl_{LP}$, are raised simultaneously, and those of the transitions between $BChl_{LA}$ and BPh_{L} also are adjusted so as to maintain a self-consistent set of energies. The computations in essence attribute all of the energy changes to $BChl_{LA}^{-}$, though they could just as well assign them to $BChl_{MP}^{+}$ or to some other component of the system.) The coefficients representing the contributions of $BChl_{MP}^{+}BChl_{LA}^{-}$ in υ and η are shown in Fig. 1B. On the left-hand sides of Figs. 1A and 1B, where the energy of $BChl_{MP}^{+}BChl_{LA}^{-}$ is relatively high, this CT transition makes a much larger contribution to η than it does to υ. Here, the energy of η varies almost linearly with the energy of the CT transitions,while the energy of υ hardly changes. One can therefore identify η with β in this region; υ can be identified with α (P*) on the basis of its energy. As the energy of $BChl_{MP}^{+}BChl_{LA}^{-}$ is decreased, η and υ come closer together in energy and the coefficient for the CT state in υ increases (Fig. 1B). On the right-hand sides of Figs. 1A and 1B the compositions of the adiabatic states are reversed, so that η now corresponds to α and υ corresponds to β.

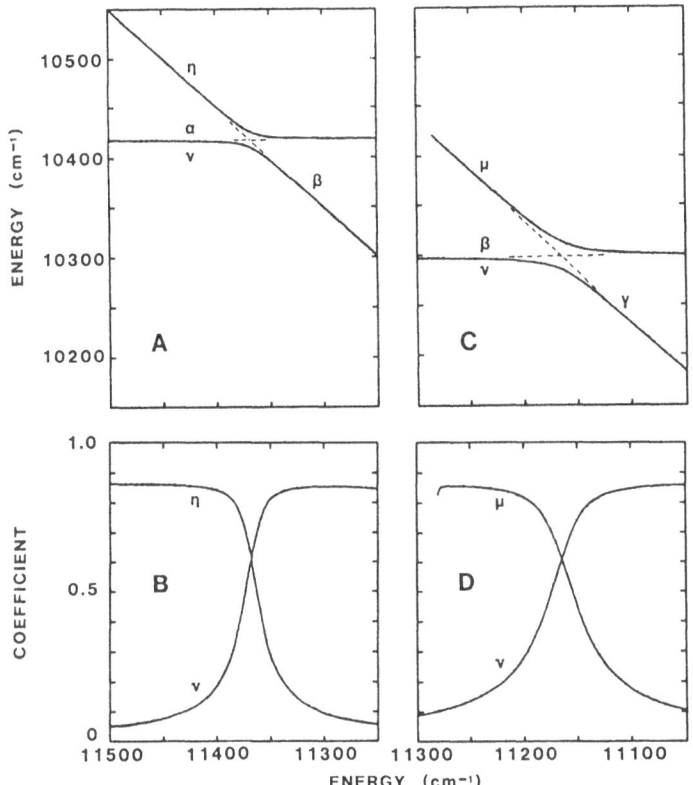

Figure 1 (A) The *solid lines* show the energies of two of the adiabatic states (η and ν) found by diagonalizing the interaction matrix **U** for the *Rps. viridis* reaction center. All parameters were as in Table I and Fig. 3 of [24], except that CT transitions involving $BChl_{LP}$ and BPh_L were included, bringing the total number of basis states in **U** to 88. The energies of the CT transitions of $BChl_{LP}$ were varied as described in the text. The abscissa gives the energy of the lowest CT basis transition in which an electron moves from $BChl_{MP}$ to $BChl_{LP}$ ($BChl_{MP}{}^+BChl_{LP}{}^-$). The straight *dashed lines* represent two diabatic states, α and β, which can be related to η and ν as discussed in the text. **(B)** Absolute values of the coefficients of $BChl_{MP}{}^+BChl_{LP}{}^-$ in η and ν. (The coefficients in η and ν have opposite signs.) **(C)** The energies of the CT transitions in which an electron moves from P to $BChl_{LP}$ are fixed at 120 cm^{-1} below the resonance point in (A), and the energies for the transfer of an electron from P to BPh_L are varied. The abscissa is the lowest energy for the formation of $BChl_{MP}{}^+BPh_L{}^-$. The *solid lines* show the energies of two adiabatic states (ν and μ); *dashed lines* represent two diabatic states (β and γ) that can be related to ν and μ. **(D)** Absolute values of the coefficients of $BChl_{MP}{}^+BPh_L{}^-$ in ν and μ. In (A) and (C), note that the energies of the adiabatic states are below the energy of the lowest CT basis transition; this is because of mixing with the higher CT transitions and with local excitations.

The energies of η and υ do not cross in the central region of Fig. 1A, but maintain a minimum splitting ($\Delta E_{\eta\upsilon}$) of approximately 15 cm^{-1}. In this region, the states of interest are not the adiabatic states η and υ, but the diabatic states represented by the *dotted lines*. The diabatic states (α and β) can be written as linear combinations of the adiabatic states. The energies of α and β do cross in the central part of Fig. 1A. At this point, the $BChl_{MP}{}^+BChl_{LA}{}^-$ CT transition makes a large contribution to an antisymmetric combination of η and υ (β), and no contribution to a symmetric combination (α). The other transitions in which an electron moves from $BChl_{MP}$ or $BChl_{LM}$ to $BChl_{LA}$ behave similarly, contributing only to β, whereas α contains contributions from the local excitations and from the CT transitions that move an electron between $BChl_{MP}$ and $BChl_{LM}$.

Electron transfer reactions occur as a result of fluctuations of dipoles in the pigments, protein and solvent, which modulate the energies of basis states and bring α and β into resonance. If the interactions among the BChls are relatively weak, the reaction center will tend to remain on the horizontal dashed line (α) as the energies fluctuate, and will have a low probability of moving to the diagonal line (β). This is a description of *nonadiabatic* electron transfer. The effective electronic interaction matrix element that mixes α and β ($H_{\alpha\beta}$) is given by $\Delta E_{\eta\upsilon}/2$, or 7.5 cm^{-1}. Inserting $H_{\alpha\beta}$ = 7.5 cm^{-1} into (3) and taking λ as about 200 cm^{-1} gives a rate constant $k_{\alpha\beta}$ of 9.3×10^{10} s^{-1}. This is reasonably close to the overall rate constant of 3×10^{11} s^{-1} that is found experimentally for electron transfer from P^* to BPh_L.

In the two-step scheme (1), state β must evolve quickly into another state in which there is an increased electron density on BPh_L. To model this second step, we assume that β first relaxes by about 120 cm^{-1} in energy below the resonance point at which it forms from α. A relaxation equivalent to the solvent reorganization energy λ is expected to occur as a result of the reorientation of dipoles around $BChl_{MP}{}^+$and $BChl_{LA}{}^-$, and one can view it as proceeding along the diagonal line in Fig. 1A. The exact magnitude of the relaxation is not critical for our present purposes, provided that it takes β sufficiently far out of resonance with α. We now seek a third diabatic state (γ) that contains a major contribution from the lowest CT transition in which an electron moves from P to BPh_L ($BChl_{MP}{}^+BPh_L{}^-$). The resonance between γ and the relaxed β is shown in Figs. 1C and 1D. The matrix element $H_{\beta\gamma}$ is found to be 14.0 cm^{-1}. Inserting this value into (3) gives $k_{\beta\gamma} \approx 3.2 \times 10^{11}$ s^{-1}. The finding that $k_{\beta\gamma}$ is about 4 times larger than $k_{\alpha\beta}$ is in accord with (1), although the ratio of the rate constants is not as great as the experimental data require.

The matrix elements $H_{\alpha\beta}$ and $H_{\beta\gamma}$ that are found in this way depend sensitively on the molecular geometry. For example, if $BChl_{LA}$ is moved 0.5 Å closer to $BChl_{LP}$ along the line connecting the N1 atoms of the two BChls, $H_{\alpha\beta}$ increases from 7.5 to 18 cm^{-1}, and $k_{\alpha\beta}$ increases to 5.3×10^{11} s^{-1}. Moving BPh_L in the direction of $BChl_{LA}$ has a similar effect on $k_{\beta\gamma}$ Displacements of this magnitude are within the 3 Å resolution of the crystal structure [2-4] and might be expected to occur as a result of vibrational motions of the protein.

The calculated rate constants $k_{\alpha\beta}$ and $k_{\beta\gamma}$ thus are of the correct order of magnitude to fit (1). However, there remains the difficulty that the two-step scheme requires the $BChl_{MP}{}^+BChl_{LA}{}^-$ CT transition to be below the BChl Q_y transitions in energy, and also below the CT transitions of P ($BChl_{MP}{}^+BChl_{LP}{}^-$ and $BChl_{MP}{}^-BChl_{LP}{}^+$). In Fig. 1A, the resonance between α and β occurs when the energy of the $BChl_{MP}{}^+BChl_{LA}{}^-$ CT transition is about 1630 cm^{-1} below that of $BChl_{MP}{}^+BChl_{LP}{}^-$. From the crystal structure of the *Rps. viridis* reaction center it is questionable whether $BChl_{MP}{}^+BChl_{LA}{}^-$ could be this low, because there are no hydrogen-bonding or charged amino acid side chains near $BChl_{LA}$. In contrast, $BChl_{LP}{}^-$ can be stabilized by a hydrogen bond to the keto group on ring V and by its strong Coulombic interaction with $BChl_{MP}{}^+$ [22, 24]. It is possible, however, that $BChl_{MP}{}^+BChl_{LA}{}^-$ is stabilized by H_2O molecules that are not seen in the

X-ray map. Calculations of such "solvation" effects are possible even without crystallographic coordinates for the H_2O molecules provided that one has the coordinates of the polypeptide chain [41], but the X-ray structure of the reaction center polypeptides has not yet been described in detail.

The superexchange mechanism can be explored by searching for the direct resonance between diabatic states α and γ. In this case, state β can lie either above or below α but the results are very sensitive to the energy difference between these two states. The mixing of α and γ depends strongly on the coefficient of $BChl_{MP}^+BChl_{LA}^-$ in α, and this coefficient decreases as $BChl_{MP}^+BChl_{LA}^-$ is increased in energy, as was shown in Fig. 1B. If $BChl_{MP}^+BChl_{LA}^-$ is placed about 30 cm^{-1} above the resonance point in Fig. 1, and γ is brought into resonance with α by varying the energy of $BChl_{MP}^+BPh_L^-$, $H_{\alpha\gamma}$ is found to be 3.0 cm^{-1} (Fig. 2A). This gives a rate constant of $k_{\alpha\gamma}$ of 1.5×10^{10} s^{-1} for the single-step superexchange pathway, approximately an order of magnitude smaller than the value of $k_{\alpha\beta}$ in the two-step mechanism. Raising the energy of β by another 50 cm^{-1} causes $H_{\alpha\gamma}$ to decrease to about 1 cm^{-1} (Fig. 2B). $H_{\alpha\gamma}$ can be increased by moving β more closely into resonance with α and γ, but it seems improbable for the three states to be simultaneously in resonance.

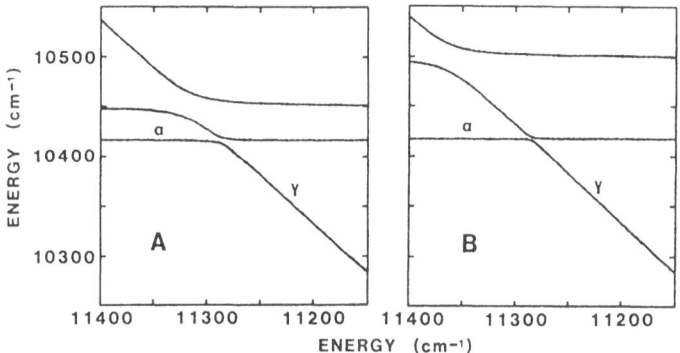

Figure 2 **(A)** The *solid lines* show the energies of three adiabatic states calculated for the *Rps. viridis* reaction center. The energies of the CT transitions in which an electron moves to $BChl_{LP}$ are fixed at approximately 30 cm^{-1} above the resonance point found in Fig. 1A, and the energies for the transfer of an electron from P to BPh_L are varied as in Fig. 1C. The abscissa is the energy of the lowest CT transition for the formation of $BChl_{MP}^+BPh_L^-$. This diagram illustrates the mixing of the diabatic states α and γ by superexchange with β. **(B)** The same as (A), except that the energies of the CT transitions in which an electron moves to $BChl_{LP}$ are fixed at about 80 cm^{-1} above the resonance point.

To evaluate the third mechanism suggested above, in which an electron moves initially from $BChl_{LA}$ to BPh_L, we consider an intermediate diabatic state β' that consists largely of $BChl_{LA}^+BPh_L^-$. The electronic matrix element $H_{\alpha\beta'}$ found for the first step in this scheme is about 1.5 cm^{-1}, which gives a rate constant of $k_{\alpha\beta'}$ of 3.7×10^9 s^{-1}. The limiting factor for this mechanism is that the excited states of $BChl_{LA}$ and BPh_L make only relatively small contributions to P*. The exact magnitudes of these contributions depend somewhat on our assumptions concerning the energies of the basis transitions, but we have not yet explored this point in detail.

Concluding remarks

The rate constants calculated for the second and third mechanisms appear to be too small to fit the experimental data on the kinetics of charge separation in the reaction center. The rate constant calculated for the first mechanism is of the right order of magnitude, but depends on a questionable value for the energy of the intermediate state. MARCUS [38] also has concluded that the superexchange mechanism has difficulty in accounting for the measured speed of the electron transfer reaction, although his estimates of the interaction matrix elements and the energy levels were based on an analysis of magnetic field effects rather than on structural information. The matrix elements that we have calculated depend strongly on the molecular geometry, and one must bear in mind that the reaction center's crystal structure is still under refinement. In addition, our calculations do not consider the possible role of nearby aromatic amino acids in mediating electron transfer from one pigment to the other. Finally, it is possible that the atomic resonance integrals that underlie the interaction matrix elements [25] fall off more slowly at large interatomic distances than we have assumed. A more detailed discussion of these points will be presented elsewhere.

Acknowledgements

This work was supported by NSF grants PCM-8303385 and PCM-8316161. We thank J. Deisenhofer and H. Michel for providing the crystallographic coordinates and D. Middendorf and A. Scherz for helpful discussion.

References

1. W. W. Parson: in *Photosynthesis*, ed. by J. Amesz (Elsevier, Amsterdam, in press)
2. J. Deisenhofer, O. Epp, K. Miki, R. Huber, H. Michel: J. Mol. Biol. 180, 385 (1984)
3. J. Deisenhofer, O. Epp, K. Miki, R. Huber, H. Michel: Nature 318, 618 (1985)
4. H. Michel, O. Epp, J. Deisenhofer: EMBO J. 5, 2445 (1986)
5. J. P. Allen, G. Feher, T. O. Yeates, D. C. Rees, J. Deisenhofer, H. Michel, R. Huber: Proc. Natl. Acad. Sci. U. S. A. 83, 8589 (1986)
6. C.-H. Chang, D. Tiede, J. Tang, U. Smith, J. Norris, M. Schiffer: J. Mol. Biol. 186, 201(1985)
7. D. Holten, C. Hoganson, M. W. Windsor, C. C. Schenck, W. W. Parson, A. Migus, R. L. Fork, C. V. Shank: Biochim. Biophys. Acta 592, 461 (1980)
8. N. W. Woodbury, M. Becker, D. Middendorf, W. W. Parson: Biochem. 24, 7516 (1985)
9. V. Z. Paschenko, B. N. Korvatovskii, A. A. Kononenko, S. K. Chamorovsky, A. B. Rubin: FEBS Lett. 191, 245 (1985)
10. J.-L. Martin, J. Breton, A. Hoff, A. Migus, A. Antonetti: Proc. Natl. Acad. Sci. U. S. A. 83, 957 (1986)
11. J. Breton, J.-L. Martin, A. Migus, A. Antonetti, A. Orsag: Proc. Natl. Acad. Sci. U. S. A. 83, 5121 (1986)
12. M. Becker, D. Middendorf, W. W. Parson: in *Ultrafast Phenomena*, ed. by G. R. Fleming and A. E. Siegman (Springer-Verlag, Berlin 1986) p. 374
13. J. Breton, J.-L. Martin, A. Migus, A. Antonetti, A. Orszag: *ibid*, p. 393
14. M. R. Wasielewski, D. M. Tiede, H. A. Frank: *ibid*, p. 388
15. W. Zinth, J. Dobler, W. Kaiser: *ibid*, p 379
16. S. V. Chekalin, Yu. A. Matveets, A. P. Yartsev: *ibid*, p. 402
17. V. A. Shuvalov, W. W. Parson: Proc. Natl. Acad. Sci. U. S. A. 78, 957 (1981)
18. V. A. Shuvalov, A. V. Klevanik, A. V. Sharkov, J. A. Matveetz, P. G. Krukov: FEBS Lett. 91, 135 (1978)
19. V. A. Shuvalov, A. V. Klevanik: FEBS Lett. 160, 51 (1983)
20. V. A. Shuvalov, L. N. M. Duysens: Proc. Natl. Acad. Sci. U. S. A. 83, 1690 (1986)
21. A. Warshel: in *Electron Transport and Oxygen Utilization*, ed. by C. Ho (Elsevier, New York 1982), p. 111

22. W. W. Parson, A. Scherz, A. Warshel: in *Antennas and Reaction Centers of Photosynthetic Bacteria* , ed. by M. E. Michel-Beyerle (Springer-Verlag, Berlin 1985), p. 122

23. J. Jortner, M. E. Michel-Beyerle: ibid, p. 345

24. W. W. Parson, A. Warshel: submitted for publ.

25. A. Warshel, W. W. Parson: submitted for publ.

26. D. deLeeuv, M. Malley, G. Butterman, M. Okamura, G. Feher: Biophys. J. 27, 111a (1982)

27. S. G. Boxer, T. R. Middendorf: Biochem., in press

28. S. R. Meech, A. J. Hoff, D. A. Wiersma: Chem. Phys. Lett. 121, 287 (1985)

29. S. G. Boxer, D. J. Lockhart, T. R. Middendorf: Chem. Phys. Lett. 123, 476 (1986)

30. S. G. Boxer, T. R. Middendorf, D. J. Lockhart: FEBS Lett. 200, 237 (1986)

31. S. R. Meech, A. J. Hoff, D. A. Wiersma: Proc. Natl. Acad. Sci. U. S. A. 83, 9464 (1986)

32. J. M. Hayes, G. J. Small: J. Phys. Chem. 90 4928 (1986)

33. R. Friesner: Proc. Natl. Acad. Sci. U. S. A., in press

34. A. Warshel: Proc. Natl. Acad. Sci. U. S. A. 77, 3105 (1980)

35. A. Warshel, J.-K. Hwang: J. Chem. Phys. 84, 4938 (1986)

36. R. Kubo, Y. Toyozawa: Prog. Theor. Phys. 13, 160 (1955)

37. R. Marcus, N. Sutin: Biochim. Biophys. Acta 811, 265 (1985)

38. R. Marcus: Chem. Phys. Lett., in press

39. T. Churg, A. Warshel: in *Structure and Motion: Membranes, Nucleic Acids and Proteins*, ed. by E. Clementi, G. Corongiu, M. H. Sarma, R. H. Sarma (Adenine, New York 1984) p. 361

40. T. Churg, T. Weiss, A. Warshel, T. Takano: J. Phys. Chem. 87, 1683 (1983)

41. A. Warshel, F. Sussman, G. King: Biochem. 25, 8368 (1986)

Femtosecond Absorption Studies of the Primary Events in Bacterial Photosynthesis and Light- and Dark-Adapted Bacteriorhodopsin

J.W. Petrich[1], *J. Breton*[2], *and J.L. Martin*[1]

[1]Laboratoire d'Optique Appliquée, Ecole Polytechnique,
 ENSTA, INSERM U275, F-91128 Palaiseau Cedex, France
[2]Service de Biophysique, CEN-Saclay, F-91191 Gif-sur-Yvette Cedex, France

1. INTRODUCTION

The ability to investigate ultrafast photophysics and photochemistry of reaction centers and antenna pigments of photosynthetic bacteria and of bacteriorhodopsin is a direct result of the advancing technology for the generation and the amplification of ultrashort light pulses [1]. The dual-beam femtosecond absorption spectrometer with which the data discussed here were obtained is described in detail elsewhere [2]. For our investigations of reaction centers and antenna pigments of photosynthetic bacteria, a tunable excitation beam was obtained by amplification [2-4] of a selected portion of a white-light continuum. For systems with several interacting chromophores, such as reaction centers, tunability of the excitation and the probe beams is equally as important as temporal resolution [3,5].

2. EXCITATION ENERGY TRANSFER AND INITIAL CHARGE SEPARATION IN REACTION CENTERS OF PHOTOSYNTHETIC BACTERIA

The remarkably efficient separation and stabilization of electric charges which constitute the key processes of photosynthesis occur in a transmembrane chlorophyll-protein complex known as the reaction center. Recently the organization of the pigments within the protein scaffold of the reaction center isolated from the photosynthetic bacterium Rps. viridis has been solved to a resolution of 3 Å [6]. The pigments exhibit an approximate C-2 symmetry with two closely interacting bacteriochlorophyll molecules forming a special pair (P). Two bacteriopheophytins (H_A and H_B) are located on either side of P, while two other bacteriochlorophylls (B_A and B_B) are arranged approximately in between P and H_A or H_B. The primary quinone electron acceptor (Q_A) lies at a greater distance on the "branch" occupied by the B_A and H_A molecules. This organization of the pigments suggests that the charge separation initially occurs between P and B_A and is followed by migration of the electron to H_A and then to Q_A. As we discuss below, however, we find no evidence of a spectroscopically observable state $P^+B^-_A$.

2.1 Excitation Energy Transfer to the Special Pair (P)

Identical risetimes for the bleaching of P at 960 nm are observed upon excitation either directly in P (930 nm) or at several wavelengths within the B or H absorption bands [3,5]. This demonstrates that the transfer of excitation energy from the B or the H molecules to the special pair occurs in less than 100 fs. The instantaneous absorbance increase observed at a variety of wavelengths is assigned to an excited state of P which we denote as P^* for simplicity [3,5]. Stimulated emission is observed within the long-wavelength band of P and is attributed to P^* whose decay kinetics can best be measured at 1050 nm (Fig. 1a), the isosbestic point of the absorption spectra of P and P^+. P^* decays with a time constant of 2.8 ps whether the reaction centers are excited in P, H, or B. The similarity of the kinetics of the absorbance changes of the 1310 nm band (Fig. 1b) corresponding to P^+ [3,5], and of the 675 nm band (Fig. 1c), corresponding to H^-_A, demonstrates that the oxidation of P occurs concomitantly with the reduction of H_A with a time constant of 2.8 ps.

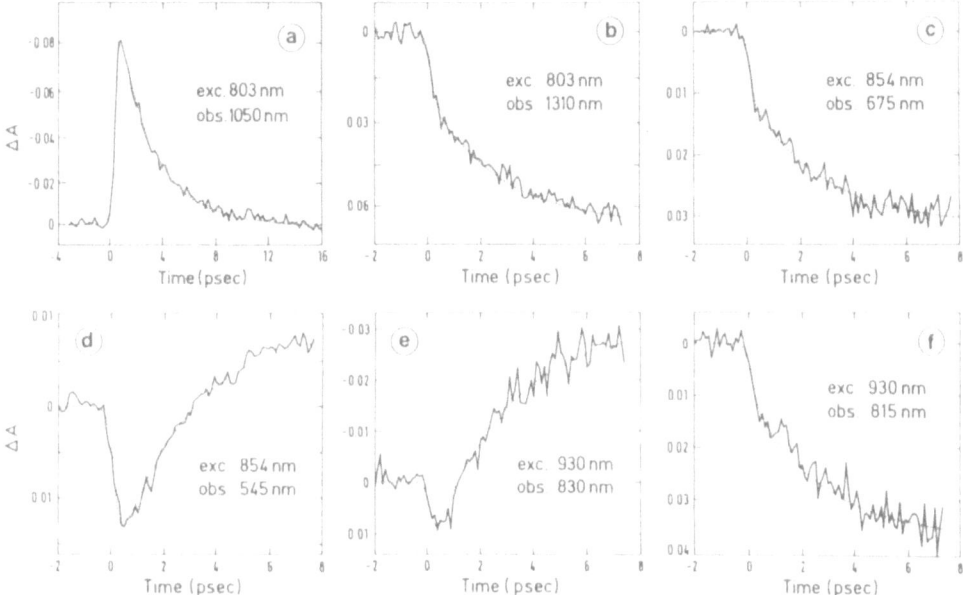

Figure 1. Transient transmission and absorbance changes in reaction centers from Rps. viridis. The best fits (·····) assume an instantaneous initial transmission (or absorbance) change and a single exponential decay (or rise) of 2.8 ps. In all cases the molecular response is convoluted with a laser pulse of 150 fs FWHM.

This is confirmed by the 2.8 ps decay component of the bleaching at 545 nm corresponding to the Q_X transition of H_A (Fig. 1d). The appearance of a bleaching with a 2.8 ps time constant at 830 nm is assigned to an electrochromic effect [3,5] of P^+ and H^-_A on the B molecules (Fig. 1e). This effect shifts the absorbance of B to the blue (Fig. 1f).

2.2 Excitation and Detection Within the 830 nm Band

Upon excitation of H or B, the excited states H^* or B^* transfer their energy in less than 100 fs to the special pair, P. It is clear from Fig. 2, however, that some of the transient bleaching of B decays more slowly, as is evident from the 400 fs component of the recovery of the bleaching. The simplest explanation for this phenomenon is to postulate the existence of two competitive processes for a given B_A molecule. In such a scheme, most of the B^*P complexes form BP^* with a time constant $\tau_1 < 100$ fs; and a small fraction of the B^*P complexes return to the ground state, BP, with a time constant, τ_2, of approximately 400 fs without the formation of P^*. It is possible, given our signal-to-noise ratio, that approximately 10% of the P^* molecules are formed in 400 fs and the remaining 90% are formed in <100 fs. Such a result would require the presence of two different types of excited B molecules which subsequently form P^* with unit quantum yield. This explanation is, however, unlikely, given the fact that the quantum yield, $\emptyset$, for the formation of P^+ in Rps. sphaeroides reaction centers has been reported to be 0.93 for excitation of B at 800 nm [7]. This result is in contrast to that obtained when P is excited directly: P^+ is formed with a quantum yield of essentially 1.0 [7]. Lastly, it is conceivable that the 400 fs decay component of the bleaching of B arises from an inhomogeneity of B molecules in the reaction centers. This is compatible with the experimental results, but it is not necessary to explain them. If the 400 fs component were due to an inhomogeneity or to sample damage, it is unlikely that identical kinetics would be observed in both the reaction centers from Rps.

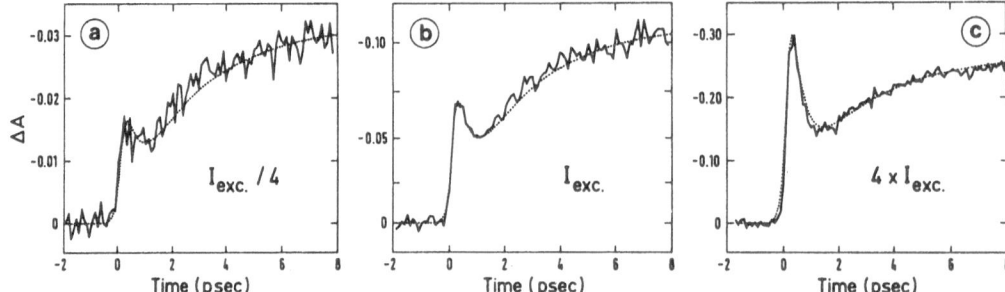

Figure 2 . (a-c) Kinetics of induced bleaching in B observed at 850 nm upon excitation at 827 nm for reaction centers of Rps. viridis. The intensity of the excitation, I_{exc}, is varied 16-fold demonstrating the change of shape due to the saturation of the formation of the state $P^+H^-_A$ which appears with a 2.8 ps time constant. The fast transient bleaching, whose amplitude is linear with excitation, is best fit (·····) with a 400 fs decay time.

sphaeroides and Rps. viridis. Furthermore, the 400 fs component persists even when the redox state of the reaction centers is modified as P^+H_AQ, $PH_AQ^-_A$, or $PH^-_AQ^-_A$.

The assumption of two competitive channels for the return of B^* to B and the observation of a quantum yield of 0.93 for the formation of P^+ upon excitation in B implies that a time-resolved experiment with sufficient temporal resolution would reveal a very fast absorption recovery for B with $\tau_1 = 30$ fs ($\tau_1 = (1-\emptyset)\tau_2/\emptyset$), where a simple branching ratio is assumed and the relevant quantities are defined above. There are three reasons why the 30 fs component is not resolved in our experiments or, put in another way, why the 400 fs contribution makes such a significant contribution in the data of Fig. 2. First, our apparatus (150 fs FWHM pulsewidth for the excitation wavelength) does not provide the necessary temporal resolution. An immediate consequence of this is that the amplitude of the bleaching of B is significantly diminished: for none of the data in Fig. 2 are we justified in fitting the absorption recovery of B to 93% of an exponentially decaying component with a 30 fs (<100 fs) time constant. This is because, as we have demonstrated in simulations, the effect of convoluting an excitation pulse of 150 fs FWHM with an exponentially decaying molecular response of 30 fs is to diminish the amplitude of the molecular response by a factor of approximately six, thus making any meaningful identification of a 30 fs component virtually impossible in Fig. 3a. Second, Fig. 2 indicates that over a 16-fold variation of excitation intensity the amplitude of the 2.8 ps process (which has been assigned to the formation of $P^+H^-_A$) saturates, while the amplitude of the bleaching of B remains linear with excitation intensity. Again, this phenomenon may be attributed to the duration of the excitation pulse. If we accept that upon excitation, B^* transfers its energy to P and returns to the ground state in approximately 30 fs for a given reaction center, then for the remainder of the 150 fs pulse, there are two possibilities: a photon can excite a B molecule in a different reaction center and produce $P^+H^-_A$, or a photon can re-excite the B molecule which has just transferred its energy to P. If energy transfer from B^* to P^* is unlikely (and since the lifetime of P^* is 2.8 ps), then B^* must return to the ground state by the channel characterized by the 400 fs time constant; and hence the amplitude of the 400 fs component is enhanced with respect to the 30 fs component. Such a scheme implies that for an excitation pulse of duration less than 30 fs, saturation of the channel producing $P^+H^-_A$ will not occur under otherwise similar conditions. Third, the amplitude of the induced absorption of P at a given wavelength should agree with the amplitude of the induced transmission of B when the respective absorption cross sections and the effect of the convolution on the absorption recovery of B is taken into account. Such an analysis, however, indicates that the amplitude of the induced transmission is larger than expected [5]. We suggest that stimulated emission from B^*

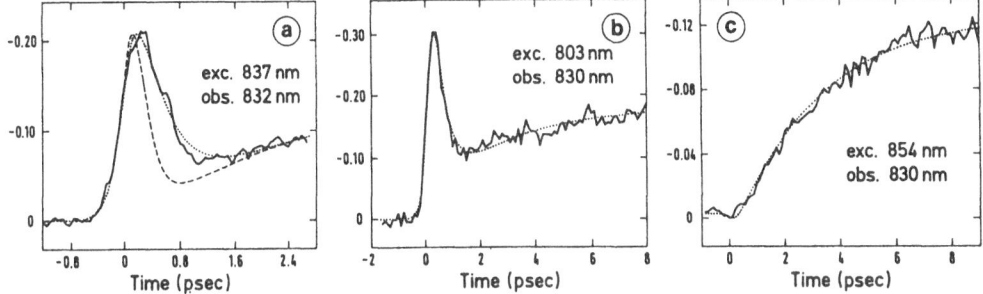

Figure 3. Transient absorption changes upon exciting and probing in B in reaction centers of Rps. viridis. (a) Kinetics of induced bleaching observed at 832 nm upon excitation at 837 nm. The bandwidth of the pump and probe pulses is approximately 5 nm. (·····): fit to the data, 400 fs decay of the transient bleaching and 2.8 ps rise time for the formation of $P^+H^-_A$; (-----): fit to the data, 100 fs decay of the transient bleaching and 2.8 ps rise time for the formation of $P^+H^-_A$. That the 400 fs component is due to stimulated emission from B^* is demonstrated by (b) and (c). When the excitation energy is higher than the probe energy (b), .the 400 fs component is observed. When, however, the situation is reversed (c), the 400 fs component is not detected. The complete absence of induced transmission at zero time can be attributed to a bleaching of B that is compensated by the absorption of P^* (Fig. 1d).

contributes to this residual amplitude. That this is the case is indicated by the following: when the probe wavelength is of lower energy than the pump wavelength, the amplitude of the induced transmission of B is small (and, in fact, appears to be negligible due to the compensating absorption of P^* (Fig. 3)).

By treating Rps. sphaeroides reaction centers with sodium borohydride, it is possible to remove the accessory B_B molecule [8,9]. Such modified reaction centers have recently been reported [10] to exhibit a fast transient bleaching in the 800 nm band upon excitation at 880 nm with pulses of 33 ps duration using only partial temporal overlap between the pump and the probe beams. This transient bleaching has been taken as evidence for the existence of a $P^+B^-_A$ state [10]. We have investigated the kinetics of such modified reaction centers (kindly provided by V. Shuvalov) with 150 fs excitation pulses at 870 nm. We observed that the kinetics at 930 nm and 805 nm are identical to those observed for unmodified reaction centers [11]. Although our studies [3,5,11] demonstrate that an intermediate state $P^+B^-_A$ cannot be observed with the resolution of our apparatus, this should not be taken as an indication that B_A plays no role in the initial electron transfer. As discussed elsewhere [12], it could lower the energy barrier between P^* and $P^+H^-_A$. It could also participate more directly in the electron transfer process: for example, the P^* state may contain, in addition to an internal charge transfer state $P^\pm$, small amount of $P^+B^-_A$ transferring very rapidly to $P^+H^-_A$ so that no detectable transient concentration of the state $P^+B^-_A$ could be measured.

3. ENERGY TRANSFER IN THE B800-850 ANTENNA CHLOROPHYLL-PROTEIN COMPLEX OF PHOTOSYNTHETIC BACTERIA

In the photosynthetic membrane of purple bacteria, most of the bacterio-chlorophyll molecules are associated with light-harvesting antenna pigment-protein complexes whose role is to capture light energy and to funnel it to the photochemically active reaction center. In Rps. sphaeroides the antenna system is composed of two different pigment-protein complexes named B800-850 and B875 corresponding to the respective center wavelengths of the main absorption bands in the near infrared. By using detergents to solubilize the membrane, these complexes have been isolated and purified. The B800-850 antenna complex has been extensively characterized by various spectroscopic techniques [13-16]. The complex consists of large aggregates of a heterodimer of polypeptides (5-6 kDa each) which holds one B800 and two B850 bacteriochlorophyll a molecules (which give rise to the 800 and 850 nm absorption bands, respectively) as well as one or two carotenoid molecules. Because of its relevance to the biological function of antenna complexes, energy transfer among the chromophores has also been extensively investigated [17-20]. Steady-state fluorescence yield data at room temperature have been interpreted in terms of a thermal equilibrium between the excitation densities in B800 and B850 with a transfer rate from B800 to B850 that is greater than 5.0×10^{11} s^{-1} [17]. Singlet-singlet exciton annihilation taking place upon excitation of B800-850 complexes with 30 ps pulses has been used to demonstrate that, in the presence of the LDAO detergent commonly used to isolate and to purify this complex, more than 600 bacteriochlorophyll molecules were connected by energy transfer [18].

Energy transfer in the isolated B800-850 antenna complex has recently been studied in two transient absorption experiments providing a time resolution of a few ps. In the first one [19], the energy transfer time from the carotenoid molecules to B850 has been estimated as 6 ± 1 ps. In the second one [20], energy transfer among B800 and B850 molecules has been probed by low intensity (below the threshold for singlet-singlet annihilation), high repetition rate, near infrared pulses. In this study the same wavelength was used to excite and to probe the system. From the decay of a polarized component observed at 800 nm, it has been deduced that energy transfer from B800 to B850 occurs on a time scale of 0.5 to 2 ps. A similar rate has also been inferred for B850 to B850 energy transfer. We have investigated the energy transfer step from B800 to B850 in the B800-850 antenna complex of Rps. sphaeroides. As shown in Fig. 4, the rise time of the bleaching at 855 nm is not resolved, indicating an ultrafast (< 100 fs) energy transfer from B800 to B850. The decay of the signal at 855 nm is fast (2.4 ps) suggesting the presence of significant singlet-singlet exciton annihilation processes at the rather high density of excitation used in the present study. Faster events, however, such as the energy transfer from B800 to B850 should be essentially unaffected by these annihilation processes. Although more detailed investigations of the absorbance changes as a function of the excitation energy and wavelength are necessary, we have found convincing evidence of an ultrafast energy transfer step from B800 to B850 which calls into question some of the conclusions drawn in a previous study [20].

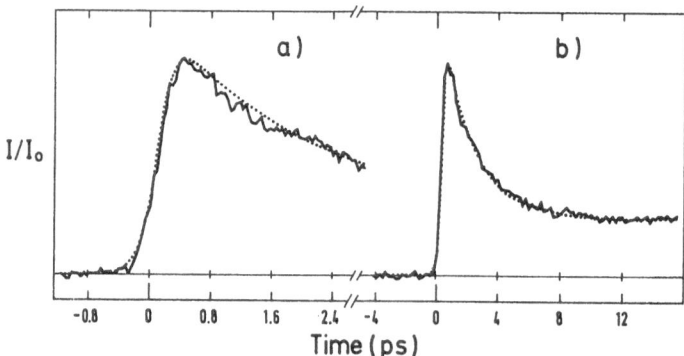

Figure 4. Induced bleaching of B850 of the B800-850 antenna complex. λ_{pump}=797 nm; λ_{probe}=855 nm; 150 fs pulsewidth. Approximately 30% of the pigments were excited. (a) The risetime of the bleaching at 855 nm cannot be resolved with our pulsewidth, and we have detemined that energy transfer from B800 to B850 is faster than 100 fs. (b) On an expanded time scale, the decay of the induced bleaching is fit well to a 2.4 ps time constant (this value was used to fit the decay in (a)).

4. LIGHT- AND DARK-ADAPTED BACTERIORHODOPSIN

The protein bacteriorhodopsin (BR) is the major constituent of the purple membrane of the bacterium Halobacterium halobium. The role of BR is to convert light into chemical energy by pumping protons across the bacterial membrane. In the membrane, BR forms trimers which comprise a two-dimensional hexagonal lattice [21]. The BR monomer is a hydrophobic protein (MW~26,000) linked from its lysine 216 to a retinal chromophore by means of a protonated Schiff base. Prolonged illumination of BR prepares the retinal in an all-trans configuration. Recent femtosecond absorption studies of the light-adapted BR have shown that subsequent to absorption of a photon, a species denoted as J is formed in approximately 0.5 ps from a proposed excited state of BR [22-24]. J has been determined to decay in 3-5 ps to the relatively long-lived (µs) bathochromic intermediate, K [22-24], whose structure and spectra have been studied by a variety of techniques [21]. In a few milliseconds, the all-trans retinal is regenerated via a pathway involving intermediates denoted as L, M, and O which have also been characterized by their absorption spectra, kinetics, and involvement in proton pumping [21]. In the dark, a thermodynamic equilibrium consisting of approximately 50% of the all-trans retinal and 50% of 13-cis, 15-cis retinal is established [25]. At low temperatures, it has been shown that upon excitation the 13-cis, 15-cis retinal also forms a bathochromic intermediate whose absorption spectrum is similar to that of K in the all-trans or light cycle. This bathochromic intermediate formed from 13-cis, 15-cis retinal does not, however, undergo subsequent transformations, but merely returns to its initial conformation with a half time of 37 ms without the pumping of protons [26]. The physiological significance of the dark cycle is not yet understood. It is now believed that the primary photophysical event in the light cycle and the dark cycle involves a trans-cis or a cis-trans isomerization, respectively [21]; and it has been shown by resonance Raman studies at 77 K that the bathochromic intermediate of the light cycle, K, has undergone an almost complete trans-to-cis isomerization about its $C_{13}=C_{14}$ bond [27]. The possibility of an accompanying isomerization about the $C_{14}-C_{15}$ bond remains controversial at this writing [28]. The isomerization is accompanied by separation of the protonated Schiff base from a negatively-charged counterion which is most likely a carboxylate group of one of four aspartic acid residues or a tyrosinate group [29].

Upon excitation of the light-adapted sample with 120 fs pulses at 612 nm, we observe two distinct phenomena [30]. The BR was bleached in less than 50 fs, and simultaneously there appeared a species with an absorption maximum at 460 nm. This absorption decayed (data not shown) with a time-constant of 500 + 40 ps to a species which has previously been designated as J. The decay of J to K (Fig. 5) is found to be 3.2 + 0.2 ps as measured by the absorption recovery at 580 nm. We obtain consistent time constants for the formation of J and K over the entire spectral range we investigated, 420 to 675 nm. Performing identical experiments on the dark- adapted BR, we observed nearly identical kinetics of formation for the counterparts of J and K in the light-adapted preparation. The only noticeable difference between the transient absorption spectra of the light and the dark-adapted BR is the amplitude of the absorption at its asymptotic value (Fig. 5). The magnitude of this absorption is consistent with that predicted from the equilibrium spectra and serves as a criterion for the integrity of the dark-adapted samples.

4.1 Photophysics of the Retinal Chromophore in Bacteriorhodopsin

The quantum yield for the bleaching of light-adapted BR has been measured to be 0.33 [31] and 0.60 [32]. It is of interest to consider the implications of a quantum yield that is less than unity for the bleaching of BR. In particular, two points must be addressed. 1) For a given quantum yield and the assumption of a simple branching ratio, one can predict a rate for the process competitive with the BR photochemistry. With sufficient time resolution, one should be able to observe this process experimentally. 2) What is the nature of this competitive process?

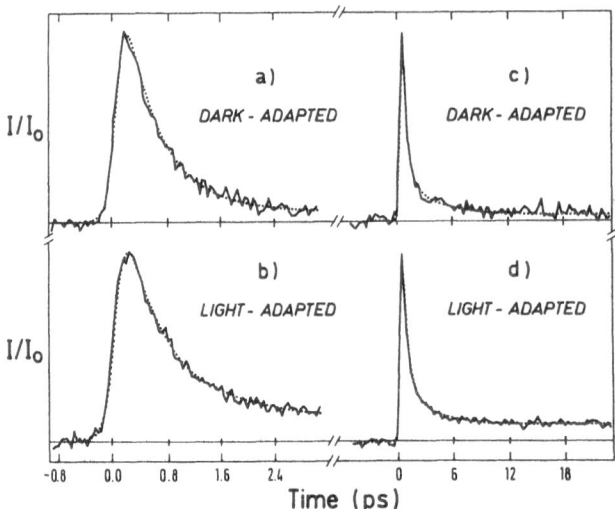

Figure 5. Transient bleaching of dark- and light-adapted BR. λ_{pump} = 612 nm, λ_{probe} = 580 nm, 120 fs pulse width. The fast recovery of the bleaching represents the formation of the first ground-state intermediate which is designated J in the light cycle. The slower recovery of the bleaching represents the formation of the ground-state intermediate designated K in the light cycle.

dark adapted: $I(t)/I_0 = 0.75\exp(-t/500 \text{ fs}) + 0.25\exp(-t/3200 \text{ fs})$
light adapted: $I(t)/I_0 = 0.70\exp(-t/500 \text{ fs}) + 0.30\exp(-t/3200 \text{ fs})$

If the quantum yield of the formation of K from J were 0.33, then, using our value of 3.2 ps for the formation of K and assuming a simple branching ratio, J would be expected to decay in 1.6 ps to another species. Similarly, if the species absorbing at 460 nm (which has been attributed [22-24] to an excited state of BR, BR*) forms J with a quantum yield of 0.33, then BR* should populate another species in 0.25 ps. Neither in the data presented in Fig. 5, nor in our data collected over the range from 420 to 675 nm, nor in the literature [22-24] is there any evidence that BR* and J are depopulated by competitive channels with time constants of 0.25 and 1.6 ps, respectively. It may, however, be that the quantum yield is closer to 0.60 [32], or to the average of the two reported values. If, for example, the quantum yield were 0.50, then the rate of formation of J is the same as that of the competitive process (~500 fs); and thus these two process cannot be distinguished by means of their kinetics. Alternatively, a quantum yield less than unity and significantly different from 0.50 for the bleaching of BR is consistent with the time-resolved measurements presented here and elsewhere [22-24] if the existence of a precursor to BR* is postulated. This precursor would be required to form BR* in less than 50 fs (the time resolution of our apparatus). Such a precursor has been suggested by Nuss et al. [22] to be a delocalized excited state of the bacteriorhodopsin trimer, in which case the quantum yield can be interpreted as the efficiency of exciton trapping by a monomer of the trimeric unit. Such an explanation reconciles the time-resolved data with the quantum yield measurements, but the existence of efficient exciton coupling [33] in the trimer has been called into question. Notably, the large absorption polarization of K has been cited by El-Sayed et al. as evidence that exciton transfer is not competitive with BR photochemistry [34]. Godfrey has analyzed this problem in terms of a very weak Förster-type excitonic coupling in the trimer [35].

4.2 The Light and the Dark Cycles: Photochemistry

Based on the similarity of the absorption spectra of J and K and the fact that K already has its retinal in a cis, or at least partly-cis, conformation [27], Nuss et al.

[22] suggested that photoisomerization has already been accomplished with the formation of J. Polland et al. [24] studied the kinetics of bacterio-opsin combined with a retinal analog modified to prohibit transformation to a 13-cis configuration. They found that such an analog has an excited-state lifetime of 10 ps, a 12-fold enhancement of its fluorescence quantum yield compared with that of native BR [36], and formed no photoproduct. This was cited as further evidence that the primary photophysical event in light-adapted BR is a rapid trans-to-cis isomerization about the $C_{13}=C_{14}$ bond. It is important to note, however, that the structure of J has not yet been determined. One possibility is that J is a vibrationally hot form of K, and resonance Raman experiments with subpicosecond resolution [4] could verify this.

For light- and dark-adapted BR and for bacterio-opsins containing retinal analogs whose equilibrium populations ranged from 30% to 85% 13-cis, Trissl and Gärtner [37] found that, within their time resolution of 70 ps, charge separation was in all cases effected with the same rate and in a direction opposite to that of proton translocation [38]. Based on these observations, they suggested that the inability of the dark cycle to pump protons could not be explained in terms of a photoisomerization of the 13-cis, 15-cis retinal in a direction opposite to that of the all-trans retinal. They concluded that the bathochromic intermediates derived from the 13-cis and the all-trans retinals resemble each other in their conformation and approach their respective conformations by two different routes. Finally, they suggested that the ability of BR to pump protons in the light cycle is more a consequence of different protonation states of potential counterions for the protonated Schiff base rather than of different geometries of the parent retinal moieties. Our transient absorption data are consistent with the conclusions and the model presented by Trissl and Gärtner [37]. A bathochromic intermediate that appears with the same time constant in both the dark- and the light-adapted preparations need not have the same structure in both of the preparations. For example, it has been shown that a bathochromic shift in the retinal absorption spectrum can be produced by either separation of the protonated Schiff base from its counterion or by protonation of the counterion [39]. It is thus conceivable that the similarities in the time constants for the formation of the bathochromic intermediates in the dark and the light cycles are not due to the acquisition of similar retinal structures but to similar displacements from their counter ions.

5. ACKNOWLEDGEMENTS

JWP was supported by an NSF Industrialized Countries postdoctoral fellowship, an INSERM poste orange, and a fellowship from La Fondation pour la Recherche Médicale. Parts of this work were funded by INSERM, ENSTA, AFME, and Le Ministère de la Recherche et de la Technologie.

6. REFERENCES

1. R.L. Fork, C.V. Shank, R. Yen, and C. Hirlimann, IEEE J. Quant. Electron. QE-19, 500 (1983); Ultrafast Phenomena V, edited by G.R. Fleming and A.E. Siegman (Springer, Berlin, 1986).
2. J.L. Martin, A. Migus, C. Poyart, Y. Lecarpentier, R. Astier, and A. Antonetti, Proc. Natl. Acad. Sci. USA 80, 173 (1983).
3. J.L. Martin, J. Breton, A.J. Hoff, A. Migus, and A. Antonetti, Proc. Natl. Acad. Sci. USA 83, 957 (1986).
4. J.W. Petrich, J.L. Martin, D. Houde, C. Poyart, and A. Orszag, J. Chem. Phys. Submitted.
5. J. Breton, J.L. Martin, A. Migus, A. Antonetti, and A. Orszag, Proc. Natl. Acad. Sci. USA 83, 5121 (1986).
6. J. Deisenhofer, O. Epp, K. Miki, R. Huber, and H. Michel, J. Mol. Biol. 180, 385 (1984).
7. C.A. Wraight and R.K. Clayton, Biochim. Biophys. Acta 333, 246 (1973).
8. P. Maroti, C. Kirmaier, C. Wraight, D. Holten, and R.M. Pearlstein, Biochim. Biophys. Acta 810, 132 (1985).

9. V.A. Shuvalov, A.Y. Shukuropatov, S.M. Kulakova, M.A. Ismailov, and V.A. Shkuropatova, Biochim. Biophys. Acta 849, 337 (1986).
10. V.A. Shuvalov and L.N.M. Duysens, Proc. Natl. Acad. Sci. USA 83, 1690 (1986).
11. J. Breton, J.L. Martin, J.W. Petrich, A. Migus, and A. Antonetti, FEBS Lett. 209, 37 (1986).
12. N.W. Woodbury, M. Becker, D. Middendorf, and W.W. Parson, Biochemistry 24, 7516 (1985).
13. J. Breton, A. Vermeglio, M. Garrigos, and G. Paillotin, in Proceedings of the 5th International Congress on Photosynthesis Vol. 3, edited by G. Akoyunoglou (Balaban International Science Services, Philadelphia, 1981), p. 445.
14. J.D. Bolt and K. Sauer, Biochim. Biophys. Acta 667, 342 (1981).
15. H.J.M. Kramer, R. Van Grondelle, C.N. Hunter, W.H.J. Westerhuis, and J. Amesz, Biochim. Biophys. Acta 765, 156 (1984).
16. A. Scherz and W.W. Parson, Photosynth. Res. In press.
17. R. Van Grondelle, H.J.M. Kramer, and C.P. Rijgersberg, Biochim. Biophys. Acta 682, 208 (1982).
18. R. Van Grondelle, C.N. Hunter, J.G.C. Bakker, and H.J.M. Kramer, Biochim. Biophys. Acta 723, 30 (1983).
19. M.R. Wasielewski, D.M. Tiede, and H.A. Frank, in Ultrafast Phenomena V edited by G.R. Fleming and A.E. Siegman (Springer, Berlin, 1986), p. 388.
20. H. Bergström, V. Sundström, R. Van Grondelle, E. Åkesson, and T. Gillbro, Biochim. Biophys. Acta 852, 279 (1986).
21. W. Stoeckenius and R.A. Bogomolni, Ann. Rev. Biochem. 51, 587 (1982).
22. M.C. Nuss, W. Zinth, W. Kaiser, E. Kölling, and D. Oesterhelt, Chem. Phys. Lett. 117, 1 (1985).
23. A.V. Sharkov, A.V. Pakulev, S.V. Chekalin, and Y.A. Matveetz, Biochim. Biophys. Acta 808, 94 (1985).
24. H.-J. Polland, M.A. Franz, W. Zinth, W. Kaiser, E. Kölling, and D. Oesterhelt, Biophys. J. 49, 651 (1986).
25. M.J. Pettei, A.P. Yudd, K. Nakanishi, R. Henselman, and W. Stoeckenius, Biochemistry 16, 1955 (1977).
26. W. Sperling, P. Carl, C.N. Rafferty, and N. Dencher, Biophys. Struct. Mech. 3, 79 (1977).
27. S.O. Smith, J. Lugtenburg, and R.A. Mathies, J. Membrane Biol. 85, 95 (1985).
28. P. Tavan and K. Schulten, Biophys. J. 50, 81 (1986).
29. M. Englehard, K. Gerwert, B. Hess, W. Kreutz, and F. Siebert, Biochemistry 24, 400 (1985); G. Dollinger, L. Eisenstein, S.-L. Lin, K. Nakanishi, and I. Termini, Biochemistry 25, 6524 (1986).
30. J.W. Petrich, J. Breton, J.L. Martin, and A. Antonetti, Chem. Phys. Lett. Submitted.
31. J.B. Hurley, T.G. Ebrey, B. Honig, and M. Ottolenghi, Nature 270, 540 (1977).
32. D. Oesterhelt, P. Hegemann, and J. Tittor, EMBO J. 4, 2351 (1985).
33. T.G. Ebrey, B. Becher, B. Mao, P. Kilbride, and B. Honig, J. Mol. Biol. 112, 377 (1977).
34. M.A. El-Sayed, B. Karvaly, and J. Fukumoto, Proc. Natl. Acad. Sci. USA, 78, 7512 (1981).
35. R.E. Godfrey, Biophys. J. 38, 1 (1982).
36. H.-J. Polland, M.A. Franz, W. Zinth, W. Kaiser, E. Kölling, and D. Oesterhelt, Biochim. Biophys. Acta 767, 635 (1984).
37. H.-W. Trissl and W. Gärtner, Biochemistry. In press.
38. H.-W. Trissl, Biochim. Biophys. Acta 723, 327 (1983).
39. U.C. Fischer and D. Oesterhelt, Biophys. J. 31, 139 (1980).

The Primary Photochemical Reactions in Systems I and II of Photosynthesis

H.J. van Gorkom and A.M. Nuijs

Department of Biophysics, Huygens Laboratory of the State University,
P.O. Box 9504, NL-2300 RA Leiden, The Netherlands

1. Introduction

In plant photosynthesis the electron transport from water to carbon dioxide is light-driven at two different sites: the photosystems I and II. In both systems the light energy is absorbed by antenna molecules, such as chlorophylls and carotenoids,and thereby converted to the energy of a singlet excited state. The excitation is then transferred among the various pigment molecules until the reaction center is reached. In the reaction center the excitation of the primary electron donor chlorophyll, called P700 in PS I and P680 in PS II, leads to electron transfer to the primary acceptor, followed by secondary electron transport steps which stabilize the charge separation. In principle the rates of electron transfer as well as the identity of the various 'early' electron acceptors can be studied by means of fast optical absorption spectroscopy, but in practice these measurements are seriously complicated by the large absorbance of antenna chlorophylls. Much of our knowledge on the primary reactions in plant photosynthesis was therefore based on indirect kinetic measurements (e.g. antenna fluorescence) and spectral studies by 'photoaccumulation' experiments. Recently the measurement of picosecond absorbance changes in systems with large antennae became possible and allowed a direct verification of some earlier conclusions. This paper summarizes the main results of recent picosecond absorption studies on relatively intact PS I and PS II from spinach (70 - 100 Chl a/reaction center), as well as on the recently isolated reaction center of PS II.

2. Photosystem I

In the reaction center of photosystem I three iron-sulfur centers, designated F_X, F_B and F_A, act as secondary electron acceptors [1]. The data on the primary electron acceptor are controversial. Picosecond absorption measurements on small PS I particles led SHUVALOV et al. to the conclusion that the primary acceptor is a chlorophyll molecule absorbing at 694 nm [2], whereas EPR and optical measurements under continuous illumination at progressively lower potentials provided evidence that two acceptors A_0 and A_1 function prior to F_X [3,4]. These data suggested that A_1^- is a semiquinone and indicated that A_0 is a chlorophyll molecule absorbing around 670 nm. Recently, picosecond absorption measurements on relatively intact PS I (70 - 100 Chl a/reaction center) were reported [5,6], which supported the earlier conclusion by SHUVALOV et al. [2].

Figure 1 shows the absorbance difference spectra of PS I at 40 ps (solid circles) and 200 ps (open circles) after the 35 ps excitation pulse at 532 nm. The primary electron donor, P700, was kept in the oxidized state by ferricyanide and continuous background illumination. Both spectra are characterized by a bleaching in the region 660 - 740 nm, flanked by shallow increases in absorption. The 40 ps spectrum is ascribed to the formation of singlet excited antenna chlorophyll, Chl^*a, and that at 200 ps to the formation of the triplet state, Chl^Ta. The kinetics at 685 nm are given in the inset. A deconvolution procedure indicated a lifetime of Chl^*a of 40 ± 5 ps.

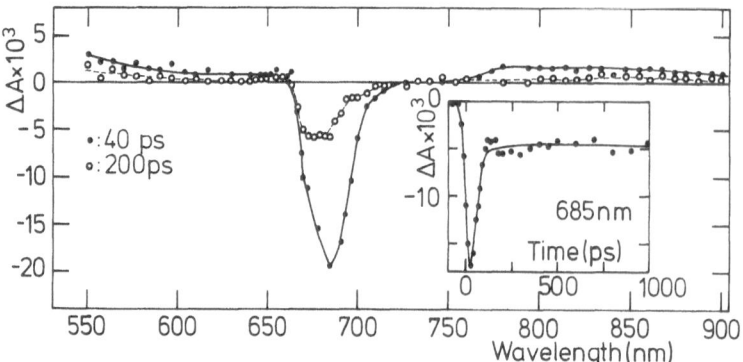

Fig. 1. Absorbance difference spectra at 40 ps (●) and 200 ps (o) after the 35 ps excitation pulse at 532 nm, in the presence of 3 mM ferricyanide and under continuous background illumination. The inset shows the kinetics at 685 nm

When the reaction centers were initially open, the difference spectrum measured at 200 ps (Fig. 2A, circles) shows minima at 701, 683 and 668 nm, and is similar to the difference spectrum of P700 oxidation induced by continuous illumination (Fig. 2A, crosses). No distinct absorbance changes due to reduction of an electron acceptor can be observed.

The difference spectrum detected at 40 ps (Fig. 2B, circles) is characterized by negative bands around 699 nm and 685 nm, which are due to the formation of $P700^+$ and Chl^*a, respectively. The spectrum can be simulated (Fig. 2B, crosses) by a sum

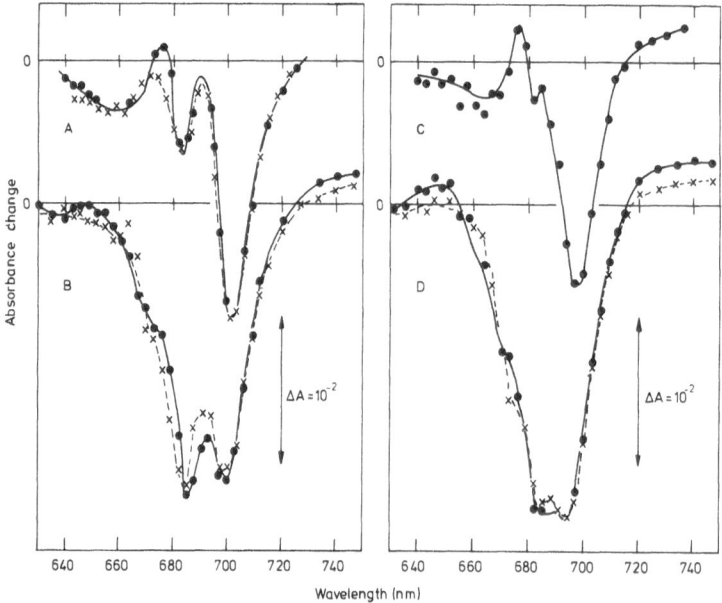

Fig. 2. A: A comparison of the ps pulse-induced spectrum of $P700^+$ formation (●) with the spectrum under continuous illumination (x). B: The 40 ps spectrum under non-reducing conditions (●) simulated by a sum (x) of the $P700^+$ and the Chl^*a spectra. C: Spectrum at 500 ps under reducing conditions. D: The 40 ps spectrum under reducing conditions (●) simulated by a sum (x) of the Chl^*a spectrum and that of the radical pair.

of the P700$^+$ and the Chl*a spectra. The differences between the spectra of Fig. 2B are due to the reduction of the primary acceptor, as will be made more clear below.

When the secondary acceptors are prereduced chemically, the difference spectrum observed at 500 ps after the flash (Fig. 2C) is different from that due to formation of P700$^+$ alone. Compared to the P700$^+$ spectrum, the main bleaching has shifted from 701 to 698 nm, and the steep increase around 690 nm has disappeared. The spectrum is ascribed to formation of the primary radical pair. At 40 ps after the flash this radical pair is also present, since the recorded spectrum (Fig. 2D, circles) can be well simulated by a sum (Fig. 2D, crosses) of the spectrum of Chl*a and that of the radical pair.

Subtracting the P700$^+$ spectrum from that of the radical pair yields the spectrum given in Fig. 3. These absorbance changes are characterized by a narrow bleaching around 693 nm, similar to that earlier observed by SHUVALOV et al. [2] and are attributed to the reduction of the primary electron acceptor. This acceptor thus proves to be a chlorophyll a species absorbing around 693 nm, and will be designated as C-693.

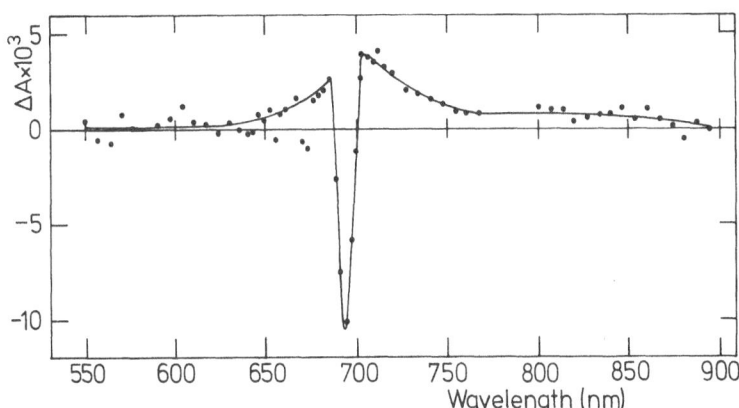

Fig. 3. Absorbance difference spectrum calculated for the reduction of the primary acceptor

When the secondary electron transport is inhibited, due to chemical reduction of the iron-sulfur centers, the primary radical pair decays in 20 - 50 ns (not shown) to the triplet state of P700 (Fig. 4) with a yield of about 30 %. The lifetime of the triplet state is about 3 μs (inset).

Figure 5A shows the kinetics at 700 nm under conditions of normal electron transport for different excitation energy densities. At low intensity (crosses) a 50 ps risetime of P700$^+$ formation is observed which reflects the process of trapping. At higher intensities a short-lived bleaching appears which is due to formation of Chl*a.

In an attempt to observe the electron transfer from C-693$^-$ to a secondary acceptor, the primary donor was excited directly at 710 nm, thus reducing antenna excitation. The kinetics at 690 nm, divided by those at 700 nm, are shown in Fig. 5B. The best simulation (dashed curve) of these kinetics was obtained using a 32 ps time constant for electron transfer from C-693$^-$ to the next acceptor.

These data thus provide evidence that the primary acceptor in PS I is a chlorophyll molecule absorbing around 693 nm. No evidence was obtained for the participation of the chlorophyllous electron acceptor A$_0$ absorbing around 670 nm [3,4]. Possibly this species is located on a sidepath. Based on various EPR and optical

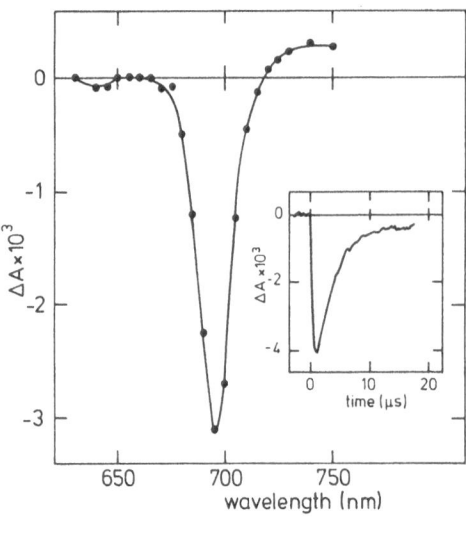

Fig. 4. Difference spectrum and decay kinetics of the triplet state of P700

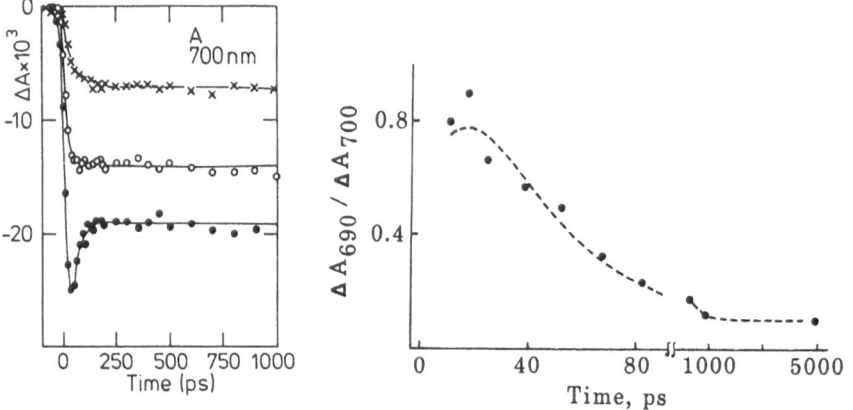

Fig. 5. A: Kinetics at 700 nm under non-reducing conditions at different excitation intensities. B: Kinetics at 690 nm divided by those at 700 nm following excitation at 710 nm. The best fit (dashed curve) yields a time constant of 32 ps

measurements under continuous illumination at low potentials, it has been postulated that yet another acceptor, A_1, would function prior to F_X [3,4]. These data suggested that A_1^- is a semiquinone. It was also argued that a reaction observed by flash spectroscopy and having a halftime of 120 µs at 10 K would represent a recombination between $P700^+$ and A_1^- [7]. Recently the absorbance difference spectrum between 240 and 525 nm has been measured for this reaction, and from a comparison with the room temperature difference spectrum for the oxidation of P700 it has been concluded that A_1 could be vitamin K_1 (phylloquinone) [8]. As a support for this hypothesis, vitamin K_1 has been found as an integral constituent of PS I [9]. Since reduction of A_1^- does not produce absorbance changes in the region above 500 nm, our ps data do not provide information on the participation of A_1 in the main chain of electron transfer after C-693. The transient reduction of A_1^- during normal PS I turnover remains to be demonstrated.

3. Photosystem II

In the reaction center of photosystem II a plastoquinone molecule, Q_A, acts as the secondary electron acceptor [10]. By analogy to the situation in purple bacteria [11] the role of pheophytin a as the primary electron acceptor was postulated on the basis of its band shifts upon Q_A reduction [12], of its reduction in photoaccumulation experiments at low redox potential [13], and of its possible participation in a radical pair generated by a ns flash in small PS II particles when Q_A was in the reduced state [14]. However, a transient reduction of Phe a during normal PS II turnover remained to be demonstrated. Direct evidence for that came from recent picosecond absorbance measurements on relatively intact PS II membranes containing about 100 Chl molecules per reaction center [15].

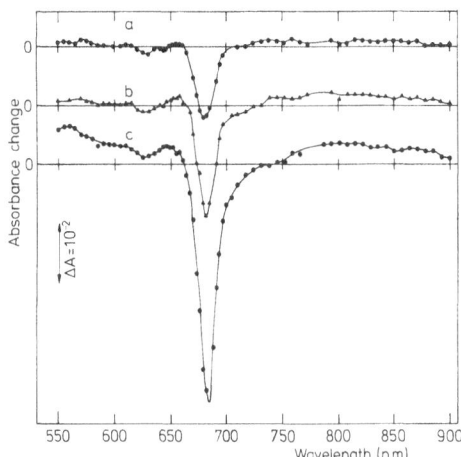

Fig. 6. Absorbance difference spectra of PS II particles in the presence of 2.5 mM ferricyanide at different delays after the 35 ps excitation pulse at 532 nm. (a) 1 ns; (b) 200 ps; (c) 0 ps. The extinction of the sample was 1.4 at 675 nm in a 2 mm cell

Fig. 6 shows absorbance difference spectra of these particles at different delay times after the flash. At 0 ps (Fig. 6C) the changes are predominantly due to formation of singlet excited antenna Chl a. At 1 ns, however, (Fig. 6A) a somewhat different spectrum is observed, in which the bleaching has shifted from 685 to 680 nm. This spectrum is ascribed to the oxidation of the primary donor, P680 [16]. This same spectrum contributes to the changes observed at 200 ps (Fig. 6B), but in addition absorbance changes are present caused by the reduction of Phe a. When the contribution by the formation of P680$^+$ is subtracted from the 200 ps spectrum, the spectrum of Fig. 7 (solid circles) remains. The spectral shape of these absorbance changes is quite similar to those obtained earlier by KLIMOV by accumulation experiments [13], and thus shows clearly the reduction of Phe a during normal PS II turnover. As expected, flash-induced reduction of this acceptor is also observed when Q_A is chemically reduced (Fig. 7, open circles).

The kinetics of Phe$^-$a reoxidation are shown in Fig. 8. The first 200 ps of the decay mainly represents the deactivation of excited antenna Chl a (cf. Fig. 6C). The residual increase in absorption is due to reduction of Phe a. During normal electron transport Phe$^-$a is reoxidized in about 250 ps (Fig. 8, open circles), whereas its lifetime is several nanoseconds under reducing conditions (solid circles).

As to the shape of the spectrum observed upon reduction of Phe a, it has been put forward by GANAGO et al. [17] on the basis of linear dichroism studies that the spectrum observed by KLIMOV et al. [13] upon reduction of Phe a under continuous illumination consists of the bleaching of a Phe a molecule absorbing at 680 nm, and of a blue shift of an other molecule, perhaps P680, together resulting in a maximal absorbance decrease at 685 nm. Since the picosecond spectra of Phe a re-

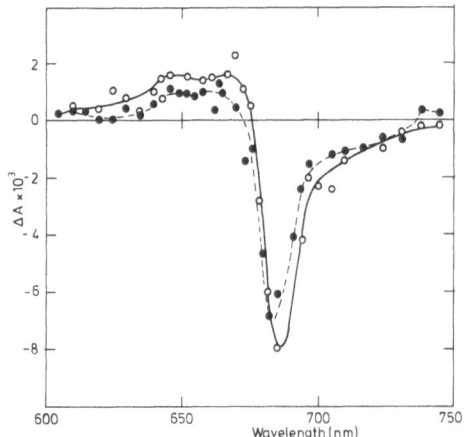

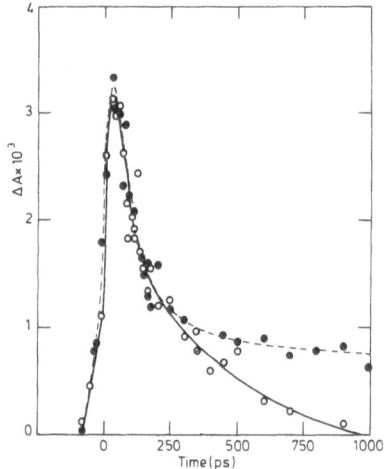

Fig. 7. Absorbance difference spectra of the PS II particles due to reduction of Phe a in the presence of 2.5 mM ferricyanide (•), and in the presence of dithionite (o)

Fig. 8. Kinetics of absorbance changes of the PS II particles at 655 nm in the presence of dithionite (•), and 2.5 mM ferricyanide (o)

duction (Fig. 6) resemble those by KLIMOV et al. reasonably well and do neither show the maximal bleaching at 680 nm, nor the zero transition at 665 nm attributed to Phe a reduction by GANAGO et al. [17], it is likely that a shift is included in the ps spectra as well. Since in the time-resolved experiments P680 was oxidized concomitantly with the reduction of Phe a, the shift cannot be due to the primary donor, but must be ascribed to another chlorophyllous molecule close to the Phe a acceptor, and absorbing around 680 nm.

The role of Phe a as an early electron acceptor in PS II is highlighted by picosecond absorption measurements on the isolated reaction center complex [18]. This complex was recently isolated by NANBA and SATOH and contains only the D1-D2 and cytochrome b_{559} polypeptides [19]. Photoaccumulation experiments showed the reversible reduction of Phe a in this preparation [19], and in ESR measurements the spin-polarized triplet state of P680 was detected, which is likely to arise from a recombination of the radical pair [20]. Since plastoquinones are lacking in this preparation, the photochemical activity of the complex should be limited to the formation of the radical pair $P680^{+}Phe^{-}a$, followed by a recombination. Evidence for these reactions came from picosecond measurements by DANIELIUS et al. [17].

Fig. 9 shows the absorbance difference spectra of the D1-D2-cyt b_{559} complex at 4 ns, and partly at 24 ns (dashed line) after the excitation pulse. The absorbance increase at 450 nm, and the negative band at 545 nm are attributed to the reduction of Phe a, similar to the changes observed when $Phe^{-}a$ is photoaccumulated [13, 19]. The oxidation of the primary donor, P680, is reflected in part of the bleachings at 430 and 680 nm, but the excitation of some unconnected chlorophyll has to be assumed as well in order to account for the large amplitude of the band at 680 nm.

Recombination of the radical pair in this preparation takes place with a time constant of 36 ns, as reflected in the kinetics at 450 nm (Fig. 10, solid circles). The kinetics at 680 nm (open circles) are more complicated. Assuming the presence of a 36 ns phase, two more phases can be extracted: one with a lifetime of 5 ns, and a constant component. The 5 ns phase probably reflects the decay of the singlet excited state of the unconnected chlorophyll, whereas the constant component is due to the formation of triplet states of both P680 and chlorophyll.

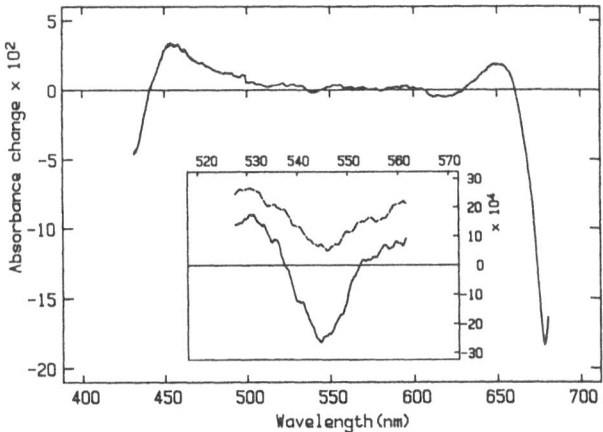

Fig. 9. Absorbance difference spectra of the D1-D2-cyt b_{559} complex at 4 ns after the 25 ps, 695 nm excitation pulse. Inset: the 545 nm region enlarged; the dashed line shows the difference spectrum at 24 ns after excitation. The absorbance of the sample was less than 0.3 at the excitation wavelength

It is noteworthy that the lifetime of the primary radical pair in more intact PS II is much shorter: when the secondary electron acceptor Q_A is in the reduced state the flash-induced absorbance changes due to $P680^+Phe^-a$ decay in only a few ns (Fig. 8, solid circles). In whole chloroplasts with Q_A in the reduced state the fluorescence lifetime of PS II chlorophyll is 2 ns and it is not yet clear to what extent the primary radical pair is formed at all in this case [21]. It seems like-ly that these shorter lifetimes are caused by the decay of singlet excited chloro-phyll, formed by recombination of the primary radical pair. This recombination may be slower in the absence of Q_A^-, and retrapping of the excitation may be faster in the absence of a chlorophyll antenna; both mechanisms may play a role in the long lifetime of $P680^+Phe^-$ in the isolated PS II reaction center.

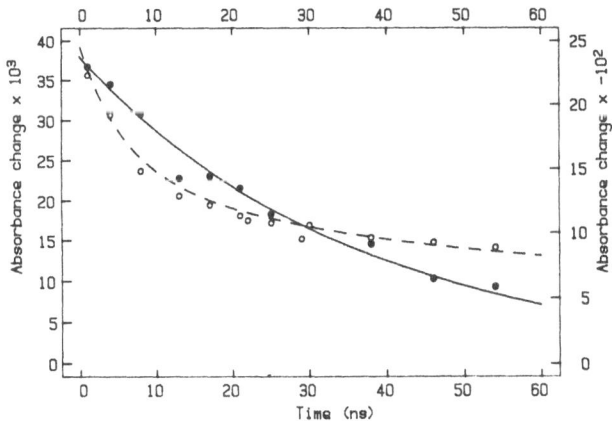

Fig. 10. Kinetics of absorbance changes at 450 nm (solid circles, right-hand scale) and 680 nm (open circles, left-hand scale). The solid line shows the best-fitting exponential decay time (1/e) of 36 ns. The kinetics at 680 nm were simulated as the sum of two exponentials, with lifetimes (and amplitudes) of 3.5 ns (37 %), and 36 ns (36 %), and a constant component (27 %)

4. Concluding Remarks

The application of sensitive picosecond absorbance difference spectroscopy to relatively intact PS I and PS II particles has provided much information concerning the rates of electron transport and the identity of the electron acceptors involved in the early steps of the charge separation.

In the reaction center of PS II the role of pheophytin a as an intermediary electron acceptor during normal turnovers has now been established. The electron is transferred to the next acceptor, which is presumably a plastoquinone, in about 250 ps. The acceptor side of PS II is thus remarkably similar to that of reaction centers from purple bacteria, in which the electron is transferred from reduced bacteriopheophytin to ubiquinone in about 200 ps [11]. Also the pigment and protein composition of the two types of reaction center are similar, as was shown after the recent isolation of the reaction center of PS II [19].

In the reaction center of PS I a chlorophyllous molecule absorbing around 693 nm has been identified as an early electron acceptor. This acceptor is reoxidized in about 30 ps, which is an order of magnitude faster than the corresponding reaction in any other type of photosynthetic reaction center. The subsequent electron acceptor could be vitamin K_1, but direct evidence for that is still lacking. Ultimately the electron is transferred to the iron-sulfur centers. The reaction center of PS I has some similarity to that of green sulfur bacteria, in which a bacteriochlorophyll c-like molecule acts as an intermediary acceptor [22-24], and iron-sulfur centers function as secondary acceptors [25]. Some evidence has been obtained that in Heliobacterium chlorum, the reaction center of which is similar to that of green sulfur bacteria [26], a quinone is involved in the electron acceptor chain as well [27].

In all the photosystems studied, both of plants and photosynthetic bacteria, direct evidence, i.e. by time-resolved absorption spectroscopy, for the reduction of the secondary electron acceptor is still lacking. Extension of the picosecond absorption technique to the blue and ultraviolet region will be needed to obtain direct evidence for the identities and kinetics of the secondary electron acceptors.

Another point for future concern is that the 'primary electron acceptors' of PS I and PS II, as they are designated here, have not really been shown to be the first acceptors. In reaction centers isolated from purple bacteria the photosynthetic charge separation has successfully been studied with much better time resolution, and was found to take 2.8 ps [28]. Until recently no reaction center preparation from plant photosystems was available in which the charge separation could be studied with a similar time resolution. Such measurements may now be possible with Ki. Satoh's PS II reaction center preparation, which contains only about 5 chlorophyll molecules per reaction center.

5. Acknowledgements

The authors are much indebted to their coworkers in the studies reviewed here: V.A. Shuvalov, H.W.J. Smit, J.J. Plijter, R.V. Danielius, Ki. Satoh, P.J.M. van Kan, L.N.M. Duysens, and to F.T.M. Zonneveld for skillful preparative work.

The investigations were supported by the Netherlands Foundations for Biophysics and for Chemical Research, financed by the Netherlands Organization for the Advancement of Pure Research (ZWO).

6. References

1. A.W. Rutherford, P. Heathcote: Photosynth. Res. 6, 295 (1985)
2. V.A. Shuvalov, A.V. Klevanik, A.V. Sharkov, P.G. Kryukov, B. Ke: FEBS Lett. 107, 313 (1979)

3. P. Gast, T. Swarthoff, F.C.R. Ebskamp, A.J. Hoff: Biochim. Biophys. Acta 722, 168 (1983)
4. R.W. Mansfield, M.C.W. Evans: FEBS Lett. 190, 237 (1985)
5. A.M. Nuijs, V.A. Shuvalov, H.J. van Gorkom, J.J. Plijter, L.N.M. Duysens: Biochim. Biophys. Acta 850, 310 (1986)
6. V.A. Shuvalov, A.M. Nuijs, H.J. van Gorkom, H.W.J. Smit, L.N.M. Duysens: Biochim. Biophys. Acta 850, 319 (1986)
7. P. Sétif, P. Mathis, T. Vänngård: Biochim. Biophys. Acta 767, 404 (1984)
8. K. Brettel, P. Sétif, P. Mathis: FEBS Lett. 203, 220 (1986)
9. E. Interschick-Niebler, H.K. Lichtenthaler: Z. Naturforsch. 36C, 276 (1981)
10. H.J. van Gorkom: Photosynth. Res. 6, 97 (1985)
11. W.W. Parson, B. Ke: In Photosynthesis. Volume 1: Energy Conversion by Plants and Bacteria, ed. by Govindjee (Academic Press, New York 1982) p. 331
12. H.J. van Gorkom, Doctoral thesis, University of Leiden (1976)
13. V.V. Klimov, A.V. Klevanik, V.A. Shuvalov, A.A. Krasnovsky: FEBS Lett. 82, 183 (1977)
14. V.A. Shuvalov, V.V. Klimov, E. Dolan, W.W. Parson, B. Ke: FEBS Lett. 118, 279 (1980)
15. A.M. Nuijs, H.J. van Gorkom, J.J. Plijter, L.N.M. Duysens: Biochim. Biophys. Acta 848, 167 (1986)
16. H.J. van Gorkom, M.P.J. Pulles, J.S.C. Wessels: Biochim. Biophys. Acta 408, 331 (1975)
17. I.B. Ganago, V.V. Klimov, A.O. Ganago, V.A. Shuvalov, Y.E. Erokhin: FEBS Lett. 140, 127 (1982)
18. R.V. Danielius, Ki. Satoh, P.J.M. van Kan, J.J. Plijter, A.M. Nuijs, H.J. van Gorkom: FEBS Lett., in the press (1987)
19. O. Nanba, Ki. Satoh: Proc. Natl. Acad. Sci. USA, in the press (1987)
20. M.Y. Okamura, Ki. Satoh, R.A. Isaacson, G. Feher: In Progress in Photosynthesis Research, ed. by J. Biggins, Vol. 1 (Martinus Nijhoff Publishers, Dordrecht 1987) p. 379
21. H.J. van Gorkom: In Light Emission by Plants and Bacteria, ed. by Govindjee, J. Amesz, D.C. Fork (Academic Press, New York 1986) p. 267
22. A.C. van Bochove, T. Swarthoff, H. Kingma, R.M. Hof, R. van Grondelle, L.N.M. Duysens, J. Amesz: Biochim. Biophys. Acta 764, 343 (1984)
23. A.M. Nuijs, H. Vasmel, H.L.P. Joppe, L.N.M. Duysens, J. Amesz: Biochim. Biophys. Acta 807, 24 (1985)
24. V.A. Shuvalov, J. Amesz, L.N.M. Duysens: Biochim. Biophys. Acta 851, 1 (1986)
25. T. Swarthoff, P. Gast, A.J. Hoff, J. Amesz: FEBS Lett. 130, 93 (1981)
26. A.M. Nuijs, R.J. van Dorssen, L.N.M. Duysens, J. Amesz: Proc. Natl. Acad. Sci. USA 82, 6865 (1985)
27. M. Brok, H. Vasmel, J.T.G. Horikx, A.J. Hoff: FEBS Lett. 194, 322 (1986)
28. J.-L. Martin, J. Breton, A.J. Hoff, A. Migus, A. Antonetti: Proc. Natl. Acad. Sci. USA 83, 957 (1986)

Phototransformation Pathway of Phytochrome

Y. Inoue

Department of Botany, Faculty of Science, University of Tokyo,
Hongo, Tokyo 113, Japan

1 Introduction

Phytochrome is a water soluble photochromic chromoprotein playing an important role
in plant photomorphogenesis as a photoreceptor pigment of the red/far-red photoreversible reaction which is generally observed in green plants [1]. The monomer
phytochrome molecule has an open tetrapyrrole called phytochromobilin (Fig. 1) as a
chromophore which binds to a cysteine residue of apoprotein with a thioether linkage
[2,3].

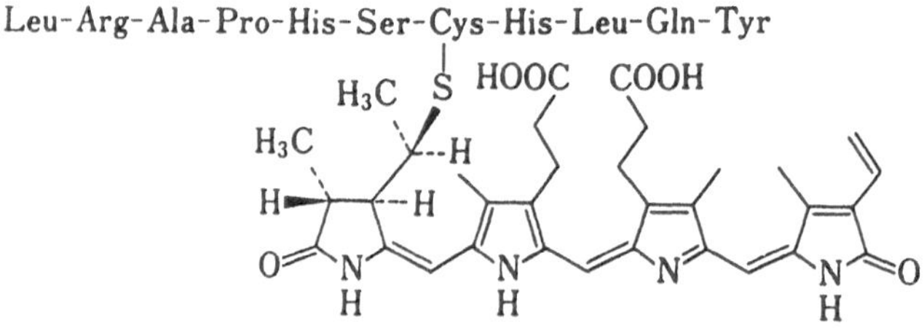

Leu-Arg-Ala-Pro-His-Ser-Cys-His-Leu-Gln-Tyr

Figure 1.   Primary structure of phytochromobiliundecapeptide [2]

The biologically inactive red-light-absorbing form named Pr converts to the
biologically active far-red-light-absorbing form named Pfr after red light irradiation. Pfr can revert back to Pr by absorbing far-red light. Since the discovery of
phytochrome [4], this photochromic property is its the most interesting characteristic, and so many studies have been done [5]. In the early days, proteolytically
degraded phytochrome having monomer molecular weight of ca. 60 kDa, called "small"
[6], was used as a sample. In later studies, partially degraded phytochromes of
molecular mass ca. 114–118 kDa, called "large" [7], were used. Nowadays, it is
believed that the molecular size of "native" phytochrome is ca. 121–125 kDa [8].

The phototransformation pathway of phytochrome between Pr and Pfr had been determined by flash photolysis using degraded "small" oat phytochrome [9-12], but nothing
had been reported about that of "large" and "native" phytochrome and phytochrome in
vivo at physiological temperature. Therefore, we constructed a multichannel transient spectra analyser, and made clear the phototransformation pathway of "large"
and "native" phytochrome and phytochrome in etiolated pea epicotyl tissue at physiological temperature (ca. 1-10°C) using this apparatus.

2 Multichannel Transient Spectra Analyser

To make clear the phototransformation pathway of phytochrome at physiological temperature, we constructed a multichannel transient spectra analyser [13] (Fig. 2),
which could record flash-induced transient absorbance spectra with microsecond time
resolution in the wavelength range from 350 to 800 nm with ca. 0.5 nm resolution.

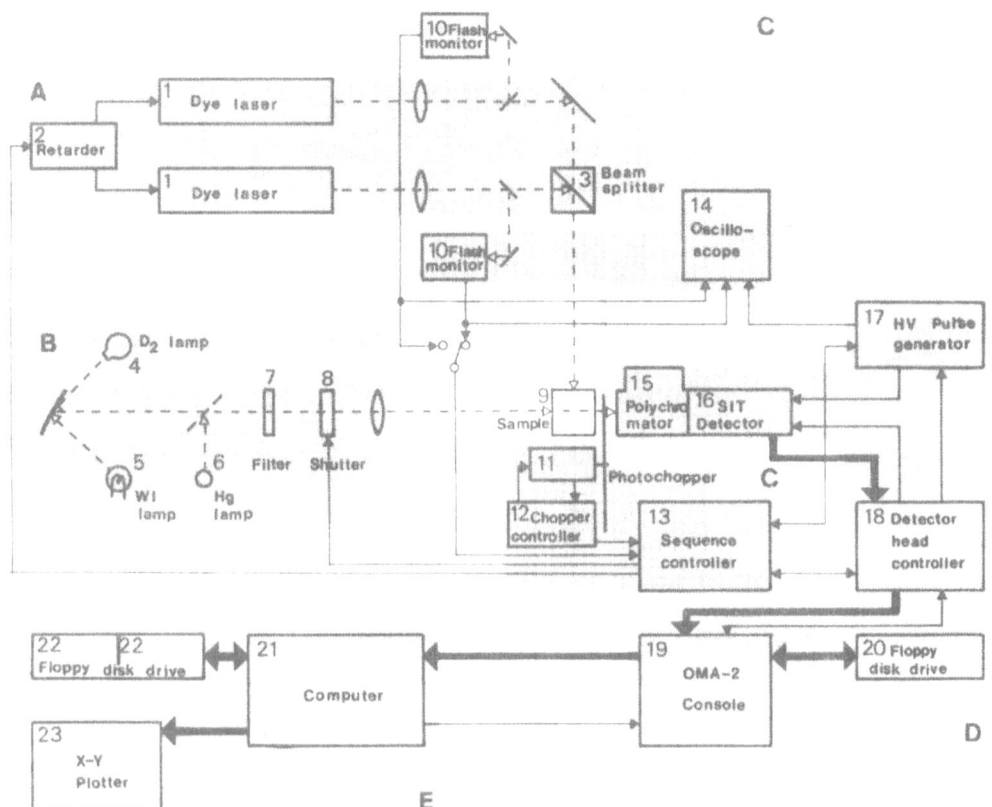

Figure 2. Schematic diagram of major features of the multichannel transient spectra analyzer [13]. A, flash excitation apparatus; B, measuring light assembly; C, photometric sequence controller; D, detector assembly; E, computer system.

The second laser-flash obtained from one of two flash-lamp-excited dye lasers (Fig. 2-1) was used to determine the photoreactivity of the phototransformation intermediates which were induced by the first laser-flash irradiation. A photomultiplier having two stages of bistable gate circuits with a digital oscilloscope was used instead of a SIT detector (Fig. 2-16) to determine decay kinetics of flash induced intermediates by analyzing the absorption change at fixed wavelength [14]. Optics consisting of a Glan Thompson prism and Fresnel rhomb retarders were used to get controlled polarization plane of laser-flashes to determine transition moments of phototransformation intermediates of phytochrome in fern protonemata [15]. Procedures for measurement of transient spectra changes have been described in previous papers [14-17,19,20]. The temperature of the sample was always kept constant by the circulation of temperature-controlled coolant inside the metal cuvette holder.

3 Phototransformation Pathway of "Large" Phytochrome

Four phototransformation intermediates of "small" oat phytochrome were originally demonstrated by flash kinetic spectroscopy on the pathway from Pr to Pfr [9].

A double-flash experiment [10] suggested that the above described pathway consisted of two kinetically identifiable stages and the final stage was composed of three parallel first-order reaction components.

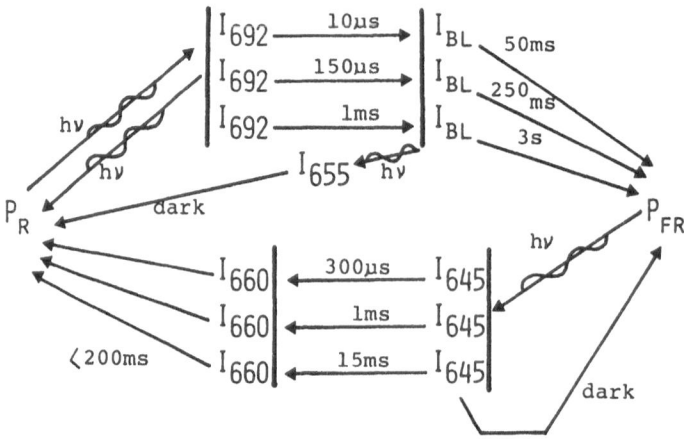

Figure 3. A general scheme showing probable phototransformation pathways and intermediates of phytochrome constructed from data in [16-20]

Phototransformation pathways were re-examined in "large" phytochrome of pea [16,17] and oat [18-20], demonstrating that the phototransformation pathway of "large" phytochrome from Pr to Pfr consisted of at least three reaction stages, as with "small" phytochrome. These included a photoinduction process of the first intermediate having absorption maximum about 692 nm (I_{692}), a decay process of I_{692} to the second intermediate having a relatively low extinction coefficient (I_{bl}) and the formation of Pfr from I_{bl} (Fig. 3). The last two processes proceeded in the dark. Decay curves of them could be resolved to three parallel first-order reaction components by kinetic analysis [16,17,19]. Transient absorption spectra changes induced by decay of three reaction components in the process of the first inter- mediate decay were similar [19], and supported the results of kinetic analysis that the decay process consisted of parallel reaction components. The second laser-flash irradiation on I_{692} induced rapid photoconversion of it back to Pr without formation of a detectable new intermediate [20]. The amount of Pfr induced by a saturation level of single red laser-flash on Pr was only about 1/3 of that induced by contin- uous red-light irradiation of Pr [16], and it was reduced when the wavelength of the laser flash was close to the peak wavelength of I_{692} [Inoue, unpublished data]. These results suggested that photoequilibrium between Pr and I_{692} was established within the period of single laser flash irradiation (ca. 600 ns). The second laser flash irradiation on I_{bl} also induced photoconversion back to Pr [20], but, in this case, at least one new intermediate (I_{655}) was formed on the way from I_{bl} to Pr. These results suggested that the conformation of apoprotein did not change so much in the phototransformation process from Pr to I_{692}, but that of I_{bl} was different from that of I_{692}. In other words, conformational change of the chromophore was mainly induced in the first reaction stage, from Pr to I_{692}, and phototransformation processes proceeding in the dark, from I_{692} to I_{bl} and from I_{bl} to Pfr, were relaxa- tion processes of apoprotein. The similarity of oscillator strength ratios between blue and red regions in intermediates calculated from their estimated absorption spectra [21] also supported this interpretation.

The phototransformation pathway of "large" pea phytochrome from Pfr to Pr [17] also consisted of at least three reaction stages, and the last two stages were composed of three parallel first-order reaction components (Fig. 3). Some parts of the first intermediate reverted back to Pfr in the dark. The photoreactivity of each component has not been made clear yet.

4 Phototransformation of Phytochrome in Pea Epicotyl Tissue

Phototransformation pathways of isolated phytochrome at physiological temperature have been relatively well characterized by flash photolysis, as described above [9-12,16-20]. However, little was known about phytochrome phototransformation in vivo, except at very low temperatures [see 5] where the spectra of phytochrome are distorted by fluorescence of chlorophyll [see 22]. The deposition of intermediate(s) in the slowest reaction step in the chain from Pr to Pfr under continuous light irradiation [23-25] and the decay of the intermediate(s) to Pfr [26,27] have only been reported at physiological temperature.

Phototransformation of phytochrome in vivo was rather suggested by the data on effects of flash light on physiological responses using the knowledge obtained in the phototransformation in vitro [28-31]. But no one knew whether the phototransformation pathway determined in vitro was applicable to that in vivo or not. Therefore we tried to determine the phototransformation pathway of phytochrome in etiolated pea epicotyl from Pr to Pfr with a flash photolysis technique [14].

Effects of a red laser flash on the induction of phototransformation from Pr to Pfr were saturated at ca. 15 mJ for flash wavelengths of both 640 and 655 nm. This level was about a half of that determined in isolated "large" pea phytochrome [16]. The amount of Pfr induced by a saturating laser pulse was ca. 50 % of that obtained at the photostationary equilibrium established under continuous red light irradiation. This value was about 1.5 times higher than that determined in isolated "large" pea phytochrome [16]. A difference spectrum measured 15 μs after the flash showed an absorbance increase at about 697 nm and a decrease at 663 nm. A difference spectrum determined 200 ms after the flash showed no such major absorbance increase. These data suggested that the intermediates, I_{692} and I_{bl}, which were observed in isolated "large" phytochrome, were also formed in the phototransformation pathway of phytochrome in vivo. Kinetic analysis of the rapid absorbance decrease at 700 and 710 nm gave one simple first-order reaction component having a rate constant of 2500 s^{-1}. Kinetics of Pfr appearance measured by absorbance increase at 750 nm was resolved into three parallel first-order reactions having rate constants of 5, 1.8, and 0.4 s^{-1} as for the case of isolated "large" phytochrome. Both intermediates were photoreactive and irradiation of these intermediates by the second flash induced the photoconversion back to Pr also like isolated "large" phytochrome.

These results suggested that the phototransformation pathway of phytochrome in vivo is in principle the same as that of isolated "large" phytochrome except for the lack of parallel reaction components in the decay of the first intermediate, and the knowledge obtained in isolated phytochrome is useful to interpret physiological data obtained in vivo.

5 Phototransformation Pathway of "Native" Pea Phytochrome

Recently, it was reported that many physico-chemical characters of "native" oat phytochrome are different from those of "large" phytochrome [see 32]. The phototransformation pathway of isolated "native" phytochrome is not yet well understood. RUZSICSKA and coworkers [33] measured the decay of the first intermediate from Pr to Pfr in both "small" and "native" oat phytochrome by flash photolysis, and reported that the kinetic parameters of both species were similar. On the other hand, EILFELD and RUDIGER [34] reported that a bleached intermediate (I_{bl}) observed in the phototransformation pathway of "small" [9] and "large" [19] oat phytochrome, was normally not formed in the phototransformation of "native" oat phytochrome determined by low temperature spectroscopy. Therefore, analysis of the entire phototransformation pathway of "native" pea phytochrome from Pr to Pfr at physiological temperature was undertaken next.

Effects of a red laser pulse on the induction of the final phototransformation product (Pfr) were saturated at ca. 30 mJ. This fluence level was equal to that of isolated "large" pea phytochrome [16] but was about 2 times higher than that of phytochrome in pea epicotyl [14]. The amount of Pfr induced by a saturating red laser pulse was about a half of that obtained at the photostationary state imposed by continuous red light irradiation. This level was equal to that of phytochrome in vivo [14], but was ca. 1.6 times higher than that obtained in "large" pea phytochrome [16]. These results suggest that the quantum efficiency of phototransformation between Pr and I_{692} in "native" phytochrome is a little different from that in "large" phytochrome. Flash induced absorption spectra changes determined at 5 µs and 2.5 ms also suggested the formation of I_{692} and I_{bl} [35]. Kinetic analysis of the first and second intermediate decay gave three parallel first-order reaction components in both steps like "large" phytochrome (Table 1).

These data suggest that the phototransformation pathway of isolated "native" phytochrome is similar to that determined in isolated "large" phytochrome.

6 Comparison between Kinetic Data of "Large", "Native" and in vivo Phytochrome

Kinetic parameters of phytochrome phototransformation from Pr to Pfr determined in isolated "large" [16,18], "native" [35] and in vivo [14] pea phytochrome have been compared (Table 1). The rate constant of the first and the second intermediate decay determined in "native" phytochrome were similar to those determined in "large" phytochrome. In particular, rate constants of the second intermediate decay to Pfr in "native" phytochrome showed good coincidence with those determined in "large" phytochrome by SHIMAZAKI et al. [16]. Data obtained in isolated "native" pea phytochrome was not only qualitatively but also quantitatively similar to that determined in isolated "large" phytochrome. This conclusion was consistent with data obtained in isolated "small" and "native" oat phytochrome [33]. These results suggested that the difference of monomer molecular weight of phytochrome had little effect on the phototransformation pathway of phytochrome, and that the photochromic nature of phytochrome mainly depended on the chromophore domain of phytochrome apoprotein.

Table 1. Comparison of kinetic parameters for the reaction monitored during the phototransformation of Pr to Pfr at ca. 2°C in several kinds of pea phytochrome [35].

Reaction	Rate constants(% amount) (s^{-1})			
	"large"[a]	"large"[b]	"native"[c]	Tissue[d]
Decay of the first intermediate				
k_{11}	46000(25)	46000(7)	42000(26)	-----
k_{12}	2500(75)	2500(59)	4100(20)	2570(100)
k_{13}	---[e]	430(34)	1500(54)	----
Decay of the second intermediate				
k_{21}	2.3 (30)	4.1 (31)	2.4 (14)	5.1 (37)
k_{22}	0.4 (35)	0.9 (69)	0.4 (69)	1.8 (50)
k_{23}	0.07(34)	---[e]	0.08(17)	0.36(13)

[a][16], [b][18], [c]Present study, [d][14], [e]not determined.

On the other hand, kinetic parameters determined in phytochrome in etiolated pea epicotyl [14] were a little different from those of isolated phytochrome. The detector system used in the in vivo study was the same as that used in the study of "native" phytochrome, but was different to that used in the study of "large" phytochrome. The monomer molecular weight of phytochrome in etiolated pea epicotyl was equal to that of isolated "native" phytochrome [36]. Therefore, the difference in detection methods and molecular size were not the causes of the difference of kinetic parameters between "native" and in vivo phytochrome. The buffer systems used in the studies of isolated "large" and "native" phytochrome were similar to each other. But these buffers probably could not entirely reproduce the micro-environment of phytochrome molecules in living cells. Slight but significant differences of kinetic parameters between the phototransformation of "native" and in vivo phytochrome were probably brought about by the different micro-environment of phytochrome molecules.

7 Effect of pH on the Phototransformation of "Native" Pea Phytochrome

A comparison of the kinetic data determined in isolated "large" and "native" pea phytochrome and phytochrome in etiolated pea epicotyl tissue (Table 1) suggests that micro-environment of phytochrome molecules has some effect on the phototransformation of phytochrome. The micro-environment of the phytochrome samples used in the previous flash photolysis studies could be regarded as similar, because the buffer conditions used in the previous reports were similar. The pH condition was conventionally always kept at 7.8 [9-12,16-20,33]. Recently, we have found that the pH condition affects the absorption spectra of isolated "native" pea phytochrome determined under continuous red-light irradiation [37]. This report suggested that the pH condition has some effect on the phototransformation of isolated phytochrome. Therefore, effects of pH on kinetics of phototransformation in isolated "native" pea phytochrome from Pr to Pfr were next examined.

In the previous studies, the temperature of the samples was always kept at 1-2°C. But in this study, the temperature was kept at 10°C. This was necessary since, in basic pH conditions, the reaction speed of the second intermediate decay was reduced too much to get entire reaction within the limit of machinery (100 s) at 2°C.

Table 2. Effect of pH on kinetic parameters for the reaction monitored during the phototransformation of Pr to Pfr at 10°C in isolated "native" pea phytochrome [35].

pH	Rate constants(% amount) (s^{-1})		
	k_1	k_2	k_3
Decay of the first intermediate			
6.8	------	13600(6)	4200(94)
7.8	------	13000(14)	3900(86)
8.8	------	9000(26)	3700(74)
Decay of the second intermediate			
6.8	40(11)	5.7(56)	1.5 (33)
7.8	32(19)	5.5(33)	1.4 (48)
8.8	12(45)	1.7(8)	0.38(47)

Kinetic analysis of the first intermediate decay brought only two parallel first-order reaction components (Table 2, upper row), because increment of the temperature accelerated the reaction speed and the decay speed of the fast component was too fast to determine the correct rate constant at any pH. Rate constants were about three times higher than the corresponding rate constants measured at 2°C. Rate constants of both components were higher in acidic conditions, but the difference was small. This result suggested that difference in pH had little effect on the decay of the first intermediate within the examined pH range.

Difference of pH had distinct effects on the rate constants of the second intermediate decay (Table 2, lower row). Rate constants of all three reaction components showed little difference between neutral and acidic pH condition. At basic pH, however, rate constants of all three components were reduced to ca. 1/3 of the corresponding rate constants determined at neutral pH. Besides the change of rate constants, the amount of the moderate-speed reaction component decreased when the pH shifted from acidic to basic, and amount of the fast-speed component increased.

As shown in Table 2, the pH condition clearly affects the kinetics of phytochrome phototransformation, especially in the final reaction step from Pr to Pfr. The absorption spectra determined at the photostationary state brought under continuous red-light irradiation showed a low Pfr peak at higher pH condition [37]. But the absorption spectra obtained 55 s after the irradiation were similar irrespective of the pH [37]. The deposition of intermediate(s) (probably I_{b1}) in the slowest reaction step in the chain from Pr to Pfr under continuous light irradiation have been reported [23-25]. Slow rate constants of the second intermediate decay at basic pH condition could explain the low absorption observed under continuous red-light at higher pH. TOKUTOMI and co-workers [38] also reported that proton uptake and release occurred during phototransformation of isolated "large" pea phytochrome. The data in Table 2 suggest that the proton uptake occurs in the final phototransformation step from Pr to Pfr.

At basic pH, rate constants of all three parallel reaction components in the decay of I_{b1} were reduced to ca. 1/3 of the corresponding rate constants determined at neutral pH. This result favors the parallel reaction model as shown in Fig. 3. In this case, the amount of the reaction components also changed when the pH shifted. This result suggests that the parallel reaction is not due to a fixed difference in the phytochrome molecules such as a difference of molecular size but is caused by a flexible difference in the phytochrome molecules such as a difference in the state of association.

8 Dichroic Orientation of Phytochrome Intermediates in the Pathway from Pr to Pfr Determined in Polarotropism of Fern Protonemata

A difference in the dichroic orientation of Pr and Pfr has been demonstrated physiologically in polarotropism of fern protonemata [39-41], in phototropism of moss protonemata [42] and in photoorientation of a chloroplast in alga [43]. A difference of the direction of the transition moment between Pr and Pfr was also indicated in isolated phytochrome [44-46]. But it is not established in which phototransformation step the direction of the transition moment of phytochrome changes. As described above, the phototransformation pathway of phytochrome in vivo is probably not so different from that of in vitro [14]. As both phototransformation intermediates, I_{692} and I_{b1}, in the phototransformation pathway were photoreactive and reverted back to Pr with a second far-red flash without formation of Pfr [14,20], the effects of a polarized far-red flash given at appropriate periods after the first inductive red flash should reflect the orientation of intermediates. Therefore, the effects of a polarized second flash on the suppression of polarotropism of Adiantum protonemata induced by the first polarized red-flash irradiation were examined [15].

A polarotropic response was effectively induced with a flash of polarized red light (640 nm) having the vibration plane of the electrical vector parallel to the protonemal cell axis. When a flash of polarized far-red light (710 nm) was given 30 s after the red flash, the red-flash-induced response was reversed by a far-red flash vibrating normal to the cell axis but not by one vibrating parallel. This result meant that dichroic orientation of Pfr was perpendicular to that of Pr, and the physiological response of protonemata to flash light was equal to the previously reported results where continuous light was used [40,41]. However, when the second flash was given 2 µs or 2 ms after the first red flash, the polarotropic response was not reversed by a polarized far-red flash vibrating normal to the cell axis but was reversed by a parallel-vibrating flash. These results meant that dichroic orientation of I_{692} and I_{b1} were parallel to that of Pr, and suggested that the dichroic orientation of phytochrome in fern protonemata changed at the final phototransformation step from Pr to Pfr, and that the direction of the transition moment of phytochrome molecule should change at this step.

9 Concluding Remarks

An outline of the phototransformation pathway of phytochrome has now become clear as described above. However, it is still an open question as to what kind of reaction occurs at the molecular level at each phototransformation process, and what is the primary action of signal transduction by phytochrome. The primary structure of phytochrome chromophore has already been determined [2,3], but little is known about its higher order structure. Probably, elucidation of the fine higher order structure of chromophore and apoprotein of Pr is one of the most urgent subjects to make clear the molecular mechanism of phototransformation. In the early days of phytochrome study, isolation of phytochrome was laborious, and phytochrome was believed to be a labile molecule. Nowadays, a simple and rapid isolation method has been developed [47], and thus isolated phytochrome can be stored for about one month in a refrigerator without proteolytic degradation. Therefore, it seems possible to apply several physico-chemical methods to get information about higher order structure of the phytochrome molecule including X-ray analysis and Raman spectroscopy.

Flash photolysis is well-suited to study the phototransformation pathway of phytochrome at physiological temperature. On the other hand, low temperature spectroscopy is good to determine the definite absolute absorption spectrum of the phototransformation intermediate. Low temperature spectroscopy has already been applied to the study of phytochrome phototransformation [5,34,48-53], and absolute absorption spectra of intermediates were roughly estimated [34,52]. However, strict fluence response experiments on the formation of intermediate to estimate the correct amount of intermediate checking the existence of an isosbestic point such as used in the study of rhodopsin [54] were not made in these previous studies. Now, these type of low temperature studies are required not only to know the absolute spectra of intermediates but to apply other spectroscopic methods on phytochrome at low temperature [55].

Utilization of chromophore analogs is a fruitful method to make clear the molecular mechanism of photoreaction as discussed by TOKUNAGA et al. and MAEDA et al. [this symposium]. In the case of phytochrome, it is not easy to adapt this method, since chromophore of phytochrome is tightly linked to apoprotein. Probably, the integration step of chromophore to apoprotein in biosynthesis is the only way to assemble chromophore analogs in phytochrome. DNA of oat phytochrome has already been isolated [56], but nothing is known about the integration of chromophore. Studies of the biosynthesis of phytochrome are also expected.

Many problems are still unsolved in the field of phytochrome phototransformation. However, one may be sure that the photochromic nature of phytochrome is so unique that further studies are worth pursuing.

Acknowledgments

This work has been made possible by collaboration with many experts, who I should like to gratefully acknowledge for their help. I am specially indebted to Prof. Furuya of Tokyo University for his continuous encouragement and advice.

1. R.L. Satter, A.W. Galston: In Chemistry and Biochemistry of Plant Pigments, ed. by T. W. Goodwin, Vol.1, (Academic Press, New York 1976) p.680
2. C.J. Lagarias, H. Rapoport: J. Am. Chem. Soc. 102, 4821 (1980)

3. W. Rüdiger, T. Brandlmeier, I. Blos, A. Gossauer, J. -P. Weller: Z. Naturforsch. 35C, 762 (1980)
4. W.L. Butler, K.H. Norris, H.W. Siegelman, S.B. Hendrick: Proc. natl. Acad Sci. U.S.A. 45, 1703 (1959)
5. R.E. Kendrick, C.J.P. Spruit: Photochem. Photobiol. 26, 201 (1977)
6. F.E. Munford, E.L. Jenner: Biochemistry, Wash. 5, 3657 (1966)
7. W.R. Briggs, H.V. Rice: Ann. Rev. Plant Physiol. 23, 293 (1972)
8. R.D. Vierstra, P.H. Quail: Proc. natl. Acad. Sci. U.S.A. 79, 5272 (1982)
9. H. Linschitz, V. Kasche, W.L. Butler, H.W. Siegelman: J. Biol. Chem. 241, 3395 (1966)
10. H. Linschitz, V. Kasche: Proc. natl. Acad. Sci. U.S.A. 58, 1059 (1967)
11. L.H. Pratt, W.L. Butler: Photochem. Photobiol. 11, 361 (1970)
12. S.E. Braslavsky, J.I. Mathews, H.J. Herbert, J. Dekok, C.J.P. Spruit, K. Schaffner: Photochem. Photobiol. 31, 417 (1980)
13. M. Furuya, Y. Inoue, Y. Maeda: Photochem. Photobiol. 40, 771 (1984)
14. Y. Inoue, M. Furuya: Plant Cell Physiol. 26, 813 (1985)
15. A. Kadata, Y. Inoue, M. Furuya: Plant Cell Physiol. 27, 867 (1986)
16. Y. Shimazaki, Y. Inoue, K.T. Yamamoto, M. Furuya: Plant Cell Physiol. 21, 1619 (1980)
17. Y. Inoue, K. Konomi, M. Furuya: Plant Cell Physiol. 23, 731 (1982)
18. M.-M. Cordonnier, P. Mathis, L.H. Pratt: Photochem. Photobiol. 34, 733 (1981)
19. L.H. Pratt, Y. Shimazaki, Y. Inoue, M. Furuya: Photochem. Photobiol. 36, 471 (1982)
20. L.H. Pratt, Y. Inoue, M. Furuya: Photochem. Photobiol. 39, 241 (1984)
21. T. Sugimoto, Y. Inoue, H. Suzuki, M. Furuya: Photochem. Photobiol. 39, 697 (1984)
22. L.H. Pratt: Photochem. Photobiol. 27, 81 (1978)
23. W.R. Briggs, D.C. Fork: Plant Physiol. 44, 1089 (1969)
24. R.E. Kendrick, C.J.P. Spruit: Nature New Biol. 237, 281 (1972)
25. R.E. Kendrick, C.J.P. Spruit: Photochem. Photobiol. 18, 139 (1973)
26. C.J.P. Spruit: Photochem. Photobiol. 35, 117 (1982)
27. R.E. Kendrick, J. Kome, P.A.P.M. Jaspers: Photochem. Photobiol. 42, 785 (1985)
28. G. Fuhr, W. Bleies, H. Göring: Plant Cell Physiol. 21, 561 (1980)
29. R. Scheuerlein: Z. Pflanzenphysiol. 109, 319 (1983)
30. M. Kraml, E. Schäfer: Photochem. Photobiol. 38, 461 (1983)
31. M. Kraml, M. Enders, N. Bürkel: Planta 161, 216 (1984)
32. P.H. Quail, J.T. Colbert, H.P. Hershey, R.D. Vierstra: Phil. Trans. R. Soc. Lond. B303, 387 (1983)
33. B.P. Ruzsicska, S.E. Braslavsky, K. Schaffner: Photochem. Photobiol. 41, 681 (1985)
34. P. Eilfeld, W. Rüdiger: Z. Naturforsch. 40C, 109 (1985)
35. Y. Inoue: In Phytochrome and Photomorphogenesis in Plants, ed. by M. Furuya, (Academic Press, Tokyo in press)
36. P.J. Lumsden, K.T. Yamamoto, A. Nagatani, M. Furuya: Plant Cell Physiol. 26, 1313 (1985)
37. S. Tokutomi, Y. Inoue, N. Sato, K.T. Yamamoto, M. Furuya: Plant Cell Physiol. 27, 765 (1986)
38. S. Tokutomi, K.T. Yamamoto, Y. Miyoshi, M. Furuya: Photochem. Photobiol. 35, 431 (1982)
39. H. Etzold: Planta 64, 254 (1965)
40. A. Kadota, M. Wada, M. Furuya: Photochem. Photobiol. 35, 533 (1982)
41. A. Kadota, M. Wada, M. Furuya: Planta 165, 30 (1985)
42. E. Hartmann, B. Klegenaberg, L. Bauer: Photochem. Photobiol. 38, 599 (1983)
43. W. Haupt, G. Mörtel, I. Winkelnkemper: Planta 88, 183 (1969)
44. C. Sundqvist, L.O. Björn: Photochem. Photobiol. 37, 69 (1983)
45. C. Sundqvist, L.O. Björn: Physiol. Plant. 59, 263 (1983)
46. N.G.A. Ekelund, C. Sundqvist, P.H. Quail, R.D. Vierstra: Photochem. Photobiol. 41, 221 (1985)
47. R. Grimm, W. Rüdiger: Z. Naturforsch. 41C, 988 (1986)
48. D.R. Cross, H. linschitz, V. Kasche, J. Tenenbaum: Proc. natl. Acad. Sci. U.S.A. 61, 1095 (1968)
49. L.H. Pratt, W.L. Butler: Photochem. Photobiol. 8, 477 (1968)

50. P.-S. Song, H.K. Sarkar, I.-S. Kim, K.L. Poff: Biochem. Biophys. Acta <u>635</u>, 369 (1981)
51. K. Manabe: Plant Cell Physiol. <u>24</u>, 1255 (1983)
52. W. Rüdiger, F. Thümmler: Physiol. Plant. <u>60</u>, 383 (1984)
53. N. Sasaki, Y. Oji, T. Yoshizawa, K.T. Yamamoto, M. Furuya: Photobiochem. Photobiophys. <u>12</u>, 243 (1986)
54. T. Yashizawa, G. Wald: Nature <u>197</u>, 1279 (1963)
55. Y. Inoue, H. Hamaguchi, K.T. Yamamoto, M. Tasumi, M. Furuya: Photochem. Photobiol. <u>42</u>, 423 (1985)
56. H.P. Hershey, R.F. Barker, K.B. Idler, J.L. Lissemore, P.H. Quail: Nucleic Acids Res. <u>13</u>, 8543 (1985)

Photochemical Holeburning and Stark Spectroscopy of Photosynthetic Reaction Centers

S.G. Boxer, D.J. Lockhart, and T.R. Middendorf

Department of Chemistry, Stanford University, Stanford, CA 94305, USA

1. Introduction

The initial step in photosynthesis involves photoexcitation of a primary electron donor and electron transfer to a primary electron acceptor. These processes take place in a membrane-protein complex called the photosynthetic reaction center (RC). The best characterized RCs have been isolated from the photosynthetic bacteria R. spheroides (R-26 mutant) and R. viridis. In both species, the complexes consist of three proteins each of approximate molecular weight 25-30kDa, four bacteriochlorophylls (BChls), two bacteriopheophytins (BPheo, BChl in which two H atoms replace the central Mg atom), two quinones and one non-heme iron [1]. In R. spheroides the pigments are a-type BChl, whereas in R. viridis the pigments are b-type. The prosthetic groups are embedded in the RC protein in vectorial fashion across the membrane, as shown by x-ray crystallography (Fig. 1) [2-4]. Combined with extensive earlier spectroscopic studies, the x-ray structure identifies the functional components with specific structural elements: the primary electron donor, often called P or the special pair, is identified as the strongly-coupled

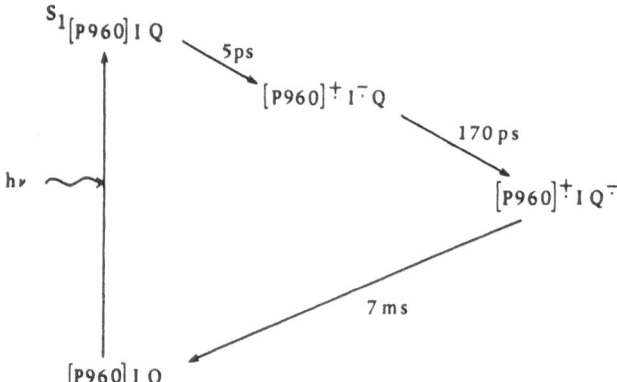

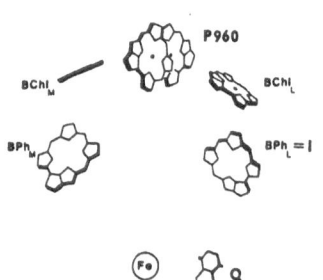

Figure 1: Kinetic scheme of the initial charge separation and recombination steps in RCs from R. viridis and the physical relationship of the reactive components adapted from the x-ray structure of Deisenhofer and co-workers [2]. P960 or P is the primary dimeric electron donor, I is an intermediate redox carrier and involves the BChl and BPheo monomers on the L-side of the structure, and Q is menaquinone. The lifetimes of the states are at cryogenic temperatures.

BChl dimer; the intermediate acceptor, often called I, is identified as the BPheo on the L-side; and the primary quinone (Q_A, ubiquinone in R. spheroides, menaquinone in R. viridis) is the quinone on the L-side. The function of the BChl situated between the dimeric electron donor and the BPheo acceptor is uncertain, though it is certainly involved in mediating electron transfer from the excited state of P to I. The role of the chromophores on the M-side of the RC, which are related to those on the functional L-side by a C_2 axis running vertically through the Fe, is unknown. Recent structural work on the R. spheroides RC suggests that the gross features of its structure are very similar to those of R. viridis [5,6].

The initial electron transfer steps in the RC are very rapid as illustrated in Fig. 1, and the charge separation has a quantum efficiency approaching unity. The absorption spectrum of the RC, especially of the dimeric primary electron donor, has been the subject of very great interest for many years. The absorption band in the near-infrared region (about 870 nm for R. spheroides, about 960 nm for R. viridis, at room temperature), due to the lowest electronic transition of the dimer, is rather broad. It is generally believed that this band is the lower exciton (dimer) band for the lowest singlet excited state [7], and that the upper exciton component is weak due to the orientation of the transition dipole moments in the monomers comprising the dimer (very similar results were obtained in a model system for the primary electron donor which has the same symmetry and orientation of the monomer transition moments [8]). This broad, near infra-red absorption is nearly completely bleached when the RC is excited with a saturating flash of white light; a free radical epr signal is observed whose intensity is proportional to the degree of bleaching and whose decay kinetics exactly correspond to the recovery of absorption for the near infra-red band [1]. On this basis, the initial photochemistry is assigned to loss of an electron from the excited state of P to form P^+. Ps transient absorption experiments demonstrate that the excited state lifetime of P is several ps [9-12] and that the excited state decay is accompanied by other changes in absorption due to the formation of I^-, where I is the BPheo on the L-side. This excited state decay rate is consistent with that estimated many years ago from the fluorescence quantum yield of P [13]. No evidence is found for the intermediate formation of the anion radical of a monomeric BChl, although, given the sensitivity of current experiments, it might go undetected if its rate of decay were more than 3-4 times its rate of formation [14].

In addition to being the lowest energy electronic absorption in the RC, the absorption of the primary electron donor is unusual in several other respects. Its linewidth is considerably greater than that due to the monomeric BChls and BPheos (this is in spite of the fact that the monomer absorption bands are due to overlapping bands from chromophores on the L- and M-sides, which can be partially resolved at cryogenic temperatures). Secondly, both the linewidth and absorption maximimum for the absorption are strongly temperature dependent for the primary electron donor, whereas both change very little for the other absorption features (the dimer absorption band shifts to the red and narrows as the temperature is lowered). Until recently there has been very little consideration of the origin of the absorption linewidths in the RC. Struck by the fact that the primary electron donor absorption band is so much broader than the monomer absorption bands, one of us [15] suggested that the greater width might be due to a very short excited state lifetime. From the relationship between linewidth and lifetime given in Table 1, it is seen that an excited state lifetime of several ps corresponds to a linewidth of a few cm^{-1}. Since the observed linewidth for a chromophore in a glassy matrix (or a protein) is typically several hundred cm^{-1}, the excited state lifetime would have to be much shorter than several ps in order for the absorption linewidth to be noticeably affected by lifetime broadening.

Table 1 Conversion between lifetime (T_1) and homogeneous linewidth in the absence of pure dephasing (Γ_0)

T_1	50fs	500fs	5ps	50ps	500ps	5ns
$\Gamma_0 (cm^{-1})$	100	10	1	.1	.01	.001

The homogeneous linewidth can be measured by a photon echo or a holeburning experiment. For holeburning experiments, the measured holewidth (FWHM) is equal to twice the homogeneous linewidth when the hole is narrow compared to the inhomogeneous line and the laser linewidth is small compared to the holewidth. In addition to the minimum linewidth determined by the excited state lifetime, rapid fluctuations in the site energy during the excited state lifetime contribute to the homogeneous linewidth. Such pure dephasing processes, characterized by time T_2^*, generally have a strong temperature dependence which is characteristic of electron-phonon coupling. The homogeneous linewidth (FWHM = Γ_h) is given by:

$$\Gamma_h = \frac{1}{\pi T_2} = \frac{1}{\pi T_2^*} + \frac{1}{2\pi T_1} . \tag{1}$$

It should be noted that Eqn. 1 is correct for a two-level system coupled to a bath. The effects of electron plus phonon transitions and transitions into congested levels are neglected.

The temperature dependence of the pure dephasing contribution to the homogeneous linewidth is very different in a crystalline and a glassy host matrix. In non-photochemical holeburning studies, we have shown that the pure dephasing contribution to the homogeneous linewidth for a chlorophyllide chromophore substituted for heme in the heme-binding pocket of apo-myoglobin is comparable to that for the same chromophore in an organic glass matrix and that its temperature dependence is very similar to that in a glassy matrix [16]. Even at the lowest temperature studied to date (1.35K), the holewidth for chlorophyllide in apo-myoglobin was found to be greater than 1GHz (.03cm^{-1}), which is greater than that due to the excited state lifetime alone (about 5 ns or 0.001cm^{-1}, Table 1). Thus, pure dephasing contributes to the homogeneous linewidth even at this very low temperature. In the context of the RC problem, however, it is seen that this linewidth is much smaller than the minimum possible linewidth for an excited state lifetime of several ps. Thus, when considering holeburning in the RC, it is safe to neglect the pure dephasing contribution to the holewidth.

2. Experimental Methods

Holeburning The method for photochemical holeburning in RC samples has been described in detail elsewhere [17,18]. Briefly, RCs in poly(vinylalcohol) (PVA) films were immersed in pumped liquid He in an optical cryostat. The temperature was varied between 1.35 and 2.2K by adjusting the pressure above the liquid He (accuracy ±0.02K). The sample was excited with pulses from a Nd:YAG pumped dye laser whose output was Raman shifted to provide near infrared excitation (laser linewidth 1-2 cm^{-1}). The change in absorption (hole spectrum) was probed with a weak probe beam passed through a monochrometer (ultimate resolution ~0.3cm^{-1}). RCs were prepared by standard methods. The film containing R. viridis RCs was a generous gift from Professor Parson; the film containing R. spheroides RCs treated with NaBH$_4$ to remove one BChl monomer [19] was a generous gift from Professor Holten.

Stark Spectroscopy The method for obtaining Stark spectra of RC samples has been described in detail elsewhere [20]. Briefly, RCs in PVA films (~0.1mm thick) were coated on both sides with semitransparent Ni electrodes by vapor deposition (thickness ~80Å). High voltage leads were connected to the Ni electrodes by springs and an AC field was applied. The samples were probed with light from a 250W tungsten-halogen lamp passed through a 1/4m monochrometer. The small AC component (ΔI) was lock-in detected at the second harmonic of the field modulation frequency and was digitized along with the large DC component corresponding to the transmitted intensity (I). For polarization measurements the probing light was passed through a Glan-Thompson polarizer to produce horizontally polarized light and the sample was rotated around a vertical axis to vary the direction between the electric vector of the light and the applied field direction (F_{ext}). The experimental arrangement and relevant angles are shown in the inset to Fig. 4. All measurements were performed at room temperature and 77K.

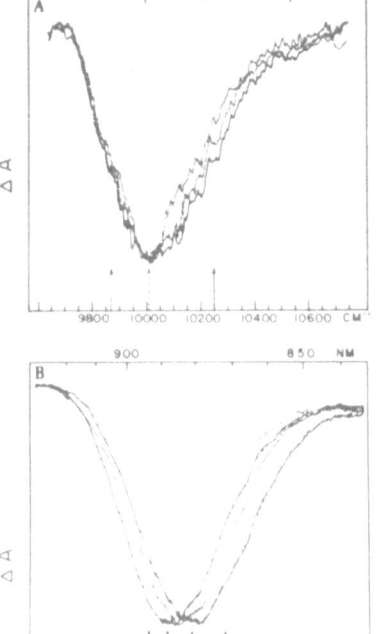

Figure 2: Photochemical holes observed in (A) R. viridis RCs at 1.4K, and (B) R. spheroides RCs at 1.5K. The burn wavelengths are indicated with vertical arrows. The amplitudes of the hole spectra were scaled to the same magnitude to facilitate comparison.

3. Results

Holeburning Spectroscopy   The changes in absorption following photoexcitation at various frequencies within the lowest energy electronic absorption band of the primary electron donor of both species of RCs are compared in Fig. 2. The holewidths and variation in hole position with excitation frequency for R. spheroides RCs treated with $NaBH_4$ were indistinguishable from those for untreated RCs. Likewise, RCs which had been incubated in D_2O-containing buffer for 1 month at 40°C showed identical results (PVA lyophylized repeatedly from D_2O solution). No variation of the holewidth was observed for either species between 1.5 and 2.2K, and the holewidths in all cases were found to be independent of excitation energy density for a bleach of less than 30% down to the weakest excitation flashes commensurate with measuring the holewidth (bleaches of ~5%).

Stark Spectroscopy   The room temperature Stark spectrum in the Q_y region of the absorption spectrum is shown in Fig. 3, along with absorption and derivative spectra. A plot of the dependence of $\Delta A(\chi)/\Delta A(\chi=90°)$ on χ for the Q_y transition of the dimer absorption band is shown in Fig. 4. Other data is presented in the discussion section.

4. Discussion

Holeburning Spectroscopy   The most striking and significant result to emerge from the photochemical holeburning studies of both species of RCs is that very broad holes are formed irrespective of burn frequency, power and temperature. Furthermore, the removal of one BChl in R. spheroides RCs has no effect on the holewidth, ruling out involvement of this chromophore (presumably the monomer BChl on the M-side) in causing the broad holes, and there is no deuterium isotope effect due to exchangeable protons. In the spirit of the background outlined in the introduction and Table 1, it is tempting to ascribe the very broad holes to a very short excited state lifetime. This is the view we expressed in a preliminary

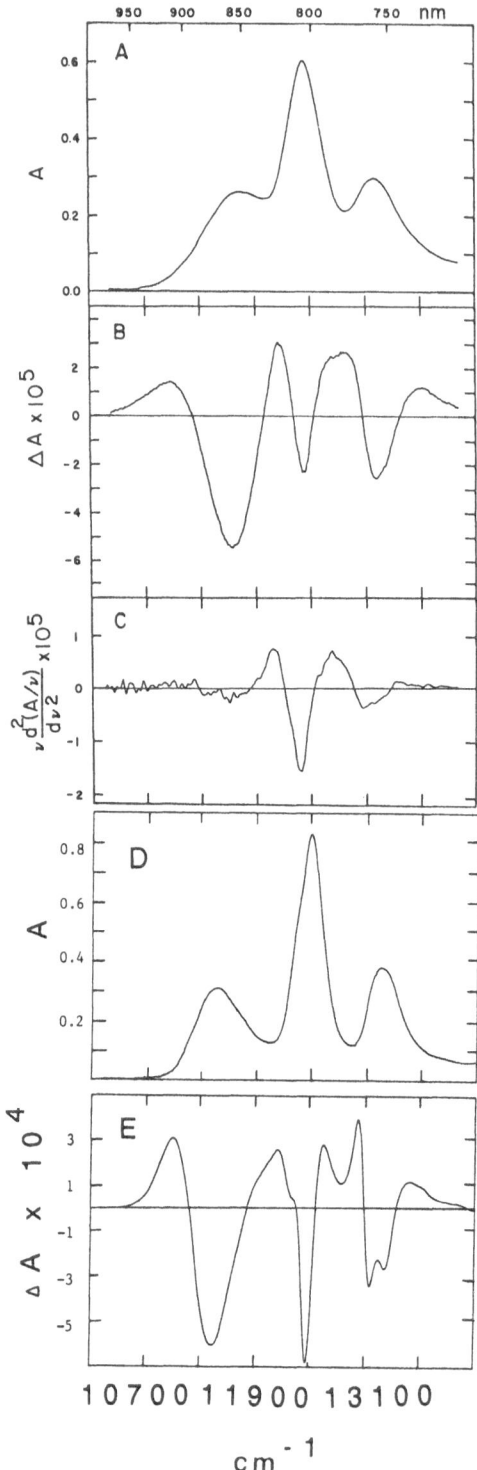

Figure 3: Absorption (A), Stark (B), and second derivative (C) spectra for the Q_y region of R. spheroides RCs (F_{ext} = 9.12x10[4] V/cm) at room temperature. Absorption (D) and Stark (E) spectra at 77K (F_{ext} = 2.59x10[5] V/cm).

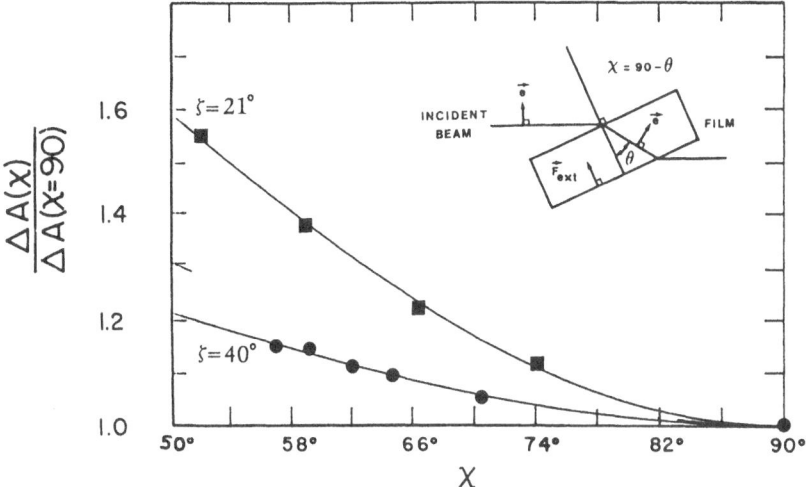

Figure 4: Plots of the calculated dependence of $\Delta A(\chi)/\Delta A(\chi=90°)$ on the experimentally variable angle χ as a function of the angle ζ between the direction of a transition dipole moment and the difference dipole moment, $\Delta\mu$. Solid curves show the best fits to the experimental data: ●●●● , special pair Q_y transition at room temperature; ■■■■ , BChl monomer Q_y transition in the RC at 77K. The inset illustrates the geometry of the sample and probe beam used for the Stark measurements.

presentation of these results [21]. Meech et al. [22] specifically endorse this picture. Even taking into account the expected presence of vibronic structure, the excited state lifetime of P would have to be much shorter than that observed in ps transient absorption experiments. Aside from the observed ps kinetics, two other observations make it unlikely that the broad holes are simply explained by a sub-ps excited state lifetime. In addition to measuring transient ps kinetics, both groups observed stimulated emission from the excited state of P, and the decay kinetics of the stimulated emission matched that of transient absorption changes [9,10]. Furthermore, it appears that the shape of the stimulated emission is similar to the spontaneous emission and that its shape does not change over the several ps that it is observed [W. W. Parson, personal communication]. Although a quantitative analysis of the intensity of this stimulated emission has not been presented, it is reasonable to argue that the excited state of P decays directly to the intermediate P^+I^- with no intermediate state formed very rapidly. The second observation is that the fluorescence quantum yield is consistent with the measured ps decay kinetics [13]; the quantum yield would have to be considerably lower if the excited state lifetime of P were much shorter. The problems with measurements of the fluorescence quantum yield for weakly emitting systems are well known, and it is possible that the observed fluorescence quantum yield is greater than the true quantum yield due to highly fluorescent, non-functional RCs. On balance, however, we feel that the evidence favors an excited state lifetime of several ps, so we must look elsewhere for an explanation of the broad holes.

As an alternative explanation, we consider the possibility that photoexcitation of P involves a substantial change in the equilibrium nuclear coordinates of P, possibly coupled to changes in its surroundings [17,18]. This could lead to complete suppression of the zero-phonon line leaving a broad (featureless) hole. Two related general classes of mechanisms can be suggested which could produce such an effect. The first is that the equilibrium structure of the excited state is substantially different from that of the ground state. Possibilities include movement of the macrocycles comprising the dimer, a well-known phenomenon for large aromatic molecules in molecular crystals (excimer formation), or movement of hydrogen bonded or liganded groups (possibilities include movement of the central Mg atom in the excited state or movement of hydrogen-bonded protons on the periphery of the macrocycle; the absence of a deuterium isotope effect provides some evidence against the latter suggestion). A second possibility is that the excited state dipole moment is very different from the ground state dipole moment, such that photoexcitation leads to a substantial shift in charge density within the dimer. Several related examples have been studied in crystals of charge-transfer complexes where a broad phonon band dominates a weak or non-existent zero-phonon line [23]. Such a change in charge density could, of course, be coupled to changes in the geometry of the environment of the dimer.

One of the well-known characteristics of molecules with a substantial change in dipole moment is that the fluorescence maximum is very sensitive to the polarity of the environment, i.e. rather large Stokes shifts are commonly observed. There is a substantial Stokes shift for the emission from RCs compared with what is observed for a BChl or chlorophyll monomer. The Stokes shift for a model pyrochlorophyllide a dimer is comparable to what is observed in the RC [8]. An analysis of the temperature dependence of the emission linewidth and position for R. viridis RCs has been presented by Scherer et al.[24]. Their analysis leads to the proposal that the excited state of P is coupled strongly to low frequency modes of average frequency 30cm^{-1}. The exact nature of this mode was not specified. We note, however, that the data used in their analysis is somewhat different from what has been obtained in other labs [25, D. Holten and C. Kirmaier, personal communication], and the analysis at very low temperature is inappropriate as high-temperature-limit equations are used. We also note that a rather different temperature dependence has been observed in our laboratory for R. spheroides RCs [L. Takiff and S. G. Boxer, to be published]. Hayes and Small [26] have accepted the analysis of Scherer et al. and have developed a model for our R. spheroides holeburning results (the analysis of Scherer et al. [25] was for R. viridis). Hayes and Small were able to simulate the very broad holes and the variation in hole position with excitation frequency which we have observed for R. spheroides [17]; it is not yet clear whether this very interesting proposal will be as successful if the analysis of the temperature dependence of the fluorescence maximum and linewidth by Scherer et al. requires revision. An alternative treatment by Won and Friesner explicitly considers strong coupling to charge transfer states involving the monomer BChl and BPheo chromophores [27].

In summary, we have argued [17,18] that a substantial displacement in equilibrium nuclear coordinate upon photoexcitation offers an alternative to lifetime broadening as an explanation for the broad photochemical holes observed in the RC. At least in part, the underlying mechanism may involve a substantial change in dipole moment which would be expected if the excited state has substantial charge-transfer character. The work of Hayes and Small [26] provides a very interesting theoretical framework for explaining the R. spheroides results, though it leaves open the question of the nature of the mode which is strongly coupled, does not explain the absence of a burn frequency dependence in the R. viridis holeburning results (Fig. 2B), and depends for numerical analysis on the data in [25]. Our suggestion that a significant degree of charge-transfer character may be important leads naturally to Stark spectroscopy as an approach to measuring the change in dipole moment between the ground and excited state.

Stark Spectroscopy Application of an electric field to an immobilized, randomly oriented sample leads to a broadening of the absorption spectrum if the ground and excited state dipole moments are different. In this case, the change in absorbance due to the difference between the ground and excited state dipole moments as a function of transition frequency is given by [28]:

$$\Delta A(\nu) = \frac{C_\chi}{30h^2} F_{int}^2 \nu \frac{d^2(A/\nu)}{d\nu^2} ,$$
(2)

where $C_\chi = 5\Delta\mu^2 + (3\cos^2\chi - 1)[3(\mathbf{p}\cdot\Delta\mu)^2 - \Delta\mu^2]$, χ is the angle between the applied electric field direction and the polarization vector of the probing beam (see inset to Fig. 4), $\mathbf{p}$ is a unit vector in the direction of the transition dipole moment being probed at frequency ν, and $\Delta\mu$ is the difference dipole moment between the ground and excited state. F_{int} is the actual field felt by the molecules under investigation, which is different from the applied electric field because of the dielectric properties of the environment. The magnitude and origin of this local field correction, $F_{int} = f \cdot F_{ext}$, can have a substantial effect on the correct quantitative analysis of $|\Delta\mu|$. We express the value of $|\Delta\mu|$ as the product of f^{-1} and the observed value of $|\Delta\mu|$ assuming $F_{int} = F_{ext}$.

$\Delta\mu$ is a vector quantity, and it is also possible to measure the angle, ζ, between $\Delta\mu$ and the direction defined by the transition dipole moment (p) for the associated electronic transition because of the $p \cdot \Delta\mu$ term in Eqn. 2. To the extent that the direction of the transition moment is known with respect to the **molecular axes** of the molecule(s) under consideration, a physical model for the charge separation pathway can be developed.

From the Stark spectrum (χ=90°) of R. spheroides RCs (Fig. 3), it is qualitatively clear that the electric field has a much greater effect on the Q_y absorption band of P than on the absorption features at 760 and 802 nm. This difference has been previously noted qualitatively by DeLeeuv and co-workers in an abstract [29]. Correcting for differences in the linewidths of each band, $|\Delta\mu|$ for the special pair Q_y band is a factor of approximately 3.7 times greater than for the monomer BChl (802 nm) Q_y band. The data in Fig. 4 demonstrate that the angle between the special pair Q_y transition dipole moment and $\Delta\mu$ is 40.4±2° at room temperature and about 35° at 77K, while that for the Q_y transition of the monomer BChl is 21° at 77K. Quantitative analysis of the Stark data is discussed elsewhere [20]; for the special pair Q_y transition $|\Delta\mu|$ is calculated to be $f^{-1}(9.3\pm0.7)$D at 20C and $f^{-1}(6.4\pm0.6)$D at 77K. Using the spherical cavity approximation and an effective dielectric constant of 2, we obtain the value $|\Delta\mu|$=7.8D at room temperature and 5.4D at 77K (assuming that f^{-1} and the sample thickness are temperature independent; the origin of the temperature dependence is unknown). We have also obtained Stark data for BChla and BPheoa monomers in polystyrene (ε = 2.6) [20]. For both molecules $|\Delta\mu|$ for the Q_y transition was found to be approximately $f^{-1}(2-2.5)$D and ζ was approximately 17°. at room temperature. We note that the values given here for room temperature correct a small error in the analysis presented in [20]. We have also obtained very similar results for $|\Delta\mu|$ and ζ for the Qy transition of the special pair in R. viridi RCs at 77K [Lockhart and Boxer, to be published].

The value $|\Delta\mu| = f^{-1}(9.3\pm0.7)$D for the Q_y transition of the special pair is considerably larger than for the monomer BChls and BPheos in the RC or for the pure monomers in polystyrene. Both $|\Delta\mu|$ and ζ are comparable for the monomeric BChl in the RC and in polystyrene. The most striking effects are the differences in both $|\Delta\mu|$ and ζ for the Q_y transition of the special pair compared to monomer BChl a in the RC or pure in polystyrene. Ignoring possible differences in f, $|\Delta\mu|$ for the special pair is almost $f^{-1} \cdot 7$D larger than for monomer BChl a, a very large difference, and $\Delta\mu$ is rotated relative to the Q_y axis in the special pair as compared with monomer BChl a. If the ground state dipole moment for the dimer were very small compared to 8D, then $|\Delta\mu|$ is dominated by the dipole moment of the excited state. A full charge separated between the monomers comprising the dimer would produce a dipole moment of about 34D. If the ground state is dipolar and its dipole moment direction is substantially different from that in the lowest excited singlet state, the magnitude of $\Delta\mu$ and the angle ζ are less straightforward to interpret.

The measured angle between the transition dipole moment for excitation of the lowest singlet state of the dimer and $\Delta\mu$ offers a basis for proposing a physical model in which charge is separated **between** the macrocycles comprising dimeric P. This is illustrated schematically in Fig. 5, where a reasonable orientation for the dimer transition moment is illustrated based on polarized single-crystal absorption data for R. viridis [30]. Using this direction for the Q_y transition of the special pair in R. spheroides, we can calculate the expected value of the angle ζ between the Q_y transition moment and a vector connecting the central Mg-atoms of the monomers comprising the special pair. This angle is approximately 30°. Thus, it is reasonable to propose that charge is separated between the monomers comprising the special pair upon excitation, although participation of the monomer BChl is not ruled out. A more definitive analysis awaits a precise determination

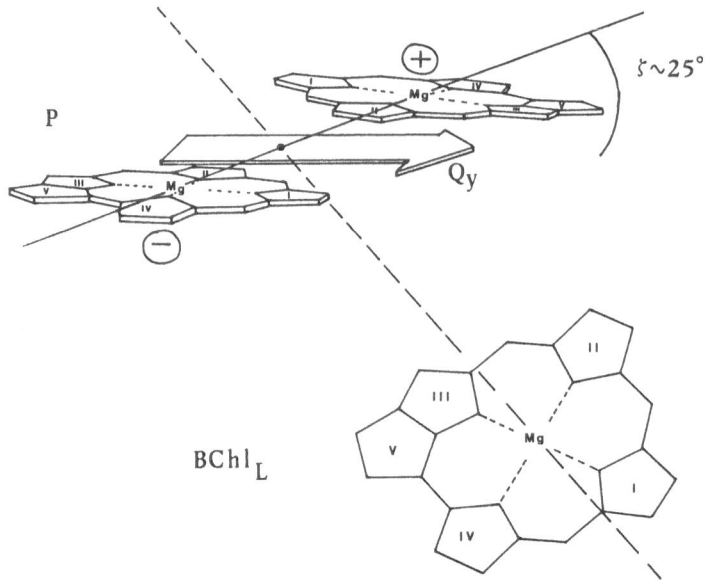

Figure 5: Schematic illustration of the angle between Δμ and the Q_y transition moment direction of the special pair obtained experimentally and a possible relationship to the structure of the special pair [2]. Also shown is the monomer BChl on the L-side.

of the direction of the Q_y transition and detailed calculations for which the Stark data provide precise values for comparison.

5. Conclusions

It is our view that a consistent description of the initial photophysical and photochemical processes in bacterial photosynthesis can be obtained by combining the holeburning and Stark data. From the Stark measurements we conclude that the state of P formed directly upon photoexcitation into the lowest energy electronic state has a substantially asymmetric charge distribution and that the line of asymmetry may be directed between the monomers composing the special pair. If the excited state of P has a much different electronic configuration than the ground state, the equilibrium nuclear configuration of P will also be different. For a large displacement between the potential energy surfaces, the Franck-Condon factor for the 0-0 transition will not be large compared to its value for transitions higher into the excited state manifold. In this case, the absorption band will be due to transitions to states above the lowest vibronic level where the density of relevant states (i.e. those coupled to the electronic transition) could be much greater. Strong coupling between the electronic transition and the environoment (electron-phonon coupling), as well as coupling to vibronic states of neighboring chromophores, can lead to further congestion of the excited state manifold. If the energy separation between the states which are coupled to the electronic transition is comparable to or less than the width of these levels, then the states are not resolvable. Under such circumstances the width of a hole burned in the absorption profile would be determined by the density of states, the distribution of Franck-Condon factors, the electron-phonon coupling strength and the rate of relaxation from the states, not the lifetime of the lowest vibronic level of the first excited singlet state. It is unlikely that the dipole moment difference of about 8D which we have observed for P at room temperature can itself account for the broad holes. Additional strong coupling to the environment and/or states of nearby chromophores [27] is probably required to ensure that the states to which the transition occurs contribute a broad band of unresolvable levels.

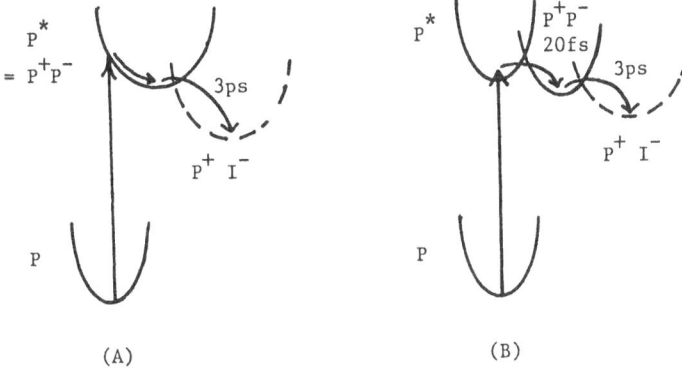

(A)   (B)

Figure 6: Schematic potential energy surfaces illustrating the difference for
a reaction scheme in which excitation of P leads directly to a highly dipolar
excited state (loosely termed P^+P^-) with a substantial diplacement of the
equilibrium nuclear coordinate relative to the ground state (A) and a scheme in
which P is excited to an essentially neutral state, followed by a rapid decay
to a charge separated state (B).   Scheme B is not expected to lead to a
substantial Stark effect.   In scheme A, vibrational relaxation in the excited
state is likely to involve solvent (protein) reorganization as well as
adjustment of the configuration of the special pair; however, given the very
short excited state lifetime of P, it is not certain whether this relaxation is
complete prior to electron transfer to I.

The physical picture which emerges is summarized in Figure 6A: P is excited to a
state whose electronic configuration is asymmetric and different from the ground
state.   This state is some linear combination of neutral and charge-transfer
states, both intradimer and involving the neighboring chromophores (see Professor
Parson's contribution to this volume for a specific and very reasonable model
[31]).   The key feature is that the excited state potential surface is
substantially displaced from the ground state surface due to the substantial
excited state dipole moment.   The Stark measurement ensures that the mixing of
charge-transfer character is strong relative to dephasing of the excited state,
rather than being weak, in which case the system would be excited to a neutral
(presumably not displaced) excited state and would subsequently evolve to an
intra-dimer charge-separated state (the latter would be a T_1 process and would not
result in a large Stark effect,as illustrated in Figure 6B).[1]

The role of solvent relaxation is not clear at this time.   The absence of any
change in the spectrum of stimulated emission during the excited state lifetime
suggests that the displaced excited state potential surface does not evolve in
time; however, insufficient data are available at the present time. A substantial
body of data suggests that a highly dipolar excited state in a fluid or disordered
solid solution leads to a substantial reorientation of solvent dipoles in response
to the newly generated excited state dipole.   In such cases the solvent is at most
weakly ordered around the neutral ground state and is far from equilibrium
immediately after excitation. Relaxation of the solvent dipoles around the excited
state dipole moment leads to a time-dependent red shift in the emission spectrum on
the time-scale of the dielectric relaxation time of the solvent.   The situation in
the vicinity of P may be quite different: the solvent in this case is the protein
matrix whose specific structure organizes the amino acid and helix dipoles. It is
possible that the protein is organized to stabilize the highly dipolar excited
state of P, a notion which follows naturally from the requirement that directional
charge separation along the L- rather than the M-branch of chromophores (Fig. 1) is
a consequence of the ordered protein environment. The time associated with protein
reorganization is expected to be comparable to or longer than the excited state
lifetime of P (it is certainly not tens of fs at cryogenic temperatures), thus it
is not clear how important the relaxation pathway along the excited state potential

surface is prior to electron transfer to I. In order to resolve this important question it will be necessary to combine the information from the temperature dependence of Stark, emission and absorption spectra with the temperature dependence of the RC structure.

ACKNOWLEDGEMENTS

This work was supported in part by NSF Grants DMB-8607799 and DMB-8352149, Gas Research Institute Contract 5083-260-0824, and the Dow Chemical Foundation.

REFERENCES

1. S.G. Boxer: Biochim. Biophys. Acta 726, 2625-292 (1983).
2. J. Deisenhofer, O. Epp, K. Miki, R. Huber & H. Michel: J. Mol. Biol. 180, 385-398 (1984).
3. J. Deisenhofer, O. Epp, K. Miki, R. Huber & H. Michel, H. Nature 318, 618-624 (1985).
4. H. Michel, O. Epp & J. Deisenhofer: The EMBO Journal 5, 2445-2451 (1986).
5. C.H. Chang, D. Tiede, J. Tang, U. Smith, J. Norris, & M. Schiffer: FEBS Lett. 205, 82-86 (1986).
6. J.P. Allen, G. Feher, T.O. Yeates, D.C. Rees, J. Deisenhofer, H. Michel, & R. Huber: Proc. Natl. Acad. Sci., U.S.A. 83, 8589-8593 (1986).
7. Parson, W.W., A. Scherz, & A. Warshel: Springer Ser. in Chem. Phys. 42, 122-130 (1985).
8. S.G. Boxer & G.L. Closs: Journal Amer.Chem. Soc. 97, 3268 (1975).
9. N.W. Woodbury, M. Becker, D. Middendorf & W.W. Parson: Biochem. 24, 7516-7521 (1985).
10. J.L. Martin, J. Breton, A.J. Hoff, A. Migus & A. Antonetti: Proc. Natl. Acad. Sci. USA 83, 957-961 (1986).
11. J. Breton, J.-L. Martin, A. Migus, A. Antonetti & A. Orszag: Proc. Natl. Acad. Sci. USA 83, 5121-5125 (1986).
12. M.R. Wasielewski & D.M. Tiede: FEBS Lett. 204, 368-372 (1986).
13. K.L. Zankel, D. W. Reed & R.K. Clayton: Proc. Natl. Acad. Sci. U.S.A. 61, 1243-1249 (1968).
14. R.A. Marcus: Chem. Phys. Letts., in press.
15. S.G. Boxer: Biochem. Biophys. Acta 726, 265 (1983).
16. S.G. Boxer, D.S. Gottfried, T.R. Middendorf & D.J. Lockhart: J. Chem. Phys., 86, 2439 (1987).
17. S.G. Boxer, D.J. Lockhart & T.R. Middendorf: Chem. Phys. Lett. 123, 476-482 (1986).
18. S.G. Boxer, T.R. Middendorf & D.J. Lockhart: FEBS Lett. 200, 237-241 (1986).
19. S.L. Ditson, R.C. Davis & R.M. Pearlstein: Biochem. Biophys. Acta 766, 623-629 (1984).
20. D.J. Lockhart & S.G. Boxer: Biochemistry 26, 664-668 (1987).
21. S. G. Boxer: Proceedings of the Vth International Congress on Energy Transfer (August 1985), J. Pantoflicek and P. Pancoska, eds, 108-115.
22. S.R. Meech, A.J. Hoff & D.A. Wiersma: Chem. Phys. Lett. 121, 287-292 (1985).
23. D. Haarer: J. Chem. Phys. 67, 4076-4085 (1977).
24. P.O.J. Scherer, S.F. Fischer, J.K.H. Hörber & M.E. Michel-Beyerle: Springer Ser. in Chem. Phys. 42, 131 (1985).
25. C. Kirmaier, D. Holten & W.W. Parson: Biochim. Biophys. Acta 810, 49-61 (1985).
26. J.M. Hayes & G.J. Small: J. Phys. Chem. 90, 4928-4931 (1986).
27. Y. Won & R.A. Friesner: Proc. Natl. Acad. Sci. USA, in press.
28. R. Mathies, & L. Stryer: Proc. Natl. Acad. Sci. U.S.A. 73, 2169-2173 (1976).
29. D. DeLeeuv, M. Malley, G. Butterman, M.Y. Okamura and G. Feher: Biophys. Soc. Abstr. 37, 111a. (1982).
30. W. Zinth, M. Sanders, J. Dobler, W. Kaiser & H. Michel: Springer Ser. in Chem. Phys., 42, 97 (1985).
31. W.W. Parson, paper in this volume; W. W. Parson and A. Warshel, J. Am. Chem. Soc., in press.

Part III

Rhodopsin, Sensory Rhodopsin, Phoborhodopsin, and Retinochrome

Photochemical Primary Process of Octopus Rhodopsin

H. Ohtani[1]*, T. Kobayashi[1], M. Tsuda[2], and T.G. Ebrey[3]

[1]Department of Physics, University of Tokyo, Bunkyo, Tokyo 113, Japan
[2]Department of Physics, Sapporo Medical College, Sapporo 060, Japan
[3]Department of Physiology and Biophysics, University of Illinois,
  Urbana, IL61801, USA

1. Introduction

A visual pigment rhodopsin is a chromoprotein with an 11-<u>cis</u> retinal. Rhodopsin is
photodecomposed to an all-<u>trans</u> retinal and an apoprotein opsin. The primary event
is the photoisomerization of retinal from 11-<u>cis</u> to all-<u>trans</u> conformation [1] followed by sequential thermal reactions [2]. Intermediates have been found by low-
temperature spectroscopy except the first intermediate [3,4] primerhodopsin (Batho'
[5] or photorhodopsin [6]), which has a bathorhodopsin-like red-shifted absorption
spectrum. There are three intermediates, bathorhodopsin, hypsorhodopsin, and primerhodopsin, in the picosecond region. The sequence among them at physiological temperature has not yet been established.

   Some time after picosecond lasers were used for time-resolved spectroscopy, the
primary process of vision was studied by Busch and others [7]. They reported that
bovine bathorhodopsin is formed within a pulse width following the excitation of
rhodopsin with the second harmonic of Nd:glass laser (530 nm) with 6-ps fwhm and
concluded that bathorhodopsin is the first intermediate in the photobleaching reaction of rhodopsin. Shichida et al. [8] observed that squid hypsorhodopsin is formed within 20 ps following the excitation of rhodopsin with 347-nm light and converted to bathorhodopsin with a time constant of 50 ps, and concluded that hypsorhodopsin is a precursor of bathorhodopsin in the photoreaction of squid rhodopsin.
Kobayashi [4] found a bathorhodopsin-like red-shifted species (here we would like
to call the intermediate primerhodopsin) by the picosecond photolysis of bovine
rhodopsin. The species was initially considered to be the lowest excited singlet
state in rhodopsin. The time constants for the species→hypsorhodopsin and hypso-
rhodopsin→bathorhodopsin processes were determined to be 15±5 ps and 50 ±20 ps,
respectively. The direct formation of bathorhodopsin from the red-shifted species
was found to be a minor pathway at room temperature. Matuoka et al. [9] proposed that hypsorhodopsin is formed by a secondary photoreaction of photorhodopsin
(primerhodopsin). However from their experiment it could not be clarified whether
or not the thermal formation of hypsorhodopsin occurs because at their excitation

---
*Present address: Hamamatsu Photonics K.K. Tsukuba Research Laboratory, Toyosato,
Ibaraki 300-26, Japan.

wavelength (532 nm) photorhodopsin efficiently absorbs the second photon to yield hypsorhodopsin photochemically.

In the present study, we excited octopus rhodopsin by a blue picosecond light pulse (461 nm) at which the absorption cross section of octopus rhodopsin is 2.7 times larger than that of primerhodopsin. The absorption spectrum of primerhodopsin is similar to that of bathorhodopsin and slightly red-shifted from the spectrum. The spectrum of bathorhodopsin is shown in Fig. 1. The excitation light (35 ± 5 μJ) was too weak to cause multiphoton events. Weak signals were measured with a highly-sensitive picosecond time-resolved spectroscopy apparatus. The formation of hypsorhodopsin was found even under the low photon density excitation (0.3 photons/molecule) by the 461-nm picosecond pulse. Octopus hypsorhodopsin is a major product at physiological temperature. On the other hand the formation yield of hypsorhodopsin was smaller than that of bathorhodopsin in the photolysis of octopus rhodopsin by continuous irradiation at 10 K.

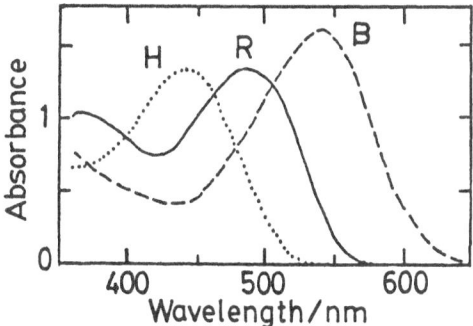

Fig. 1. Absorption spectra of octopus rhodopsin (R), hypsorhodopsin (H), and bathorhodopsin (B) at 10 K. Spectra of B and H are reconstructed from the spectra of the photo-stationary state at 10 K [10].

2. Experimental

2-1. Preparation of Octopus Rhodopsin

Octopus (Mizudako, Paroctopus defleini) rhodopsin was prepared by a method similar to that described previously [10], and solubilized by 2 % digitonin and 2 % L-1690 (lauryl ester of sucrose) [12]. The resulting solution was concentrated by centrifuge. Deuterated photoreceptor membranes were prepared by suspending the microvillar membranes in D_2O. Deuterated glycerol and digitonin were used as the solvents for deuterated samples. The absorbance of samples for the picosecond experiments was 1.2 at the excitation wavelength 461 nm (2-mm light path length).

2-2. Picosecond Photolysis of Octopus Rhodopsin at Physiological Temperature

Figure 2 shows the block diagram of the picosecond spectroscopy apparatus used in the present study. A light source was a hybridly mode-locked Nd:YAG laser (1064 nm). An excitation light source was the first anti-Stokes Raman scattering light at 461 nm which was generated by focusing the second harmonic (532 nm) of the Nd: YAG laser into acetone and amplified by an amplifier (coumarine 440 in methanol) pumped by the third harmonic (355 nm) of the mode-locked Nd:YAG laser. A probe light source was a picosecond continuum light generated by focusing the 532-nm light into CCl_4 or by focusing the 1064-nm light into D_2O. The probe light was detected by a multichannel photodiode array (512 channels) coupled with a grating polychromator (f=20 cm, 600 grooves/mm). Thirty to forty pairs of excitation and nonexcitation data were averaged by the micro- and mini-computer system [13]. The temperature of the sample was kept constant at 8°C with a semiconductor cooling device. Acid metarhodopsin formed by the irradiation of the excitation light pulse is stable below 15°C [11]. Rhodopsin was recovered from acid metarhodopsin by irradiation with a tungsten lamp through a cut-off color filter (O58) after every excitation laser exposure.

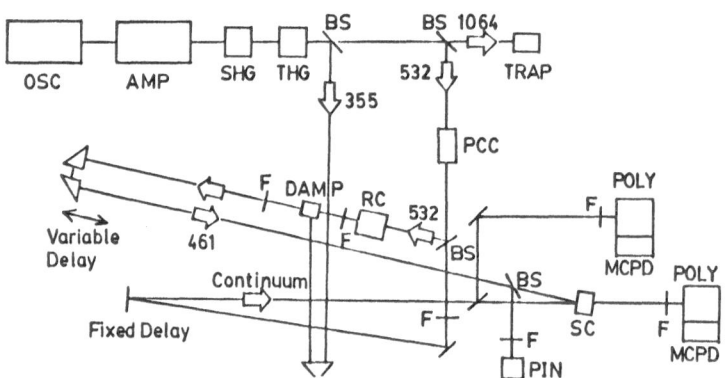

Fig. 2. Block diagram of picosecond time-resolved absorption spectroscopy apparatus. OSC: mode-locked Nd:YAG laser, AMP: Nd:YAG amplifier, SHG: KDP crystal for the second harmonic generation, THG: KDP crystal for the third harmonic generation, BS: beam splitter, PCC: picosecond continuum generation cell, RC: Raman cell, DAMP: dye amplifier, F: filter, PIN: PIN photodiode for excitation energy monitor, SC: sample cell, POLY: polychromator, MCPD: multichannel photodiode array.

2-3. Photolysis of Octopus Rhodopsin by Continuous Light at Low Temperature

Microvillar membranes in H_2O or D_2O were sonicated and mixed with glycerol or deuterated glycerol (1:3 v/v). The mixture in a 2-mm path-length cell was placed in a dewar. Absorption spectra were measured with a Cary 14 spectrophotometer.

3. Results and Discussion

3-1. Formation of Primerhodopsin

The primary photoprocess of rhodopsin is initiated by the rapid photoisomerization
of retinal. The absorption spectrum of rhodopsin in the excited singlet state has
not been measured because of its short lifetime. The fluorescence lifetime of bovine rhodopsin is on the order of 0.1 ps [14]. In the present study, we observed
the formation of the octopus deeply red-shifted precursor, primerhodopsin, at 8°C.
The lifetime of octopus primerhodopsin was estimated to be smaller than the resolution time of our apparatus (36 ps), which is consistent with the reported lifetime
of bovine primerhodopsin (15 ± 5 ps [4]).

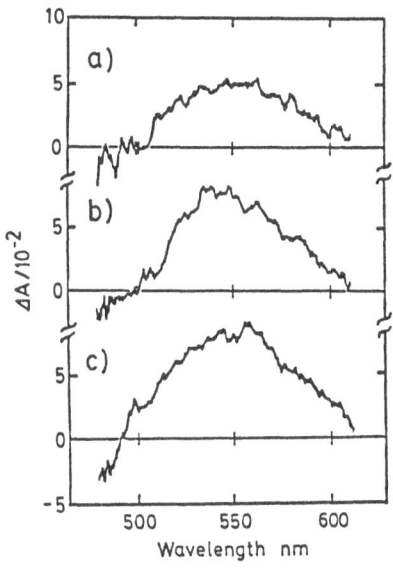

Fig. 3. Picosecond difference absorption spectra of octopus rhodopsin in
H_2O suspension following 461-nm excitation at 8°C. (a) -22 ps, (b) 100
ps, and (c) 1 ns. The excitation density is $(1.9\pm0.3)\times10^{16}$ photons/cm^2.

Figure 3a exhibits a transient spectrum 22 ps before excitation. The absorbance
change observed is due to the overlap of the tailing part of the probe pulse (30-ps
fwhm) and the leading edge of the excitation pulse (20-ps fwhm). The spectrum at
-22 ps is not attributed to bathorhodopsin but to a deeply red-shifted precursor,
primerhodopsin. Kobayashi [4] found a deeply red-shifted precursor in the photolysis of bovine rhodopsin with 530-nm excitation light (6-ps fwhm) at room temperature, and attributed it to the lowest excited singlet state in rhodopsin. However,
the lifetime (15 ± 5 ps [4]) is longer than the upper limit of the fluorescence lifetime estimated from the fluorescence quantum yield (1×10^{-5}) [14]. Figure 4 shows
the difference absorption spectra following the 461-nm excitation of rhodopsin in
both the H_2O and D_2O suspensions at 8°C. The large absorbance change around 460 nm

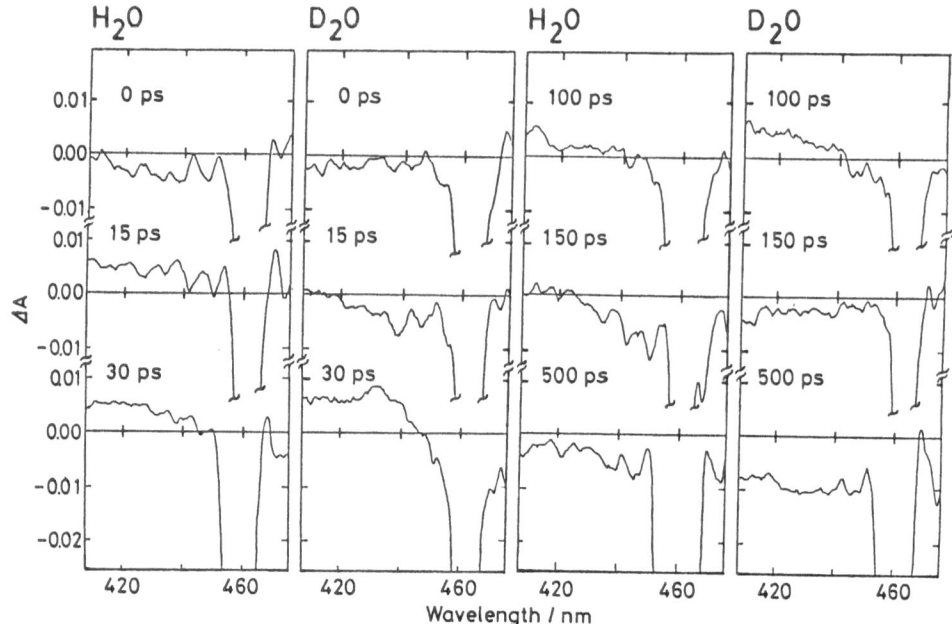

Fig. 4. Picosecond difference absorption spectra of octopus rhodopsin in H_2O and in D_2O suspensions following 461-nm excitation at 8°C. The excitation photon density is $(9.3 \pm 1.3) \times 10^{15}$ photons/cm^2.

was due to the scattering of the excitation light. The negative absorbance change in the 415–445 nm region at 0 ps in both the H_2O and D_2O suspensions shows that molar extinction coefficient of primerhodopsin in this region is smaller than that of octopus rhodopsin. Kobayashi [4] found that the difference spectrum of prime-rhodopsin minus rhodopsin is red-shifted from the difference spectrum of bathorhodopsin minus rhodopsin by about 20 nm. Such a large spectral shift from octopus bathorhodopsin to primerhodopsin was not found in the present study.

3-2. Formation of Hypsorhodopsin

The spectra 30 ps after excitation indicate clearly the formation of hypsorhodopsin in both the H_2O and D_2O suspensions (Fig. 4). The formation of hypsorhodopsin in the D_2O suspension is slower than in H_2O. The spectral changes between 30 ps and 500 ps are caused by the hypsorhodopsin → bathorhodopsin conversion. Figure 3c shows the difference spectrum at 1-ns delay time with a maximum absorbance change at 550 nm. This is in agreement with the wavelength of the bathorhodopsin-minus-rhodopsin difference spectrum obtained from Fig. 1.

Open circles in Fig. 5 show the dependence of the absorbance change measured at 430 ± 10 nm 30 ps after excitation on the excitation photon density. Solid lines are

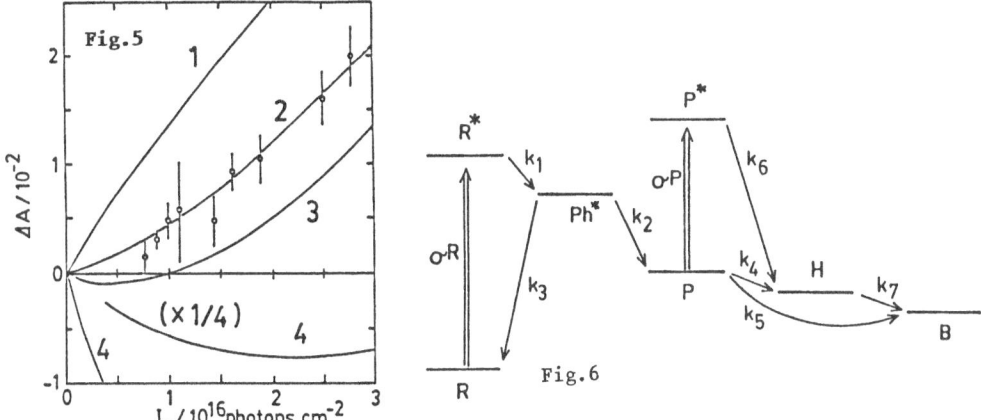

Fig. 5. Excitation energy dependence of the transient absorption 30 ps after excitation monitored at 430±10 nm. The efficiency for the hypsorhodopsin formation by the thermal decay of primerhodopsin, ϕ_H [$=k_4/(k_4+k_5)$], is varied from 1 (curves 1), 0.8 (curve 2), 0.7 (curve 3), to 0 (curve 4). Lifetimes of primerhodopsin and hypsorhodopsin are set at 15 and 70 ps, respectively. The quantum yield for the formation of primerhodopsin is set at 0.5. The absorption cross sections of rhodopsin (σ^R) and primerhodopsin (σ^P) at 461 nm are set at 1.08×10^{-16} and 0.96×10^{-16} cm^2.

Fig. 6. Kinetic scheme of the primary process of rhodopsin for numerical calculations. Rhodopsin, primerhodopsin, bathorhodopsin, and hypsorhodopsin are denoted by R, P, B, and H, respectively. Rhodopsin in the excited singlet state with 90° twisted C_{11}-C_{12} bond (phantom state) is denoted by Ph^*. The rate constant and the absorption cross section at excitation wavelength are denoted by k and σ, respectively.

obtained by numerical calculations based on a model shown in Fig. 6. Hypsorhodopsin is thermally and photochemically formed from primerhodopsin. Bathorhodopsin is thermally formed from primerhodopsin.

The quantum yield for the formation of primerhodopsin, η_P, was taken to be same as that for the bleaching of octopus rhodopsin 0.5 [15]. The lifetime of primerhodopsin, τ_P [$=1/(k_4+k_5)$], was set at that of the lifetime of bovine primerhodopsin, 15 ps [4]. The lifetime of hypsorhodopsin, τ_H ($=1/k_7$), was taken to be 70 ps from the formation time of bathorhodopsin monitored at 560 nm. The absorption cross section of primerhodopsin (σ^P) in the 410–480 nm region is smaller than that of octopus rhodopsin (Fig. 4). The σ^P value at excitation wavelength (461 nm) was set to be twice that of bathorhodopsin ($\sigma^P = 2\sigma^B = 0.87\sigma^R$). Curves 1-4 in Fig. 5 show the calculated absorbance change versus excitation photon density for several values of ϕ_H [$=k_4/(k_4+k_5)$], the efficiency for the formation of hypsorhodopsin from primerhodopsin. A parametrization, $\phi_H=0$, corresponds to the following case: hypso-rhodopsin is formed only by a photochemical reaction of primerhodopsin, and batho-rhodopsin is directly formed from primary rhodopsin in the thermal process. An intense bleaching at 430 nm is predicted for an excitation photon density lower than 3×10^{16} photons/cm^2 as shown by curve 4. This result completely derivates from the experimental results shown by open circles. Therefore the thermal conversion from primerhodopsin to hypsorhodopsin must be taken into account for the explanation of the experimental results. Curves 1, 2, and 3 were calculated for $\phi_H=1.0$, 0.8, and 0.7, respectively. Curve 2 agrees well with the experimental results. The lower limit of ϕ_H at 8°C was estimated to be 0.6. This value is consistent with the previously reported value of ϕ_H (>0.75) in bovine rhodopsin at room temperature [4].

A considerable amount of hypsorhodpsin was formed in the photolysis of octopus rhodopsin with a low excitation photon density at 461 nm (7.7×10^{15}photons/cm^2, 0.3 photons/molecule). The molar fractions of hypsorhodopsin formed thermally (f_t) and photochemically (f_p) from primerhodopsin are estimated to be 0.9 and 0.1, respectively. The present study shows that both hypsorhodopsin and bathorhodopsin are thermally formed from primerhodopsin. The primary photoprocess of octopus rhodopsin is given as the following "thermal" reactions initiated by the photoisomerization of retinal.

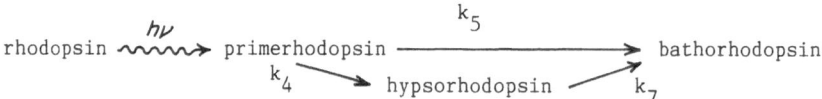

3-3. Temperature and Isotope Effects on the Decay of Primerhodopsin

The thermal formation of hypsorhodopsin was also found using 480-nm continuous irradiation of octopus rhodopsin at 10 K. Hypsorhodopsin is formed in the initial stage of irradiation before the accumulation of any species which could be photo chemically converted to hypsorhodopsin. The photochemical formation of hypsorhodopsin from primerhodopsin could not occur because of the low excitation power and the short lifetime of primerhodopsin (29$\pm$2 ps for bovine Batho' at 10 K [3]). The ϕ_H value is about 0.25 at 10 K, which is smaller than that at 8°C. Therefore it is concluded that the rate constant k_4 depends more sensitively on temperature than k_5.

The formation of hypsorhodopsin was found to be slower in D_2O than in H_2O. Peters et al. [3] reported that the lifetime of bovine primerhodopsin is about 7 times longer in a D_2O suspension than in H_2O. These two results show the isotope effect on the decay rate of primerhodopsin (k_4 and/or k_5). Any definitive isotope effect on the formation yield of hypsorhodopsin at 8°C was obtained within experimental error. On the other hand resonance Raman studies [16] show that the molar fraction of hypsorhodopsin in the photo-steady state at 12 K is 1.4–1.8 times larger in a D_2O suspension than in H_2O. The increases in the lifetime of primerhodopsin and formation yield of hypsorhodopsin in D_2O suspension indicate that the rate of the decay of primerhodopsin to bathorhodopsin (k_5) more efficiently than that to hypsorhodopsin (k_4) by H_2O/D_2O exchange. The rate-determining step in the primerhodopsin → bathorhodopsin conversion may be a proton transfer as speculated by Honig et al. [5]. The decrease in k_5 in D_2O causes the increase in the conversion efficiency from primerhodopsin to hypsorhodopsin, ϕ_H. Bathorhodopsin formation is the main decay process of primerhodopsin ($k_5 > k_4$) at 10 K. On the other hand, k_5 is smaller than k_4 at 8°C. The H_2O/D_2O exchange effect in ϕ_H is smaller at 8°C than at 10 K. No isotope effect was found in the rate of the hypsorhodopsin → bathorhodopsin conversion. This suggests that the rate-determining step for the hypsorhodopsin → bathorhodopsin conversion may be a conformational change in opsin which affects the interaction between an amino acid residue and retinal.

The detailed results and discussion will be published elsewhere together with the experimental results of the stationary absorption spectrum of octopus rhodopsin at low temperatures [17].

Acknowledgment

This work was partly supported by a Grant-in-Aid for Special Distinguished Research (56222005) and a Special Research Project (60115004) to T.K. and a Grant-in-Aid (60115001) to M.T. from the Ministry of Education, Science, and Culture in Japan. It was also supported partly by the Toray Science and Technology Foundation and Kurata Science Foundation to T.K. and the JSPS-NSF Japan-US Cooperative Science Program to M.T. and T.G.E. The authors thank Mr. Koshihara for his help in the early stage of picosecond experiment and Dr. Uchiki and Mr. Yoshizawa for their help in constructing the data-analyzing system of micro- and mini-computers.

1. R. Hubbard, A. Kropf: Proc. Natl. Acad. Sci. USA., 44, 130 (1958);
   T. Yoshizawa, G. Wald: Nature, 197, 1279 (1963).
2. T. Yoshizawa: in Handbook of Sensory Physiology VII/1 (H.J.A. Dartnall ed.)
   pp. 146-179, Springer-Verlag, Berlin (1972).
3. K. Peters, M.L. Applebury, P.M. Rentzepis: Proc. Natl. Acad. Sci. USA., 74,
   3119 (1977).
4. T. Kobayashi: FEBS Lett., 106, 313 (1980); Photochem. Photobiol., 32, 207
   (1980).
5. B. Honig, T. Ebrey, R.H. Callender, U. Dinur, M. Ottolenghi: Proc. Natl. Acad.
   Sci. USA., 76, 2503 (1979).

6. Y. Shichida, S. Matuoka, T. Yoshizawa: Photobiochem. Photobiophys., 7, 221 (1984).

7. G.E. Busch, M.L. Applebury, A.A. Lamola, P.M. Rentzepis: Proc. Natl. Acad. Sci. USA., 69, 2802 (1972).

8. Y. Shichida, T. Yoshizawa, T. Kobayashi, H. Ohtani, S. Nagakura: FEBS Lett., 80, 214 (1977); Y. Shichida, T. Kobayashi, T. Yoshizawa, H. Ohtani, S. Nagakura: Photochem. Photobiol., 24, 335 (1978).

9. S. Matuoka, Y. Shichida, T. Yoshizawa: Biochim. Biophys. Acta, 765, 38 (1984).

10. M. Tsuda, F. Tokunaga, T. G. Ebrey, K.T. Yue, J. Marque, L. Eisenstein: Nature, 287, 461 (1980).

11. M. Tsuda: Biochim. Biophys. Acta, 545, 537 (1979).

12. K. Nashima, M. Mitsudo, Y. Kitô: Biochim. Biophys. Acta, 536, 78 (1978).

13. J. Iwai, M. Ikeuchi, Y. Inoue, T. Kobayashi. In Protochlorophyllide Reduction and Greening (C. Sironval, M. Brouers eds.) pp 99–112, Martinus Nijhoff/Dr. W. Junk Publishers, Hague (1984).

14. A.G. Doukas, M.R. Junnarkar, R.R. Alfano, R.H. Callender, T. Kakitani, B. Honig: Proc. Natl. Acad. Sci. USA., 81, 4790 (1984).

15. M. Tsuda, B. Mao, T.G. Ebrey, unpublished results.

16. A.J. Pande, R.H. Callender, T.G. Ebrey, M. Tsuda: Biophys. J., 45, 573 (1984).

17. H. Ohtani, T. Kobayashi, M. Tsuda, T.G. Ebrey: Biophys. J., (1986), submitted for publication.

Phototaxis and the Second Sensory Pigment in
Halobacterium halobium

T. Takahashi, H. Tomioka, Y. Nakamori, N. Kamo, and Y. Kobatake

Faculty of Pharmaceutical Sciences, Hokkaido University, Sapporo 060, Japan

1 Introduction

Halobacterium halobium is an archaebacterium well known for its unique light energy-transducing apparatus, bacteriorhodopsin.   Usually the bacterium is energized by respiration. When the cells grow so dense that the supply of oxygen becomes insufficient, the cells produce a large amount of bacteriorhodopsin and utilize the energy of visible light. Therefore, bacteriorhodopsin appears to be an emergency apparatus [1]. The cell growing in such a dense culture is attracted to visible light. The cells accumulate at a spot of green-yellow light. On the other hand, UV and violet light are avoided by the cells [2]. The behavioral response of the cells to light is similar to that of enteric bacteria to chemical substances [3], namely, the frequency of the change in swimming direction is increased or decreased when a repellent stimulus or an attractant one is introduced to the cell, respectively. The difference lies in the way the change of swimming direction is brought about. Halobacterium halobium has flagella in both poles of the cell and swims indifferently in either direction. Therefore, the bacterium simply reverses the swimming direction by reversing the direction of flagellar rotation when repellent light is applied (step-up photophobic response) or attractant light is taken away (step-down photophobic response).   The photoattractant and the photorepellent response are controlled by different photosystems. The photorepellent system that shows maximal response of the cells to 370nm light was named photosystem 370 (PS370) [2].

At first, sensory photoreceptor for photoattractant response was thought to be bacteriorhodopsin. Bacteriorhodopsin is a membrane bound protein. When illuminated, it ejects protons from the cell and generates an electrochemical potential difference of protons across the cell membrane. Within 10ms after absorption of light, the bacteriorhodopsin molecule undergoes sequential reactions through several metastable intermediates and finally returns to its original state. Like rhodopsin in animal eyes, bacteriorhodopsin contains the chromophore retinal, which is covalently linked via a protonated Schiff base.

Recently, the second and the third retinal-containing proteins were also found in Halobacterium halobium membrane. Halorhodopsin is the second rhodopsinlike protein to be found. It introduces chloride ions into the cell [4]. The third rhodopsinlike protein was named sensory rhodopsin [5].   Although this protein is bound tightly to the cell membrane, it does not show any pumping activity which affects the membrane potential of the cells.   The two rhodopsinlike proteins also show cyclic photoreaction, but the regeneration time of sensory rhodopsin is two orders of magnitude slower than that of bacteriorhodopsin and halorhodopsin.

The name sensory rhodopsin stems from the fact that the pigment is a strong candidate for the receptor protein for photobehavioral response of the cell [5]. Several pieces of evidence have been accumulated: (1) Mutant strains with defects in both bacteriorhodopsin and halorhodopsin also show photobehavioral responses [5,6].   (2) Photorepellent response which is maximal in 370nm light is strongly dependent on background light [5,9]. This is consistent with the idea that a long-lived photointermediate of sensory rhodopsin is the receptor for PS370. The photointermediate absorbs maximally 373nm light. (3) Violet background illumination converts attractant light into a repellent stimulus [5,8]. This suggests that the long-lived photointermediate functions as a photoreceptor, absorbing violet background light as soon as it has been generated by orange attractant light.

The evidence obtained by photobehavioral experiments and discrepancies between them will be outlined here. The latter led us to the discovery of a new photoreceptor protein. This fourth rhodopsinlike protein in <u>Halobacterium halobium</u> was previously overlooked because the content is too small and the wavelength of its absorption band is too short to be detected with the routine method of flash-photolysis experiments. The appropriate wavelength of the flash was in fact predicted by the photobehavioral study [14].

The fourth rhodopsinlike pigment seems to have an important physiological role as well as sensory rhodopsin.

2 Methods

Several years ago, we developed a microcomputer-linked method for the measurement of the reversal frequency of the swimming direction of <u>Halobacterium halobium</u> [9]. It is a convenient and reproducible method for quantitative measurement of the photobehavioral response of this bacterium. The hardware has now been modified and improved so that a real-time measurement can be performed with high accuracy. A block diagram of the improved hardware is shown in Fig. 1. All traces of the cells recorded in frame memories by an A/D converter are tracked by computer. Whether the cells have reversed or not is judged by software. The system minimizes the errors due to human subjectivity.

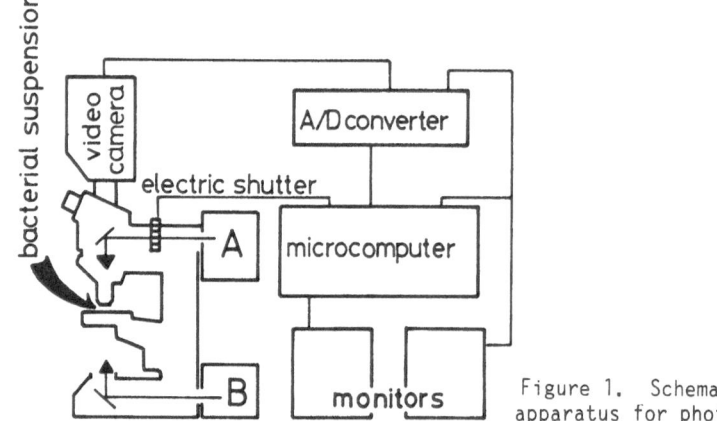

Figure 1. Schematic diagram of the apparatus for photobehavioral measurement

Note that two light beams are passed through the bacterial suspension. One is the actinic light (stimulating light: denoted A in Fig. 1) focused through an incident beam illuminator. The other is a background light (B in Fig. 1), which is an important factor in the photobehavioral experiment on <u>Halobacterium halobium</u>.

3 Photoreceptor of PS370

Photorepellent system PS370 develops as the culture of the cells are grown into the stationary growth phase. The behavioral response was inhibited when the synthesis of retinal was blocked by addition of nicotine to the culture medium, and restored soon after addition of all-trans retinal [10]. The photoreceptor is obviously a retinal-containing protein. No receptor candidate for this photosystem had been known until sensory rhodopsin was found.

Sensory rhodopsin was found in a bacteriorhodopsin-deficient mutant strain [11] and in the mutants which are deficient in both bacteriorhodopsin and halo-rhodopsin [6]. The latter were isolated with an elaborate technique developed by SPUDICH and SPUDICH [12], which is a suitable method for the selection of ion-flux mutants of Halobacterium halobium. A simplified scheme of the photoreaction cycle of sensory rhodopsin is shown in Fig. 2.

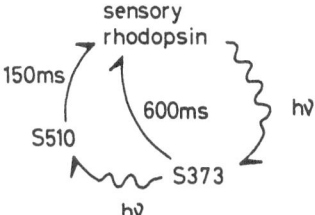

Figure 2. Simplified scheme of the photoreaction cycle of sensory rhodopsin

The lifetime of the photointermediate S373 is quite long. Therefore, it is possible that S373 acts as photoreceptor for 350-400nm repellent light. Steady-state concentration of S373 can be estimated using the photocycling rate of sensory rhodopsin along with the intensity of background light. The background light is usually present when we can see the bacterium. A capture cross section of a pigment is related to a molecular extinction coefficient:

$$q = 0.38 \times 10^{-18} \varepsilon \ mm^2. \tag{1}$$

If the molecular extinction coefficient of sensory rhodopsin in its original state and the quantum efficiency of the process from sensory rhodopsin to S373 are assumed to be 50,000 and 0.5, respectively, about half the sensory rhodopsin molecules are in the S373 state under 10^{14} photons/mm^{2}s of the orange (587nm) light. The appropriate number of receptor molecules for the response is presuma-bly dependent on the extent of signal amplification inside the cell. Neverthe-less, it is obvious that the photorepellent response should not be observed without the orange background illumination, as long as S373 is the receptor for the photorepellent response.

The effect of orange background light on photorepellent response was carefully examined by SPUDICH and BOGOMOLNI [5] and by us [7]. Both groups demonstrated a strong dependence of the photorepellent response on the intensity of the orange background light. Wavelength dependence of the response under intense yellow background light and that under dim background light are shown in Fig. 3. This is one piece of evidence for the photorepellent receptor S373. However, we can also see that the photorepellent response is still observed under dim background light [7]. This observation led to the discovery of another photorepellent recep-tor.

Further support for the idea that S373 is a receptor for PS370 is obtained from experiments with blue background illumination. Details will be discussed in the next section.

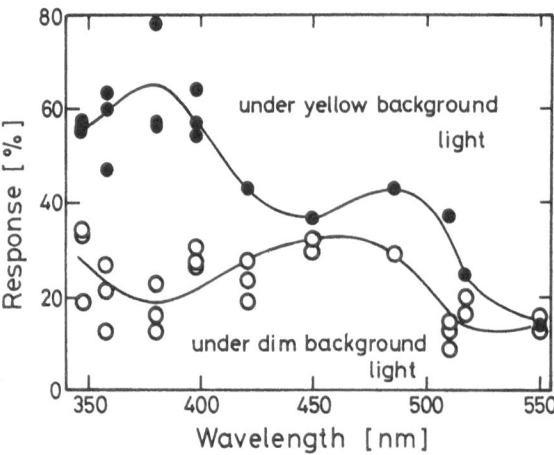

Figure 3. Wavelength dependence of reversal response of the cells to the step-up stimulus of repellent light under dim background light (○), or under intense yellow background light (●)

Possible Models of Signal Production

Based on the photoreaction cycling of sensory rhodopsin, two alternative models are possible for the photorepellent response of Halobacterium halobium. In our first model [7], photointermediate S510 produces the repellent signal. The signal is produced as long as the receptor molecule is in the S510 state, but the amount is gradually reduced according to an adaptation term. The probability of the reversal of the cell is assumed to be proportional to the amount of the signal. Using this model along with a few assumptions which are consistent with our independent observation, the behavior of the cell was explained by the photoreaction of the receptor molecule [7].

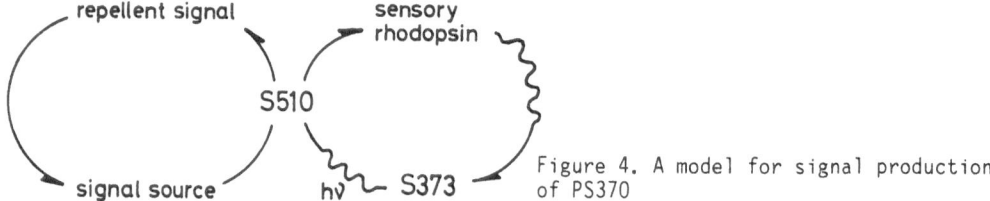

Figure 4. A model for signal production of PS370

A second possible model is that the decrease in S373 produces the repellent signal. This model also seems to explain photoattractant behavioral response. In this case, the increase in S373 may produce an attractant signal. The step-down photophobic response of the cells to attractant orange light corresponds to the decrease in S373.

Experiments with a combination of violet and orange light can test the two alternative models. Let us consider the change in the population of sensory rhodopsin molecules in the states S373 and S510 when orange and violet light are applied simultaneously but with different durations (see Fig. 5). The light with longer duration corresponds to the background light in our photobehavioral experiments. We normally observe the behavioral response of the cells to violet repellent light under orange background illumination. In this case, the populations of the S373 and S510 states change oppositely when the violet actinic light is turned

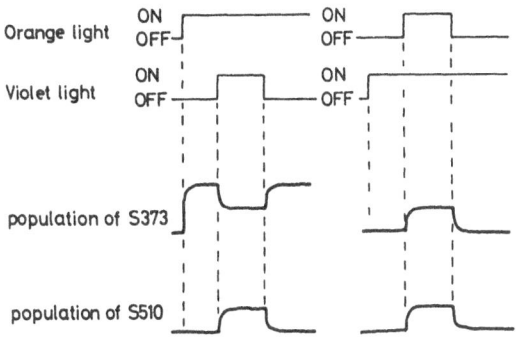

Figure 5. Changes in the population of sensory rhodopsin in intermediate states

on (left-hand side of Fig. 5). When the background light is violet, however, S373 as well as S510 increases upon turning on the orange actinic light (right-hand side of Fig. 5). Therefore, we can say which photointermediate is more likely to produce a repellent signal when we know at which stage -- upon turning on or turning off the orange actinic light under violet background illumination -- the cells show a photorepellent response.

The cells showed reversal response upon turning on the orange light under intense violet background illumination [5,8]. The first model is obviously more likely. Fortunately, the first model is quantitative. The behavioral response of the cells was successfully simulated using reaction rate constants in a photocycle of sensory rhodopsin obtained by a flash-photolysis experiment [7]. Therefore, we concluded that S373 is the photoreceptor of PS370.

4 Receptor for Photoattractant Response.

Sensory rhodopsin is also a candidate for the receptor of photoattractant light. The attractant response is maximal in 560–590 nm light. First, bacteriorhodopsin was suggested. This may be true, but at least one other photoattractant receptor is necessary because a photoattractant response was observed in mutant strains that lacked bacteriorhodopsin. A classical strategy for the identification of the photoreceptor is to determine the action spectrum.

For a single photoreceptor, the photobehavioral response is a function of the number of excited photoreceptor molecules:

$$R(\lambda, I) = f(n\phi q(\lambda)I),\tag{2}$$

where n is the number of all the photoreceptor molecules contained in the cell, ϕ is the quantum yield, q the capture cross section, λ the wavelength, and I the intensity of actinic light. When we choose the intensity of actinic light at each wavelength so that the response has a certain constant value, $n\phi q(\lambda)I(\lambda, R)$ should also be constant as long as f is monotonic. Therefore, the spectrum of the reciprocal of such light intensities will be exactly similar to the absorption spectrum of the receptor pigment.

The action spectrum of the photoattractant response obtained by such a method resembled the absorption spectrum of sensory rhodopsin [13]. However, at wavelengths shorter than 580nm, the action spectrum was distorted due to the photorepellent response. The photorepellent system is different from PS370 [14]. Under conditions where the photorepellent system was suppressed, the wavelength dependence of the photoattractant response looked like the absorption spectrum of sensory rhodopsin (Fig. 6). The spectrum in Fig. 6 was obtained with equal intensity at each wavelength. Note that the action spectrum obtained with methods other than that

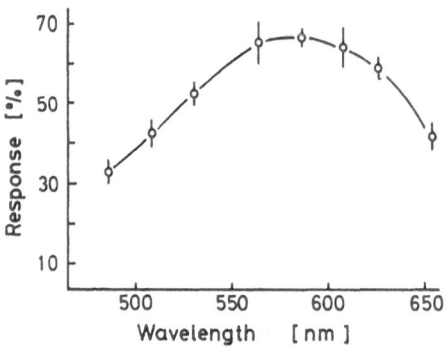

Figure 6. Wavelength dependence of photoattractant response obtained with equal-intensity method

described above should not be compared directly with an absorption spectrum of a receptor pigment.

In a growing culture of <u>Halobacterium halobium</u>, PS370 seemed to appear about 10 hours earlier than the photoattractant response [2,15]. This is inconsistent with the idea that the photoattractant response as well as PS370 is mediated by sensory rhodopsin. Several action spectra of the photoattractant response showed the maximum at 565nm. From these, DENCHER [16] mentioned that there may be un-identified photoreceptor pigments which may be called P565 and P370, in addition to the three rhodopsinlike pigments which had been found at that time.

5 A New Photoreceptor Pigment for Photorepellent Response

The action spectrum of PS370 sometimes shows a shoulder at nearly 500nm. An antenna function of carotenoids was suggested [10]. However, a new photorepellent system with its sensitivity maximum at around 500nm also explains this. As mentioned above, this new photorepellent system is consistent with the fact that the photorepellent response (PS370) can be observed even under dim background light. A mutant which does not show both photoattractant response and PS370 clearly demonstrated this new photorepellent system [14]. The sensitivity maximum of the system is at 480nm (PS480). The mutant strain named Flx30N1 (formerly called Flx3-12) contains less retinal at an early stage of the growth. When retinal was added to the cells in the early growth stage, the behavioral response was enhanced. Therefore the photoreceptor pigment in PS480 appeared to be a rhodopsinlike protein. The candidate for the photoreceptor protein was identified flash-spectrophotometrically [14]. The next step is to obtain evidence that this fourth rhodopsinlike protein is the photoreceptor of PS480.

The strain Flx30N1 contains sensory rhodopsin. The apparent lack of both PS370 and the photoattractant response is probably due to a defect in the sensory transduction pathway. Recently, we have isolated two mutant strains that also lack PS370 and the photoattractant response. One is called Flx30N-P, the derivative of Flx30N1. It contains trace amounts of sensory rhodopsin. The strain is suitable for flash-photolysis experiments because, like Flx30N1, it is rich in the fourth rhodopsinlike protein. The photoreaction scheme of the pigment shown in Fig. 7 was determined using this strain [17]. The other mutant strain is Flx3a1 which contains no sensory rhodopsin as determined by flash-photolysis experiments. Surprisingly, this strain contains a very small amount of the fourth rhodopsinlike protein whereas the sensitivity of PS480 seemed to be about 5 times higher than that of Flx30N1. This suggests that the defect in the signal trans-duction pathway in Flx30N1 also affects the sensitivity of PS480. A defect in the chemosensory transduction was also suggested in this strain [18].

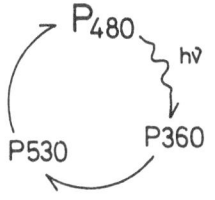

Figure 7. Photoreaction cycle of the fourth rhodopsinlike protein

PS480 and the fourth rhodopsinlike protein appear to be present in many strains of Halobacterium halobium. PS480 is detected under dim background light by comparing the sensitivity of the photorepellent response at 370nm with that at 480nm. As far as we examined, all strains, including wild-type strains, mutants that lacked bacteriorhodopsin, and a carotenoid-deficient one, showed higher photorepellent sensitivity to 480nm light than to 370nm light under dim background illumination. Also, all these strains showed the flash-induced absorbance change due to the fourth rhodopsinlike pigment. The pigment content in each strain corresponds to the photorepellent sensitivity to 480nm light, supporting the idea that the pigment is the photoreceptor of PS480.

Further support for the photoreceptor of PS480 is the action spectrum. For direct comparison with the absorption spectrum, the action spectrum was determined by the method described in Sect. 4. As we could not obtain the absorption spectrum of the fourth rhodopsinlike protein, the action spectrum was compared with the flash-induced difference spectrum. In Fig. 8, the difference spectrum of the fourth pigment obtained with Flx3ON-P, the action spectrum of the photo-repellent response of Flx3ON1, and that of Flx3a1 are shown. These are in good agreement with each other except at wavelengths shorter than 430nm, where the difference spectrum may not fit the absorption spectrum of the pigment. From these results, we conclude that the fourth rhodopsinlike protein is the photoreceptor of PS480.

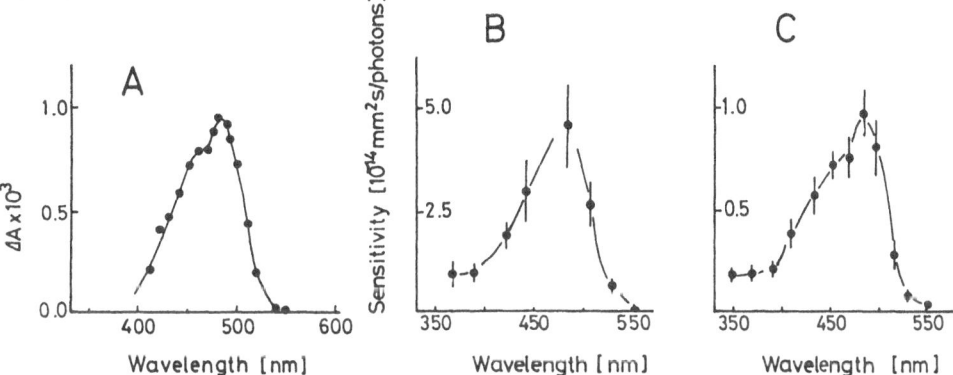

Figure 8. Absorption difference spectrum of the fourth rhodopsinlike pigment (A), action spectrum of the repellent response of Flx3a1 (B) and that of Flx3ON1 (C)

6 Discussion

Our results of photobehavioral experiments clearly showed that the receptor pigment of PS370 is S373, a long-lived photointermediate of sensory rhodopsin. An important role of the other photointermediate S510 has been suggested. In our opinion, S510 is a signal messenger. The relatively long lifetime of S510 is suitable for amplification of the repellent signal (excitation process of the behavioral response). During its lifetime, S510 can produce much repellent signal in collision with transformer molecules within the membrane of the cell. There-fore, the long-lived photointermediate seems to have a memory function. In the photoattractant response, this may be the case in S373. When the attractant signal is produced by S373, irradiation with an attractant light under violet background illumination does not cause the attractant response because S373 is immediately converted to S510 by the background light. This explains why the attractant orange light was so easily converted into a repellent stimulus under violet background illumination. The photoattractant receptor seems to be sensory rhodopsin.

Further support for the photoattractant receptor is provided by the strain Flx3a1, which contains no sensory rhodopsin and shows neither photoattractant response nor repellent response due to PS370. Similar results were obtained in other strains which contain no sensory rhodopsin [19]. In these strains, however, no photorepellent response was observed. Flx3a1 is the first strain which demonstrates that PS480 is irrelevant to sensory rhodopsin. This also indicates that the fourth rhodopsinlike protein is independent of sensory rhodopsin. Therefore, we suggest the name phoborhodopsin [17] for the fourth pigment. Recently, WOLFF et al. [20] reported a similar result to our first paper [14] on the fourth pigment. However, they did not find the mutant strain which apparently lacks PS370 but has PS480.

Strain Flx3a1 indicates that PS480 has a high sensitivity. Flx3a1 contains a very small amount of phoborhodopsin, i.e. less than one tenth of that in strain Flx3ON1. Comparing it with the amount of S373 under normal background light, we can see that the signal amplification of PS480 is as much as that of PS370. This may be the reason why phoborhodopsin was overlooked before and was detected first in a mutant strain with a defect in sensory function. It might be harder to detect the receptor of PS370 than to detect sensory rhodopsin, if the receptor is not a photointermediate. It is plausible that strain Flx3ON1 produces a large amount of phoborhodopsin because of its imperfect amplification of the photorepellent signal. PS480 seems to have an important physiological role. A hint for a physiological role of PS480 is the change in the sensitivity of PS480 during growth of the culture. Photorepellent sensitivity of the cells to 480nm light is higher at an early stage of the growth (logarithmic growth phase) whereas that of PS370 and the photoattractant system is higher in the stationary growth phase [13-15]. The amount of phoborhodopsin is also maximum in the cells in the logarithmic growth phase [17]. In the logarithmic growth phase, cells are growing at full speed. The respiration rate is maximum in this phase. In the cells in the logarithmic growth phase, we sometimes found a decrease in membrane potential under intense light. Usually, the membrane potential reflects the level of energy of the cell. Intense light may inhibit respiration of the cell, or may affect other physiological functions of the cell [21]. Thus, the avoidance of intense light with the help of PS480 may be of benefit to the cell. Indeed, strain Flx3ON1 was obtained from a long continuous culture in a liquid medium under illumination.

Increased sensitivity of PS480 in the logarithmic growth phase explains the discrepancy of the appearance of PS370 and photoattractant system reported by HILDEBRAND and SCHIMZ [15]. Phoborhodopsin seems to be a constitutive equipment [14,17]. On the other hand, if we regard bacteriorhodopsin as an emergency apparatus for restricted oxygen supply, sensory rhodopsin is also equipment for an emergency. Sensory rhodopsin guides the cells to the place filled with useful visible light but free from harmful UV light. In fact, sensory rhodopsin or retinal seemed to be induced in the stationary growth phase [13].

7 Acknowledgement

The parent strain of flx3ON1, Flx3ON-P and Flx3a1 was kindly provided by J. L. Spudich. The authors thank him. This work was supported by grants from the Ministry of Education, Science and Culture of Japan.

8 References

1. M. Sumper, H. Reitmeier, D. Oesterhelt: Angew. Chem. Int. Ed. Engl. 15, 187 (1976)
2. E. Hildebrand, N. Dencher: Nature 257, 46 (1975)
3. D. E. Koshland, Jr.: In Bacterial Chemotaxis as a Model Behavior System (Raven Press, New York, 1980)
4. J. K. Lanyi: Annu. Rev. Biophys. Biopys. Chem. 15, 587 (1986)
5. J. L. Spudich, R. A. Bogomolni: Nature 312, 509 (1984)
6. R. A. Bogomolni, J. L. Spudich: Proc. Natl. Acad. Sci. USA 79, 6250 (1982)
7. T. Takahashi, M. Watanabe, N. Kamo, Y. Kobatake: Biophys. J. 48, 235 (1985)
8. T. Takahashi, Y. Mochizuki, N. Kamo, Y. Kobatake: Biochem. Biophys. Res. Commun. 127, 99 (1985)
9. T. Takahashi, Y. Kobatake: Cell Struct. Funct. 7, 183 (1982)
10. N. Dencher: In Energetics and Structure of Halophilic Microorganisms, eds. S. R. Caplan and M. Ginzburg, (Elsevier/North Holland, Amsterdam, 1978) p.67
11. M. Tsuda, N. Hazemoto, M. Kondo, N. Kamo, Y. Kobatake, Y. Terayama: Biochem. Biophys. Res. Commun. 108, 970 (1982)
12. E. N. Spudich, J. L. Spudich: Proc. Natl. Acad. Sci. USA, 79, 4308 (1982)
13. H. Tomioka, T. Takahashi, N. Kamo, Y. Kobatake: Biochim. Biophys. Acta 884, 578 (1986)
14. T. Takahashi, H. Tomioka, N. Kamo, Y. Kobatake: FEMS Microbiol. Lett. 28, 161 (1985)
15. E. Hildebrand, A. Schimz: Photochem. Photobiol. 38, 593 (1983)
16. N. Dencher: Photochem. Photobiol. 38, 753 (1983)
17. H. Tomioka, T. Takahashi, N. Kamo, Y. Kobatake: Biochem. Biophys. Res. Commun. 139, 389 (1986)
18. E. N. Spudich, S. A. Sundberg, D. Manor, J. L. Spudich: Proteins (in press)
19. S. A. Sundberg, R. A. Bogomolni, J. L. Spudich: J. Bacteriol. 164, 282 (1985)
20. E. K. Wolff, R. A. Bogomolni, P. Scherrer, B. Hess, W. Stoeckenius: Proc. Natl. Acad. Sci. USA 83, 7272 (1986)
21. I. E. D. Dundus: Adv. Microbiol. Physiol. 15, 85 (1977)

Photocycles of Sensory Rhodopsin

T. Kobayashi[1], H. Ohtani[1]*, and M. Tsuda[2]

[1]Department of Physics, University of Tokyo, Bunkyo, Tokyo 113, Japan
[2]Department of Physics, Sapporo Medical College, Sapporo 060, Japan

1. Introduction

Four retinoid proteins have been found in the plasma membrane of Halobacterium
halobium. They are bacteriorhodopsin [1], halorhodopsin [2], sensory rhodopsin
[3,4], and phoborhodopsin [5]. Bacteriorhodopsin and halorhodopsin are light-
driven proton pump [6] and chlorideion pump [7], respectively. Sensory rhodopsin
[4,8] and phoborhodopsin are photoreceptors for the phototaxis of H. halobium [5].
    Sensory rhodopsin was formerly called third-rhodopsin-like pigment [3] or slow
rhodopsin [4]. It has an absorption maximum around 590 nm [4,9]. Hereafter we
denote it sR_{590}. The chromophore of sR_{590} is an all-<u>trans</u> retinal [10]. The following photocycles of sensory rhodopsin have been reported (see Fig. 1). There
are red-shifted and blue-shifted intermediates (sR_{680} and sR_{370}, respectively) in
the photocycle of sR_{590} [4]. The absorption spectrum of sR_{680} has not been measured. The maximum of the absorbance change in the sR_{680}-minus-sR_{590} difference
spectrum is located at 680 nm [4]. Hereafter we denote it sR_K by analogy with
the photocycle of the light-adapted bacteriorhodopsin. sR_K decays to sR_{370} with
a half-life of 20 μs [4]. sR_{370} is a long-lived species ($\tau_{1/2}$=0.8 s [4]) with 13-
<u>cis</u> retinal [10]. The formation yield of sR_{370} depends on temperature [11]. There
may exist a secondary pathway for the recovery to sR_{590} without forming sR_{370}. The

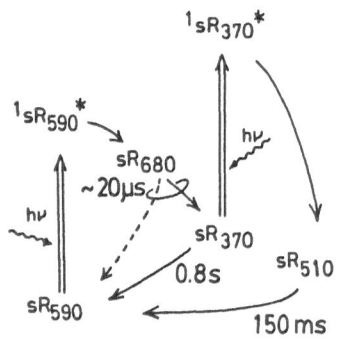

Fig. 1. Photocycles of sensory rhodopsin. Lifetimes ($\tau_{1/2}$) measured at room temperature are
given in the figure.

---

*Present address: Hamamatsu Photonics K.K. Tsukuba Research Laboratory, Toyosato,
Ibaraki 300-26, Japan

intermediate has been found to be a photoreceptor for the repellent phototaxis of
H. halobium [12]. By absorbing near UV light, sR_{370} is photochemically converted
to an intermediate sR_{510}, which yields thermally sR_{590} [10,13,14].

In the present work [15], we studied the photocycle of sR_{590} with the aid of
nanosecond time-resolved spectroscopy and compared it with that of the light-adapted bacteriorhodopsin. We found a slightly blue-shifted intermediate formed between
sR_K and sR_{370} and tentatively called it sR_L. We also found the recovery process
from sR_L to sR_{590} and that the conversion efficiency from sR_L to sR_{370} decreases
with temperature. We also studied the photocycle of sR_{370} and found that sR_{510}
intermediate is formed within 100 ns.

2. Experimental

2-1. Preparation of Sensory Rhodopsin
The calotenoid-free strain which is a mutant of H. halobium Flx3 was grown and was
harvested by standard methods [3,16]. Envelope vesicles were washed with NaCl solution and prepared by sonication methods. The vesicles were suspended with Tween-
20 to remove the membranes solubilized by Tween-20. The suspension was centrifuged
at 60000 g for 1 hour and washed with NaCl solution. The suspension was sonicated
and centrifuged at 20000 g for 10 minutes to reduce turbidity. The protein concentration was 4-5 mg/ml.

2-2. Nanosecond Time-Resolved Spectroscopy
Figure 2 shows the block diagram of the nanosecond spectroscopy apparatus used in
the present study. An excitation light source for the photolysis of sR_{590} was the
second harmonic (532 nm, 5-ns fwhm) of a Q-switched Nd:YAG laser or a dye laser
(630 nm, 15-ns fwhm) pumped by the second harmonic of the Nd:YAG laser. The third
harmonic (355 nm, 5-ns fwhm) of the Nd:YAG laser was used for the photolysis of

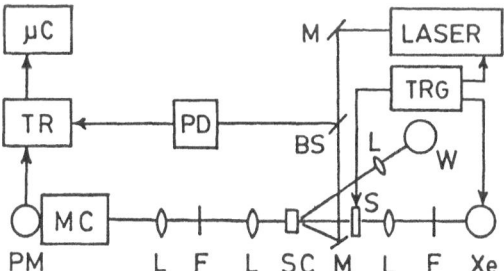

Fig. 2. Block diagram of the nanosecond time-resolved spectroscopy apparatus.
TRG: trigger circuit, W: tungsten lamp for cw-irradiation of sR_{590}, Xe: probe
light, L: lens, F: filter, M: mirror, S: shutter, BS: beam splitter, SC: sample
cell, MC: monochromator, PM: photomultiplier, TR: transient recorder, μC: micro
computer, PD: photodiode.

sR$_{370}$ which was accumulated by the continuous irradiation of sR$_{590}$ sample by a tungsten lamp with an O58 cutoff filter. A probe light source was a xenon lamp or a tungsten lamp. The probe light was detected by a photomultiplier coupled with a grating monochromator (f=20 cm, 1350 grooves/mm). The output signal of a photomultiplier was digitized by a transient recorder connected with a micro-computer [17]. The temperature of the sample was kept constant ($\leq \pm 0.5$°C) with a water bath.

3. Results and Discussion

3-1. Intermediates in the photocycle of sR$_{590}$

Figure 3 shows the transient difference absorption spectrum 2 μs after excitation of sR$_{590}$ at 22.5 and 3.5°C (open and closed circles, respectively). The spectrum measured at 22.5°C are nearly identical with that at 3.5°C. The positive maximum of the absorbance change is located around 680 nm. This result is consistent with the reported result that a red-shifted intermediate sR$_{680}$ (sR$_K$) is formed in the early stage of the photocycle of sR$_{590}$ [4]. The difference spectrum shows bleaching due to the disappearance of sR$_{590}$ and the formation of sR$_K$. sR$_K$ was formed within the resolution time (10 ns) of the apparatus in the 3.5–33.0°C temperature region.

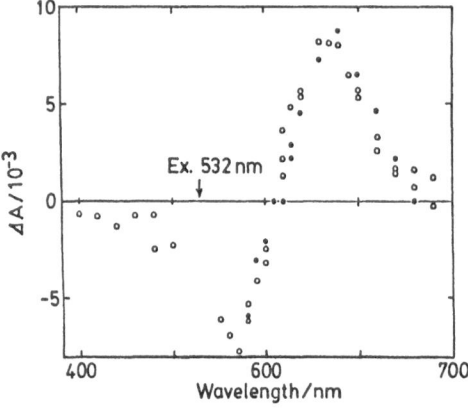

Fig. 3. Time-resolved difference absorption spectrum 2 μs after the 532-nm excitation of sR$_{590}$ at 22.5°C (open circles) and 3.5°C (closed circles).

Figure 4 shows the temperature dependences of the absorbance change at 680 nm just after 532-nm excitation of sR$_{590}$ (ΔA_{680}, open circles) and the rate constant (k, closed circles) for the decay of the absorbance change. The formation yield of sR$_K$ (η_K) was found to be insensitive to temperature in the 3.5–33°C range. The decay rate of sR$_K$ decreased with temperature. The Arrhenius plot is composed

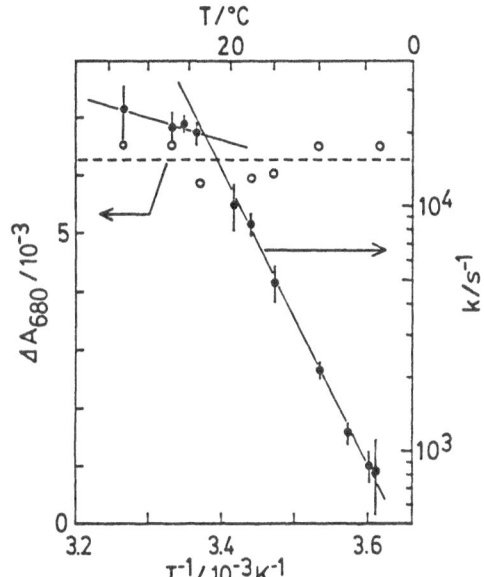

Fig. 4. Temperature dependence of the absorbance change (ΔA_{680}) immediately after excitation monitored at 680 nm (open circles) and the rate constants for the decay of sR_K monitored at 680 nm (closed circles).

of the two lines above and below 23.5 °C with different slopes. The temperature (23.5 °C) of the slope continuity is higher than that observed for sR_{370} decay (15 °C [11]). Activation energies were obtained to be 18 ± 4 kJ/mol and 114 ± 3 kJ/mol for temperature regions above and below 23.5 °C, respectively.

Figure 5 shows the kinetics of the absorbance change following the excitation of sR_{590} with 532-nm pulsed light. The formation of sR_{370} (time constant $\tau_{1/e} = $ 220 µs at 19.5 °C, 1.7 ms at 3.8 °C) slower than the decay of sR_K ($\tau_{1/e} = $ 69 µs at 19.5 °C, 840 µs at 3.8 °C) was also found both at 19.5 and 3.8 °C. These results show that sR_{370} is not directly formed from sR_K. A rapid increase ($\tau_{1/e} < 200$ µs at 19.5 °C, < 0.7 ms at 3.8 °C) and a slow decrease ($\tau_{1/e} > 200$ µs at 19.5 °C, > 1.5 ms at 3.8 °C) of the absorbance change was observed at 500 nm. The former and the latter qualitatively correspond to the decay of absorbance change at 680 nm and the rise at 385 nm, respectively. Such increase and decrease in absorbance were observed in the 500–540 nm region. It is concluded that a intermediate exists between sR_K and sR_{370}. The newly found intermediate has an absorption spectrum slightly more blue-shifted than that of sR_{590} (discussed later).

The following sequence of the intermediates has been found. The photocycle is similar to that of the light-adapted bacteriorhodopsin bR_{568}.

$$sR_{590} \xrightarrow{h\nu} sR_K \longrightarrow sR_L \longrightarrow sR_{370} \longrightarrow sR_{590}$$

$$bR_{568} \xrightarrow{h\nu} K_{610}(\text{or } KL_{596}) \longrightarrow L_{543} \longrightarrow M_{412} \longrightarrow bR_{568}$$

The rate of the $sR_L \to sR_{370}$ conversion was found to be strongly dependent on tem-

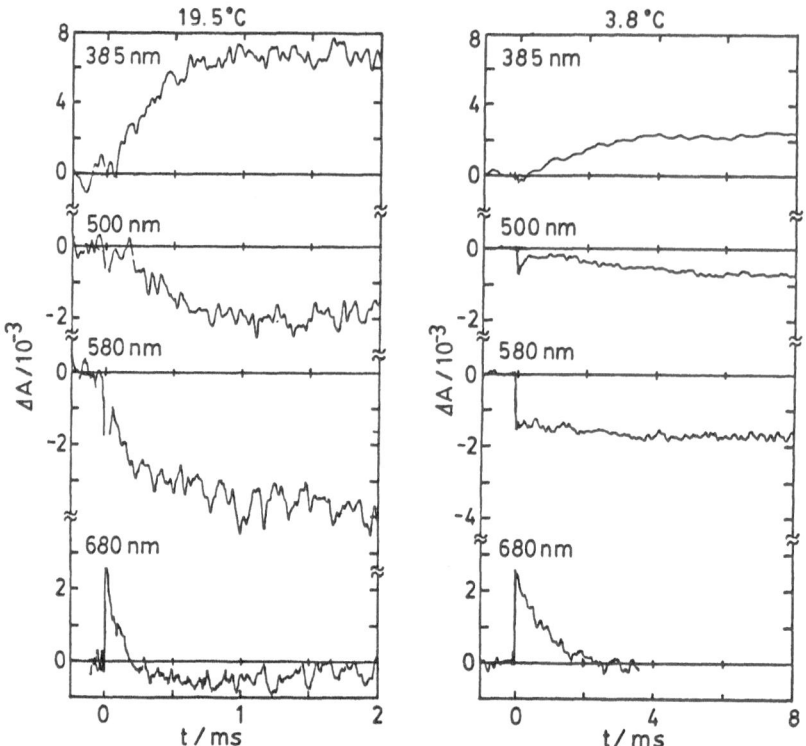

Fig. 5. Kinetics of the absorbance change following the 532-nm excitation of sR$_{590}$ at 19.5 and 3.8°C.

perature as well as that of the sR$_K$→sR$_L$ process. The activation energy was determined to be 75±13 kJ/mol.

3-2. Branching process in the photocycle of sR$_{590}$

The formation yield of sR$_{370}$ (η) decreases with temperature, for example η(40°C)/ η(5°C) = 0.25 [11]. The quantum yield for sR$_K$ formation is insensitive to temperature (see open circles in Fig. 4). There may exist the recovery pathway from sR$_K$ and/or sR$_L$ to sR$_{590}$ without formation of sR$_{370}$. The conversion efficiencies from sR$_K$ to sR$_L$ (ϕ_{K-L}) at 36.5, 19.5, and 3.8°C were estimated.

$$\phi_{K-L}(36.5°C) : \phi_{K-L}(19.5°C) : \phi_{K-L}(3.75°C)$$

$$= (1.0\pm0.1) \quad : \quad (0.8\pm0.2) \quad : \quad (1.2\pm0.3)$$

The conversion efficiency ϕ_{K-L} seems to be insensitive to temperature change. Therefore, the decrease in the conversion efficiency from sR$_L$ to sR$_{370}$ (ϕ_{L-M}) with temperature mainly contributes to the decrease in the quantum yield of sR$_{370}$ (η =

$\eta_K \Phi_{K-L} \Phi_{L-M}$). There exists a recovery process from sR_L to sR_{590}, which is similar to the recovery from L to bR_{568} in the photocycle of bR_{568} [18].

3-3. Absorption Spectra of sR_K and sR_L

The absorption spectra of sR_K and sR_L shown in Fig. 6 were obtained assuming that the efficiency for the $sR_K \rightarrow sR_{370}$ conversion is unity at 40°C. The absorption maximum (λ_{max}) of sR_K (595$\pm$5 nm) is slightly shifted from that of sR_{590} (587 nm [9]). The molar extinction coefficient at λ_{max} (ε_{max}) is smaller than that of sR_{590}. sR_L has an absorption spectrum with a maximum in the 540–570 nm region. sR_K and sR_L have spectra similar to KL_{596} [19] and L_{543} [18,19], respectively, in the photocycle of the light-adapted bacteriorhodopsin.

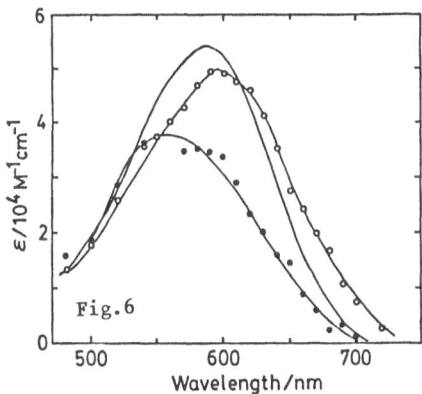 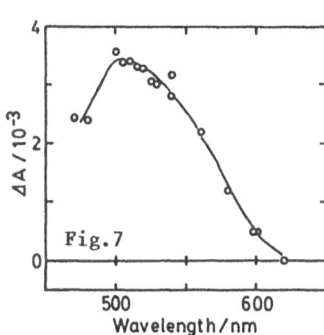

Fig. 6. Absorption spectra of sR_{590} (solid line), sR_K (open circles), and sR_L (closed circles).

Fig. 7. Time-resolved absorption spectrum 600 ns after the 355-nm excitation of sR_{370} at 23°C.

3-4. Photocycle of sR_{370}

sR_{510} intermediate is formed by the photolysis of sR_{370} with blue light [10,13,14]. In the present study, sR_{370} was formed by the cw-irradiation of sR_{590} with red light (>580 nm) and excited with the third harmonic (355 nm) of a Q-switched Nd:YAG laser. sR_{510} was formed within 100 ns. The spectrum of sR_{510} is shown in Fig. 7. It is not clarified whether or not sR_{510} is the first intermediate of the photocycle of sR_{370}. Further experiments are in progress.

Acknowledgments

The authors thank Mr. Masayuki Yoshizawa and Dr. Hisao Uchiki for their assistance in the construction of the data-analysis system and Mr. S. Koshihara for his help in the operation of the dye laser sytem. This work was supported in part by a Grant-in-Aid for Special Distinguished Research (56222005) and Special Research Project (60115004) to T.K. and by a Grant-in Aid (6058022) to M.T. from the Ministry of Education, Science, and Culture.

References

1. D. Oesterhelt, W. Stoeckenius: Nature New Biol., 233, 149 (1971).

2. A. Matsuno-Yagi, Y. Mukohata: Biochem. Biophys. Res. Commun., 78, 237 (1977); Y. Mukohata, Y. Kaji: Arch. Biochem. Biophys., 206, 72 (1981).

3. M. Tsuda, N. Hazemoto, M. Kondo, N. Kamo, Y. Kobatake, Y. Terayama: Biochem. Biophys. Res. Commun., 108, 970 (1982).

4. R.A. Bogomolni, J. L. Spudich: Proc. Natl. Acad. Sci. USA, 79, 6250 (1982).

5. T. Takahashi, H. Tomioka, N. Kamo, Y. Kobatake: FEMS Microbiol. Lett., 28, 161 (1985).

6. D. Oesterhelt, W. Stoeckenius: Proc. Natl. Acad. Sci. USA, 70, 2853 (1973).

7. B. Shobert, J.K. Lanyi: J. Biol. Chem., 257, 10306 (1982).

8. T. Takahashi, M. Watanabe, N. Kamo, Y. Kobatake: Biophys. J., 48, 235 (1985).

9. J.L. Spudich, R.A. Bogomolni: Biophys. J., 43, 243 (1983).

10. M. Tsuda, B. Nelson, C.-H. Chang, R. Govindjee, T. G. Ebrey: Biophys. J., 47, 721 (1985).

11. N. Hazemoto, N. Kamo, Y. Terayama, Y. Kobatake, M. Tsuda: Biophys. J., 44, 59 (1983).

12. T. Takahashi, Y. Mochizuki, N. Kamo, Y. Kobatake: Biochem. Biophys. Res. Commun., 127, 99 (1985).

13. J.L. Spudich, R.A. Bogomolni: Nature, 312, 509 (1984).

14. H. Tomioka, N. Kamo, T. Takahashi, Y. Kobatake: Biochem. Biophys. Res. Commun., 123, 989 (1984).

15. This work was partly reported: H. Ohtani, T. Kobayashi, M. Tsuda: Photobiochem. Photobiophys., 13, (1987) in press.

16. E.N. Spudich, J.L. Spudich: Proc. Natl. Acad. Sci. USA, 79, 4308 (1983).

17. J. Iwai, M. Ikeuchi, Y. Inoue, T. Kobayashi: In Protochlorophyllide Reduction and Greening (C. Sironval and M. Brouers, eds.) pp. 99–112. Martinus Nijhoff/Dr W. Junk Publishers, Hague, Boston, Lancaster (1984); H. Ohtani, T. Kobayashi, T. Ohno, S. Kato, T. Tanno, A. Yamada: J. Phys. Chem., 88, 4431 (1984).

18. T. Iwasa, F. Tokunaga, T. Yoshizawa: Biophys. Struct. Mech., 6, 253 (1980).

19. Y. Shichida, S. Matuoka, Y. Hidaka, T. Yoshizawa: Biochim. Biophys. Acta., 723, 240 (1983).

Flash Photolysis Study on Sensory Rhodopsin and Phoborhodopsin

N. Kamo, H. Tomioka, T. Takahashi, and Y. Kobatake

Department of Biophysics, Faculty of Pharmaceutical Sciences,
Hokkaido University, Sapporo 060, Japan

At least four retinylidene-proteins have been discovered so far in the membrane of Halobacterium halobium. They are bacteriorhodopsin (BR), halorhodopsin (HR), sensory rhodopsin (SR) and phoborhodopsin (PR). The first to be discovered was bacteriorhodopsin (BR), which functions as a light-driven H^+-pump (for a review see [1]). On illumination, BR extrudes H^+ from the inside to the outside and creates the so-called proton motive force, which is a driving force for ATP synthesis, transport of amino acids and other energy-requiring processes.

The pigment which was discovered second is halorhodopsin (HR). In 1977, Matsuno-Yagi and Mukohata [2] reported that an apparently BR-free H. halobium strain showed H^+-uptake by cells on illumination. It has been shown that addition of uncouplers increases both the extent and rate of H^+-uptake, revealing that the H^+-uptake is not active but passive [3,4]. This second-discovered pigment was, at first, considered to be a light-driven outwardly directed Na^+ pump, but later it was confirmed that it functions as a light-driven inwardly directed Cl^- pump [5]. (Cl^- can be substituted by Br^- and I^- [6,7] but physiologically this pigment works as a Cl^- pump.) Mukohata and Kaji [8] named this pigment halorhodopsin (HR). A flash photolysis study on HR was first done by Weber and Bogomolni [9] using a BR-deficient mutant. They found two photo-intermediates absorbing at 380 nm and 500 nm in the millisecond range. Later, it was shown that the photo-intermediate absorbing at around 380 nm belongs to another pigment which is now called sensory rhodopsin (SR).

Spudich and Spudich [10] have selected strains lacking both BR and HR by applying a technique developed by them for isolating ion transport mutants. A mutant isolated, called Flx3, showed no generation of membrane potential or H^+ movement on illumination. Spectroscopic analysis of mutant membranes investigated by Spudich and Bogomolni [11] has revealed a third retinal protein. This third pigment has its absorption maximum at around 590 nm (at present, 587 nm [12]) which is close to those of the other two photo-energizing pigments. The photo-intermediate in the millisecond range is absorbing at 373 nm and it is identical to that absorbing at around 380 nm found by Weber and Bogomolni previously. This intermediate is now designated S373 while SR is written as SR587 to show the maximum wavelength of absorption. The photo-intermediate S373 is transformed to the original pigment SR587 with a half lifetime of as long as 650 to 800 ms. The rate of photocycling is very slow and so, this pigment was at first named slow-cycling rhodopsin (or s-rhodopsin). They proposed the hypothesis that this third s-rhodopsin serves as a receptor of positive phototaxis and that the long-lived intermediate (S373) is a receptor of negative phototaxis whose action maximum is located at 370 nm. Since the decay of S373 is slow, this intermediate can be accumulated in sunlight. Much evidence supporting this proposal has been reported. So, at present, it is called sensory rhodopsin (SR)

Independently of Spudich and Bogomolni, we presented data showing that there are two pigments (photocycles) in a BR-deficient mutant and that the rate of one of the photocycles is very slow [16,17]. The pigment having the slow photocycle was called the third rhodopsin-like pigment, TR. The following is a summary of results obtained. Results for the slow photocycle are the properties of SR and those for the fast one are the properties of HR.

(1) The kinetic trace for the recovery or the absorbance at 590 nm is obviously bi-phasic when flash photolysis experiments are done for a BR-deficient mutant. The half-time of the recovery for one component is about 10 ms and the other is as long as 0.65 - 0.8 s. The wavelengths of the photo-intermediates were about 500 nm and 380 nm, which were the same as those reported by Weber and Bogomolni. The half-time of the decay of the intermediate absorbing at 500 nm was about 10 ms and that of the other photo-intermediate was about 0.8 s. These values agree with those observed at 590 nm.

(2) On addition of cholate to the suspension of envelope vesicles, the slow photocycle disappeared immediately while the fast one was stable for a long period. High sensitivity of SR to detergents was later described by Steiner and Oesterhelt [18].

(3) After the sample was subjected to extremely low or high pH, the slow photocycle was found to be inactivated irreversibly, while the fast photocycle was recovered.

(4) At pH 10, the wavelength of the depletion maximum of the pigment having the slow photocycle was shifted from 580 to 550 nm, while that of fast photo-cycle was not changed. Since the absorbance depletion in the flash-induced difference spectrum corresponds to the absorption of the original pigment (assuming no effect of photo-intermediates), this fact suggests that the absorption maximum of the slow component is shifted in alkaline solution.

(5) When the ionic concentration of the medium of the envelope vesicles was lowered, the flash-induced absorbance changes were decreased gradually, and it is stressed that the time course of the decrease of these two pigments was different for each of them. The pigment having the slow photo-cycle was weaker in low ionic solutions.

(6) The ratio of flash-induced absorbance change at 370 nm and 500 nm differed from one strain to another.

(7) The flash yield at 500 nm of the fast photocycle depended strongly on halogen (except F^-) concentration in the medium, while the yield at 380 nm of slow photocycle was relatively independent of the ionic species in the medium [6].

The fourth pigment is phoborhodopsin (PR). We isolated a mutant which lacks both the photo-repellent response whose action maximum is at 370 nm and the photo-attractant response [19]. It is worth noting that this mutant shows the negative phototaxis whose action maximum is located at 470-480 nm. The negative phototaxis peaked at 370 nm depends strongly on the intensity of background red light which excites SR587, since the amounts of the receptor S373 increase with increase of red background light [13,14]. On the other hand, the negative taxis centered at 480 nm does not depend on the intensity of background light. We detected the fourth pigment and named it phoborhodopsin (PR), since this pigment mediates the negative phototaxis [20].

Here, we want to discuss the following three topics connected with SR and PR.
(1) Glycerol, sucrose and other sugars prevent the inactivation of SR in weak ionic solutions.
(2) The photochemical intermediate originated from S373 by blue (near UV) light. It is noted that the negative phototaxis whose action maximum is located at 370 nm is initiated by the excitation of S373 by blue light.
(3) The flash photolysis study of PR.

1. Glycerol, sucrose and other sugars stabilize SR even in a weak ionic medium

As is described above, SR is inactivated irreversibly when the ionic concentration of the medium is low. The data have already been reported in Fig. 3-b in [17]. The amounts of photo-active SR were considered to be proportional to the amplitude of absorbance change immediately after a flash (flash duration 500 μs). With a decrease of the ionic concentration, the rate of inactivation becomes larger. It is interesting that the values extrapolated to the time zero are not unity (note that the scale is logarithmic), showing that the inactivation is bi-phasic (also see Fig. 1). Some SR loses its activity instantly when the ionic concentration of the medium is lowered and some is inactivated gradu-

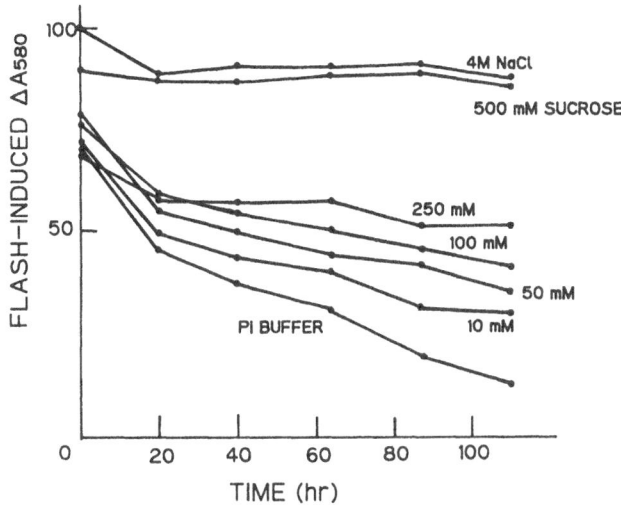

Fig. 1. Sucrose prevents the inactivation of SR in weak salt solution. Medium contains 170 mM NaCl buffered by 10 mM phosphate (pH 7.0). Envelope vesicles of Flx3 (~2 mg protein/ml) were used. The vertical values are relative to that in 4 M NaCl at time zero. The sucrose concentrations are shown in the figure. Pi buffer means 10 mM phosphate solution (pH 7.0).

ally. A similar bi-phasic character was also observed for inactivation by detergents. The reason is not clear at present and the results of further study are awaited.

It is known that sucrose and glycerol stabilize the protein [21]. We have found that the presence of such chemicals stabilizes SR even in weak ionic solutions. Results obtained are shown in Fig. 1. Essentially the same results were obtained when the absorbance change at 380 nm which monitors S373 was plotted. In 170 mM NaCl solution containing 10 mM phosphate buffer, the flash-induced absorbance change decreases gradually. On the other hand, presence of sucrose prevents the inactivation of SR in the concentration-dependent manner. Other sugars such as glucose and mannose are also effective. The half-times of the recovery of SR587 or the decay of S373 were not changed during the incubation in the solution containing 500 mM or more concentrated sucrose solution. Similar results were obtained when glycerol was used instead of sucrose. The half-times of decay of S373 or recovery of SR587 were shorter in the presence of sucrose and longer in the presence of glycerol than those in 4 M NaCl solution. When sucrose or glycerol was added 2 days after the incubation of membrane fractions in weak ionic solutions, we could not observe the effect of such chemicals. But, when the period of incubation in low salt solutions is 1 or 2 h, these chemicals exert their effect.

The turbidity of the sample is greatly reduced when membrane fractions are subjected to weak ionic solutions and so, using this finding, we are now preparing a clear sample to reduce the flash artifacts. Needless to say, purification of SR is a very important project but, unfortunately, no reports have been published describing successful purification. Quite recently, McGinnis [22] reported the partial purification of SR which yields only a few Coomassie blue-staining bands on SDS polyacrylamide gels.

2. Photochemistry when S373 is irradiated by blue light

As described above, S373 is a receptor of negative phototaxis by blue or near UV light. What happens when this intermediate is irradiated with blue light? To make the process clear one must understand the molecular mechanism of the negative phototaxis whose action maximum is 370 nm. This problem has been tackled

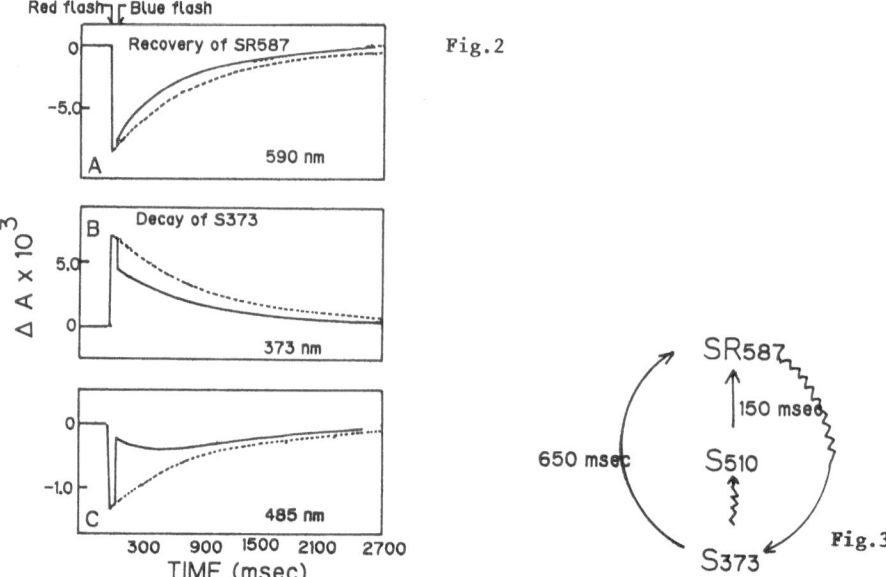

Fig. 2. Typical red flash (>600 nm) induced kinetic traces with (solid line) and without (broken line) subsequent blue flash (398±15 nm). The interval between the two flashes is 50 ms. The measuring wavelengths are 590 nm (A), 373 nm (B) and 485 nm (C). Temperature was 20° C. The curves shown in this figure were smoothed and the original data were published in Fig. 1 of [24].

Fig. 3. Photocycle of SR. Half-times shown are the values obtained at pH = 7.0, 20 °C in 4 M NaCl.

so far by three groups, Spudich and Bogomolni [14], Tsuda et al. [23], and us [24]. The first two groups used constant red illumination to generate S373, whereas we used double flashes. With a red flash S373 is formed, which is subsequently excited with a blue flash. This method has the advantage that the kinetic constants can be determined precisely.

Figure 2 shows the kinetic trace at typical wavelengths used to examine the phenomena caused by excitation of S373 with a blue flash. The membrane fragments were obtained from Flx3 which contains only SR (the content of PR is negligible in comparison with that of SR). The broken line in Fig. 2A is the kinetic trace of absorbance change at 590 nm caused only by a red flash which excites SR587. At this wavelength, the change in SR587 is monitored. The analysis of this curve gave the half-time of 650 ms (see Fig. 3). The solid line in this figure is the kinetic trace at 590 nm when red and subsequent blue flashes were applied. The interval of two flashes was 50 ms, which is much shorter than the half-time of the decay of S373 (650 ms). Excitation of SR587 with a red flash produces S373, which is excited by the subsequent blue flash. On excitation with the blue flash, the kinetic curve becomes steep, showing that the blue flash accelerated the regeneration of SR587. Analysis of this trace by a nonlinear least squares method has revealed that it can be resolved into two kinetic components with half-times of 150 ms and 650 ms. Therefore, a blue flash forms a new pathway for regeneration of SR587. Figure 2B shows the kinetic trace at 373 nm which monitors S373. Immediately when the blue flash was applied, the disappearance of S373 was observed. Figure 2C shows the kinetic trace at 485 nm. When the blue flash was applied, a rapid absorbance change was observed, indicating the formation of an intermediate. Moreover, we can see the decay of the intermediate. Nonlinear least squares analysis of the kinetic trace after the blue flash resolved it into two components with half-times of 110 ms and 650 ms. This value of 110 ms is very close to that obtained as the half-time of regeneration of SR587 by the blue flash. The magnitude of the rapid change induced by the blue flash was plotted as a function of wavelength. This plot gave the peak at 510 to 530 nm with use of envelope vesicles. When a sample solubilized with CHAPS (3-[(3-cholamido-propyl)dimethylammonio]-1-propanesulfonate) was used, the peak was determined to be 510 nm [25]. This sample is clear and stable for a long period. CHAPS does not change the wavelength of SR587 and S373 although the rate of photocycling is slowed down. Based on these observations, we present the scheme shown in Fig. 3. On irradiation of S373 with blue light, an intermediate of S510 forms and then goes back to SR587 with a half-time of 150 ms. With our instrument, we could not get information about the change between S373 and S510. To confirm this scheme, we measured the rate constant under various conditions. Figure 4 shows the temperature dependence of rate constants of decay of S510 and re-generation of SR587. Since the blue flash used could not excite all of S373, the kinetic curve measured at 590 nm consisted of two components as shown in Fig. 2A. One is S373 - SR587, which is the pathway without the blue flash, and the other is S373 - S510 - SR587. The rate constants of the former pathway are represented by open circles and are equal to those obtained only with the red flash (open triangle data). The latter, i.e. the rate constants of regenera-tion of SR587 via S510, are represented by filled circles and are equal to the rate constants obtained from the decay of S510 (filled square data). These results support the above scheme. Results of varying pH (pH 5.5 to 8.0) also support the scheme. The pH dependence of the rate constants of the pathway by blue light is relatively small.

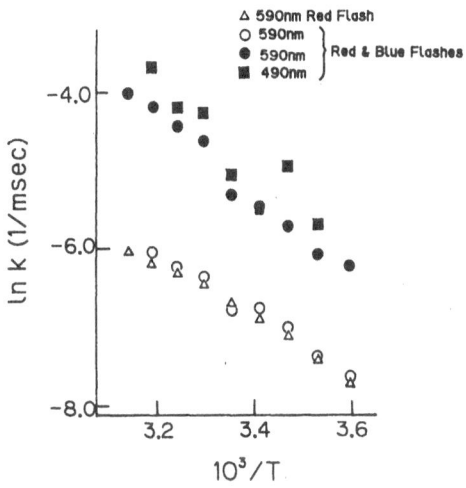

Fig. 4. Temperature dependence of rate constants of decay of S510 and re-generation of SR587. △: excitation only by red flash, monitoring wavelength was 590 nm. Other data were obtained with red and subsequent blue flashes. Moni-toring wavelengths were: ○, 590 : ●, 590 : ■, 490 nm.

3. Photocycle of PR
We isolated a mutant, named ON-P which has almost only PR. The photo-behavioral response of this and related strains and the relationship between behavior and PR are described by Takahashi et al. in this book. Using the envelope vesicles from this strain, we determined the photochemical cycle of PR [20]. The flash-induced difference spectrum is shown in Fig. 5. The open circles are the difference spectrum 70 ms after the flash and the filled circles are that 420 ms after the flash. We can see one absorbance depletion maximum around 480 nm and two absorbance increase maxima around 530 and 350 nm. The absorbance depletion is due to the flash-induced depletion of the original pigment and the bands of absorbance increase at 530 and 350 nm show the exist-

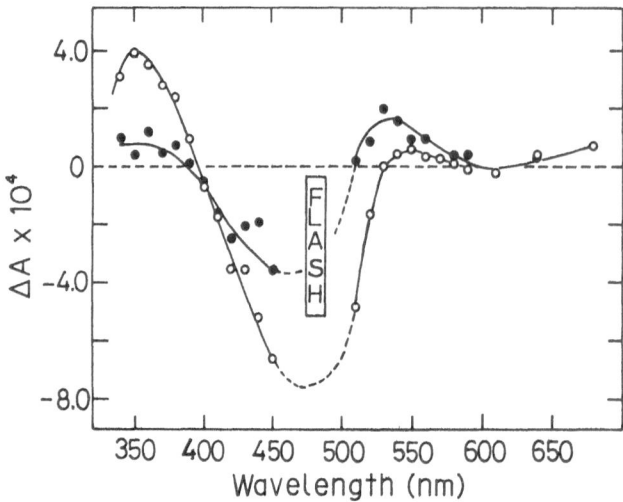

Fig. 5. Flash-induced absorbance changes in ON-P membrane fraction. Flash (486 ± 6 nm) was obtained by use of interference filter and short-cut filter. The membrane was suspended in 4 M NaCl at pH 7.0 (20 mM PIPES/NaOH). Five μl of 1.76 mM all-trans retinal was added to the 2 ml of membrane suspension (2 mg/ml). In the early logarithmic phase, retinal is not synthesized but apoprotein of PR is already present. Temperature was 20 °C.

ence of two photo-intermediates absorbing at these wavelengths. They are referred to as P530 and P350, respectively.

At 450 nm which monitored the original pigment, a kinetic component has a half recovery time of about 310 ms. At 370 nm, an absorbance increase relaxed with a half time of about 140 msec. The decay of P350 does not match the regeneration of PR480, suggesting the existence of a photo-intermediate between P350 and PR480. As is indicated in the flash-induced difference spectrum (see Fig. 5), an absorbance at 530 nm rose to a maximum with time and then declined. The kinetic curve at this wavelength is resolved into two components by a nonlinear least squares method: One had a half recovery time of about 140 ms, which is the same value as that at 370 nm, and the other had a half-time of about 300 ms, which is close to that at 450 nm. The simplest scheme which accounts for these observed absorbance changes is a cyclic reaction involving three spectrally distinct species: PR480, P350 and P530 (see Fig. 6). The numbers represent the wavelength at maxima in the difference spectrum.

Figure 7. illustrates the temperature dependence of the rate constants. The logarithm of the rate constants of various steps is plotted against the reciprocal of the absolute temperature. The filled circles represent the time constant of the decay at 530 nm and the filled squares are that of formation at 480 nm. These two agree well over temperature range examined. The activation energy is calculated to be 15.9 kcal/mol. The triangles are the time constant of the decay at 370 nm and the open circles represent the time constant of the formation at 530 nm. These two again agree fairly well over temperature range

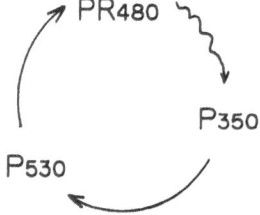

Fig. 6. Photocycle of PR. Wavy line indicates the light reaction.

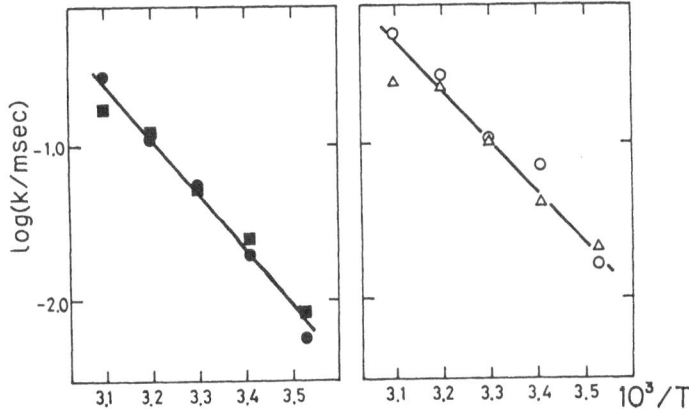

Fig. 7.   Temperature dependence of rate constants of PR.

examined.   The activation energy of this process is 13.9 kcal/mol.   These
results support our scheme.   The photo-intermediate before P350 will be the
subject of a future investigation.
    Recently, a paper [26] from Stoeckenius and Hess's laboratory appeared,
stating that this fourth pigment has a photocycle having a photo-intermediate
only at 360 nm.   In other words, P530 is not found by them.   The strain used
by them contains SR whereas the SR content of our strain is negligible.   This
may be the reason why they could not find P530.
    We found that the rate is slowed down in alkaline solution.   In addition, in
an alkaline medium, the photo-intermediate at 530 nm is not observed.   When
experiments were done in an acidic medium, we could clearly see the photo-
intermediate of P530.   These properties of P530 are similar to the O-inter-
mediate of BR.   P350 is supposed to be a deprotonated Schiff base.   Together
with the fact that P530 and the rate of photocycling are sensitive to the
acidity in the medium, it is supposed that proton movement may occur within the
PR molecule.   Since PR is not electrogenic, the proton movement is limited to
within the molecule and does not include proton transport through the membrane.
    PR differs from the other three pigments in its absorption maximum.   The
other three pigments have an absorption maximum between 560 – 590 nm, but PR has
an absorption maximum at around 480 nm.   It is nearer to that of vertebrate
rhodopsin than that of the other three pigments.   This fourth pigment has
another interesting property.   The total PR content slightly decreases as a
cell grows in culture, while the amount of the other three pigments in cell
membranes increases.   The isomeric composition of retinal of BR is all-trans in
the light-adapted state and 50 % 13-cis ,50 % all-trans in the dark-adapted
state.   Dark adaptation is also found in HR [27].   HR in the dark-adapted state
contains 70 % 13-cis and 30 % all-trans [28].   Any illumination increases the
all-trans form.   SR has all-trans retinal which isomerizes to 13-cis on illumination [23].   Therefore, the isomeric composition of PR is quite interesting.
    Quite recently, a very interesting paper [29] appeared, describing a photo-
active yellow protein in Ectothiorhodospira halophila, which is a retinoid
protein and shows a photocycle whose time constant is about 300 ms.   Yellow
protein shows that the maximum wavelength of this pigment is located at 480 nm.
These properties are quite similar to PR.   Although the physiological role of
this yellow pigment is not yet known, it is not unreasonable that this pigment
may act as a receptor for the phototaxis of this organism.   Another interesting
paper is one by Bivin and Stoeckenius [30], who show that there are two photocycles in haloalkaliphilic bacteria.   One of them has a fast photocycle and
depends on the chloride concentration.   This is similar to halorhodopsin.
Another pigment has an absorbance maximum around 500 nm and a slow photocycle of
a half-time of 500 ms.   This is similar to PR.   Thus, PR-like pigments are
widespread among the bacterial population.   The structure and function of these

123

PR-like pigments observed in various bacteria promises to be an interesting further study.

4. References
1. W. Stoeckenius, R.A. Bogomoli: Annu. Rev. Biochem. 52, 587 (1982)
2. A. Matsuno-Yagi,Y. Mukohata: Biochem. Biophys. Res. Commun. 78, 237 (1977)
3. R.E. MacDonald, R.V. Greene, R.D. Clark, E.V. Lindley: J. Biol. Chem. 254, 11831 (1979)
4. R.V. Greene, J.K. Lanyi: J. Biol. Chem. 254, 10986 (1979)
5. B. Schobert, J.K. Lanyi: J. Biol. Chem. 257, 10306 (1982)
6. N. Hazemoto, N. Kamo, Y. Kobatake, M. Tsuda and Y. Terayama: Biophys. J. 45, 1073 (1984)
7. T. Ogurusu, A. Maeda, N. Sasaki, T. Yoshizawa: Biochim. Biophys. Acta 682, 446 (1982)
8. Y. Mukohata, Y. Kaji: Archiv. Biochem. Biophys. 206, 72 (1981)
9. H.J. Weber, R.A. Bogomolni: Photochem. Photobiol. 33, 601 (1981)
10. E.N. Spudich, J.L. Spudich: Proc. Natl. Acad. Sci. USA 79, 4308 (1982)
11. J.L. Spudich, R.A. Bogomolni: Proc. Natl. Acad. Sci. USA 79, 6250 (1982)
12. J.L. Spudich, R.A. Bogomolni: Biophys. J. 43, 243 (1983)
13. T. Takahashi, M. Watanabe, N. Kamo, Y. Kobatake: Biophys. J. 48, 235 (1985)
14. J.L. Spudich, R.A. Bogomolni: Nature (London) 312, 506 (1984)
15. H. Tomioka, T. Takahashi, N. Kamo, Y. Kobatake: Biochim. Biophys. Acta 884, 578 (1986)
16. M. Tsuda, N. Hazemoto, M. Kondo, N. Kamo, Y. Kobatake, Y. Terayama: Biochem. Biophys. Res. Commun. 108, 970 (1982)
17. N. Hazemoto, N. Kamo, Y. Terayama, Y. Kobatake, M. Tsuda: Biophys. J. 44, 59 (1983)
18. M. Steiner, D. Oesterhelt: EMBO J. 2, 1379 (1983)
19. T. Takahashi, H. Tomioka, N. Kamo, Y. Kobatake: FEMS Microbiol. lett. 28, 161 (1985)
20. H. Tomioka, T. Takahashi, N. Kamo, Y. Kobatake: Biochem. Biophys. Res. Commun. 139, 389 (1986)
21. M.L. Schelanski, F. Gaskin, C.R. Cantor: Proc. Natl. Acad. Sci. 70, 765 (1973)
22. K. McGinnis, P. Scherrer, R.A. Bogomolni: Biophys. J. 51, 418a (1987)
23. M. Tsuda, B. Nelson, C-H. Chang, R. Govindjee, T. Ebrey: Biophys. J. 47, 721 (1985)
24. H. Tomioka, N. Kamo, T. Takahashi, Y. Kobatake: Biochem. Biophys. Res. Commun. 123 989 (1984)
25. M. Satoh, H. Tomioka, N. Kamo: unpublished results
26. E.K. Wolff, R.A. Bogomolni, P. Scherrer, B. Hess and W. Stoeckenius: Proc. Natl. Acad. Sci. USA 83, 7272 (1986)
27. N. Kamo, N. Hazemoto, Y. Kobatake, Y. Mukohata: Archiv. Biochem. Biophys. 238, 90 (1985)
28. J.K. Lanyi: J. Biol. Chem. 261, 14025 (1986)
29. D.E. McRee, T.E. Meyer, M.A. Cusanovich, H. E. Parge, E.D. Getzoff: J. Biol. Chem. 261, 13850 (1986)
30. D.B. Bivin, W. Stoeckenius: J. Gen. Microbiol. 132 2167 (1986)

Photochemistry of Retinochrome Studied by Nanosecond and Picosecond Spectroscopy

T. Kobayashi[1], K. Ogasawara[1,2], S. Koshihara[1], K. Ichimura[1], and R. Hara[3]

[1]Department of Physics, University of Tokoy, Bunkyo, Tokyo 113, Japan
[2]Hamamatsu Photonics K.K. Research Div., Hamamatsu, Shizuoka 435, Japan
[3]Department of Biology, Osaka University, Toyonaka, Osaka 560, Japan

1 INTRODUCTION

The cephalopod retina contains two photosensitive chromoproteins, rhodopsin and retinochrome. The various properties of retinochrome have been studied by HARA and HARA [1-9], HARA et al. [10], OZAKI et al. [11, 12], HAMDORF [13], SPERLING and HUBBARD [14], and SEKI [15]. They are summarized as follows.

(1) Rhodopsin is located entirely in the outer segment membranes of the visual cells, while retinochrome is distributed in both inner and outer segments [1-8].

(2) The two photopigments, rhodopsin and retinochrome, are both retinoid proteins but they differ from each other in the stereoisomeric form of their chromophore retinal, which is 11-cis in rhodopsin and is all-trans in retinochrome [2-15]. The all-trans retinal in retinochrome is bound to the protein moiety (aporetinochrome) with protonated Schiff base as in rhodopsin [5]. However retinochrome is, unlike rhodopsin, affected by the addition of hydroxylamine, sodium borohydride, formaldehyde and glycerol [4-8], thus it is concluded that the secondary interaction between retinal and aporetinochrome is far weaker than that between retinal and apoprotein in rhodopsin.

(3) The wavelength of the absorption maximum (λ_{max}) of retinochrome is 495 nm at pH6.5 as shown in Fig.1(a) and it is longer by 15 nm than that of rhodopsin. The peak wavelength depends on the concentrations of hydrogen ion and salts.

(4) On irradiation with green light at room temperature (23±3°C), retinochrome is bleached to a photo-product, metaretinochrome, which has an absorption maximum at 470 nm as shown in Fig.1(b) [1-12].

(5) Metaretinochrome exists in two forms, with loose and tight coupling of the chromophore to aporetinochrome [9].

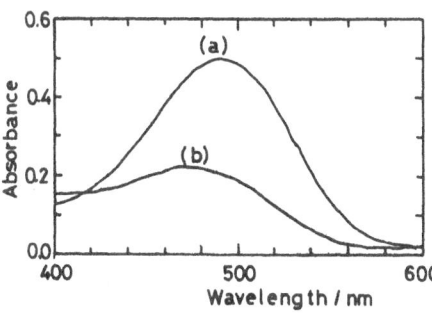

Fig.1: The absorption spectra of (a) retinochrome (Ret) and (b) metaretinochrome (M-Ret) at room temperature (23°C) (pH 6.5).

(6) Retinochrome at weakly acid pH exists in two tautomeric forms, R_1 and R_2, with their own λ_{max} around 495 nm and 476 nm, respectively, depending on a salt concentration [11]. Under the salt-free condition, retinochrome takes R_1 form.

(7) On irradiation with green light (546 nm) at liquid-nitrogen temperature, retinochrome is converted to an intermediate, lumiretinochrome $(\lambda_{max}=475$ nm), which is stable in the temperature region from $-190°C$ to about $-20°C$. Above $-20°C$, it is further converted thermally into metaretinochrome [10]. The difference absorption spectrum of lumiretinochrome minus retinochrome at liquid-nitrogen temperature is shown in Fig.2 (solid line(a)) with that of metaretinochrome minus retinochrome at room temperature (Fig.2, dashed-and-dotted line(b)).

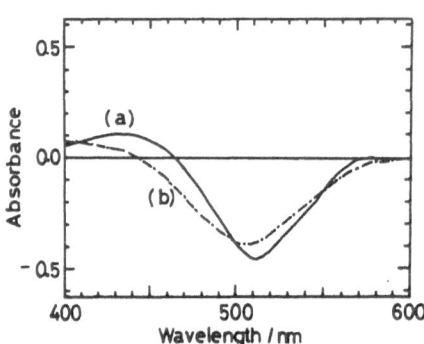

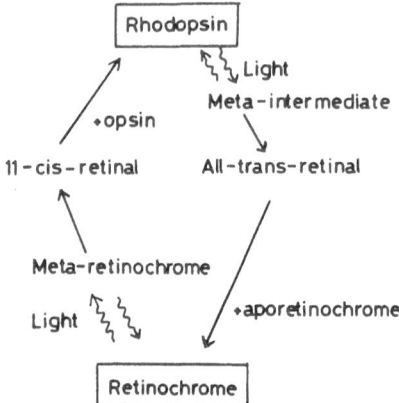

Fig.2: Difference absorption spectrum of lumiretinochrome (L-Ret) minus retinochrome (Ret) at liquid nitrogen temperature (solid line, curve (a)) and that of metaretinochrome (M-Ret) minus retinochrome (Ret) at room temperature (23°C)(dashed-and-dotted line, curve (b)).

Fig.3: Schematic representation of the dual system of photo-pigments, showing the cyclization by cis-trans isomerization of ' their chromophore retinal. Transfer of the retinal($\rightarrow$) is conducted by the aid of a retinal-binding protein.

From some of the above summarized properties of retinochrome, HARA and collaborators concluded that retinochrome contributes to the synthesis of squid rhodopsin in the visual cells as shown in Fig.3 [4-11].

In this paper, we have studied the photo-bleaching process of retinochrome at room temperature ($23\pm3°C$) by time-resolved absorption spectroscopy using both mode-locked and Q-switched Nd:YAG lasers in the time region from picoseconds to seconds for the first time. We could observe the formation processes of lumiretinochrome and metaretinochrome.

Retinochrome, lumiretinochrome, and metaretinochrome are hereafter abbreviated as Ret, L-Ret, and M-Ret, respectively.

2 EXPERIMENTAL

2.1 Sample

The squid was captured during the night and decapitated in the dark. The enu-cleated eyes were immediately frozen and stored at $-20°C$. This prior refriger-ation of the fresh eyeball facilitated separation of the inner and outer seg-ments of the visual cells. All the operations for preparing retinochrome ex-tracts were carried out under dim red light and, unless otherwise specified, at temperatures as low as 4°C.

The frozen eye was dissected to remove its anterior part with the lens and vitreous humor. The remaining eyecup was then shaken upside down in three successive portions of 67 mM phosphate buffer at pH 6.5 so as to detach the outer segments carrying the rhabdomeres and black pigment. The eyecup retained the retinochrome-bearing retina consisting of the inner segments and nerve plexus. The eyecup was spread out on a sheet of filter paper and the periphery cut away by scissors to remove different black pigment remaining there. The outer segment-free retina was peeled out of the eyecup with forceps, cut in fine pieces with scissors, homogenized in a small amount of the same phosphate buffer with a glass-Teflon homogenizer, and centrifuged.

The pellet was then suspended in a 43 %(w/v) solution of sucrose in neutral phosphate buffer, and centrifuged at 12000 rpm for 15 minutes to obtain the supernatant and the precipitate. The supernatant was withdrawn , and the remaining precipitate was mixed again with the same sucrose solution and centrifuged to yield the second supernatant. The two supernatants were pooled together, diluted with an equal volume of phosphate buffer, and centrifuged at 12000 rpm for 20 minutes to obtain a residue of cell fragments, which abounds in the myeloid bodies. All subsequent washings were carried out each time by shaking with water or solvents and by centrifuging at this speed for about 20 minutes.

This residue was then transferred to 0.5 % Na_2CO_3 for mild digestion and immediately centrifuged. The swollen cell fragments were washed repeatedly with water to release pale brown impurities that bear no relation to retinochrome, washed with phosphate buffer, and immersed in Weber-Edsall's solution (0.6 M KCl, 0.04 M $NaHCO_3$ and 0.01 M Na_2CO_3) for 30 minutes at 30°C to extrude soluble proteins. The residue after centrifuging was washed with water and phosphate buffer, and then lyophilized. The dried material was extracted with three successive portions of petroleum ether to remove lipids. After removing the solvent completely, the residue was washed once with water, mixed with 1-2 ml 2 % solutions of digitonin in 67 mM phosphate buffer at pH 6.5, and extracted by gentle shaking in a refrigerator for 20 minutes. The retinochrome extract from digitonin was obtained as a clear supernatant solution by centrifuging. Furthermore, the residue was mixed with 1-2 ml 12 mM solution of CHAPS, and extracted by gentle shaking at room temperature for 3 hours. The retinochrome extract from CHAPS was obtained as a supernatant solution by centrifuging. Retinochrome is, unlike rhodopsin, readily broken by the addition of detergents such as CTAB and Emulphogene, thus these could not be applied to extract retinochrome.

2.2 Time dependence of absorbance changes and time-resolved absorption spectrum

For the measurement of the time-resolved difference absorption spectrum in the picosecond region, the second harmonic (532 nm) of a mode-locked Nd:YAG laser (Quantel YG472, pulse width (fwhm)=35 ps, pulse energy=0.4 mJ) was used for the excitation of the sample. The details of the experimental apparatus are shown in Fig.2 of ref.16.

The time dependence of the absorbance change induced by nanosecond laser excitation in the time region from nanosecond to second was measured with the same experimental apparatus as that shown in Fig.1 of ref.16 with small modification. The second harmonic of a Q-switched Nd:YAG laser (Quanta-Ray DCR-1A, pulse width (fwhm)=5 ns, pulse energy=0.5 mJ) was used as an excitation light pulse and it was also used for the measurement of the time-resolved absorption spectrum. Figure 4 shows the experimental apparatus used for the measurement of the time-resolved absorption spectrum in the region from nanosecond to second. The absorption spectrum was measured with a combined system of Xe flash (Sugawara NP-1A, pulse width (fwhm)=200 ns), polychromator,

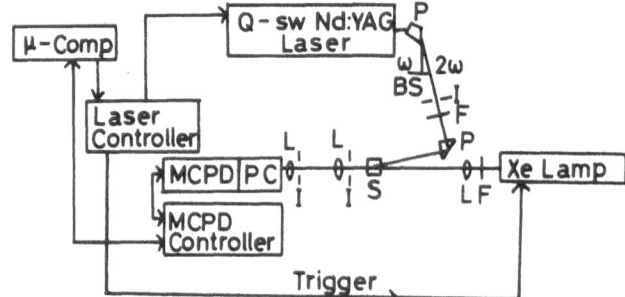

Fig.4: Block diagram of experimental apparatus used for the measurement of the time-resolved absorption spectra: BS (beam-stop), F (filter), I (iris), L (lens), MCPD (multi-channel photodiode), P (prism), PC (poly-chromator), and μ-Comp (microcomputer).

and multichannel photodiode (Unisoku USP-450). Time resolution was 200 ns. The multichannel photodiode was interfaced to an on-site microcomputer (NEC PC9801).

3 RESULTS AND DISCUSSION

3.1 Absorbance changes in the time region from nanosecond to second

The time dependences of the absorbance changes induced by the irradiation of a retinochrome sample by the nanosecond laser at 532 nm were measured at various wavelengths. The results obtained for four wavelengths ((a)440 nm, (b)500 nm, (c)620 nm, and (d)640 nm) are shown in Fig.5 as examples. Time resolution of the experimental apparatus was 10 ns and data were obtained by accumulating data taken by 32 laser shots.

The absorbancies at 440 nm (Fig.5(a)) and 500 nm (Fig.5(b)) change within resolution time and they are constant until the delay time (t_d) of 9 μs. No absorbance change could be detected at 620 nm (Fig.5(c)) and 640 nm (Fig.5(d)).

The apparent negative absorbance peaks observed at all wavelengths just after excitation are due to the strongly scattered laser pulse (532 nm) by sample and probably also due to the fluorescence of the sample. The experiment of fluorescence spectrum and decay kinetics of retinochrome is being planned.
The linear and logarithmic plots of the absorbance change at 460 nm between t_d= -200 ms and 1.8 s are shown in Figs.6(a) and (b), respectively. The data was taken with resolution time 2 ms. The results shown in Fig.6 were obtained by averaging data taken by 96 laser shots. The absorption intensity at 460 nm increases instantaneously just after excitation and decreases slowly to -0.05 with two (slow and fast) time constants (τ) which were obtained to be τ_{fast}=80±15 ms and τ_{slow}=290±30 ms by the least-squares fit.

Figure 7 shows the time-resolved difference spectra at t_d=-1.6 μs(a), 200 ns(b), 7 ms(c), and 1.6 s(d). The spectra at t_d=200 ns((b)) and 7 ms((c)) are much the same. The difference absorption spectrum (cf. Fig.2 and ref.9) of L-Ret minus Ret at -190 °C (dashed-and-dotted curve in Fig.7(b)) and that of M-Ret minus Ret at 23°C (dashed-and-dotted curve in Fig.7(d)) agree well with the time-resolved difference spectra at the delay times of 200 ns (Fig.7(b)) and 1.6 s (Fig.7(d)) after the excitation of Ret, respectively, except small discrepancy at longer wavelength than 520 nm. There was no absorbance change detected in the longer than 580 nm wavelength region.

3.2 Absorbance change in picosecond region

The time-resolved difference spectra in picosecond region are shown in Fig.8. The spectra (a), (b), (c), and (d) were observed at t_d=-100 ps, 0 ps, 60 ps, and 1 ns, respectively. The time resolution of the experimental apparatus was 35 ps.

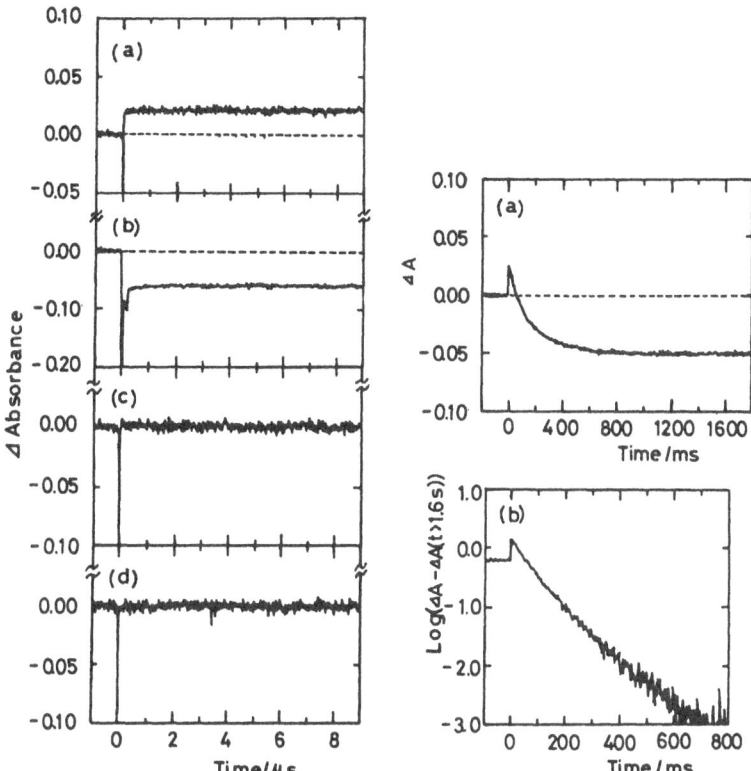

Fig.5: The time dependence of the absorbance change between $t_d= -1\,\mu$s and 9 μs at (a)440 nm, (b)500 nm, (c)620 nm and (d)640 nm. Time resolution of the measurement was about 10 ns.

Fig.6: (a)The time dependence of the absorbance change at 460 nm between $t_d= -200$ ms and 1.8 s, and (b) logarithmic plot. Time resolution of the experimental apparatus was 2 ms.

The absorbance at 460 nm decreases at t_d =0 ps and increases with time for the order of 100 ps (see Figs.8(b) and (c)) and there is no detectable difference among the absorption spectra at t_d=0 ps (Fig.8(b)), 60 ps (Fig.8(c)), and 1ns (Fig.8(d)) between 480 nm and 580 nm. The time-resolved spectrum at t_d=1 ns (Fig.8(d)) agrees with the difference absorption spectrum of L-Ret minus Ret at liquid nitrogen temperature which is shown in Fig.8(d) by a dashed-and-dotted line (cf. Fig.2 and ref.9). In the wavelength region longer than 580 nm, no absorbance change was observed in the picosecond region.

3.3 The bleaching process of retinochrome (Ret)

HARA and her coworkers observed the change in the absorption spectrum of L-Ret at liquid nitrogen temperature (cf. Fig.1) and proposed the bleaching process of Ret as follows from the results of the photochemistry at low temperature [10].

The experimental results described above are summarized and discussion on the bleaching process of Ret is made as follows.

(1) In the picosecond region at room temperature (23±3°C), absorbance decrease within an excitation pulse width(35 ps) and the formation of the intermediate of L-Ret with the time constant of the order of 100 ps were observed. The formation time of about 100 ps may correspond to the decay time

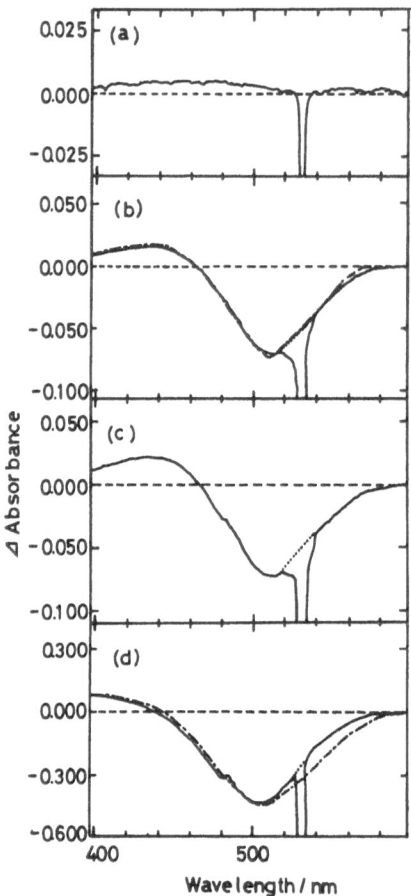

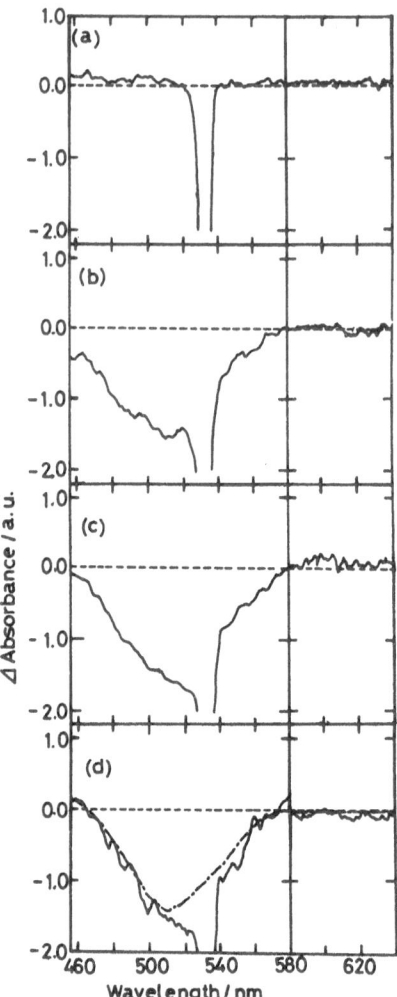

Fig.7: The time-resolved
absorption spectra at
(a) t_d= -1.6 μs, (b) 200 ns,
(c) 7 ms and (d) 1.6 s.
Time resolution of the
experimental apparatus was
200 ns. The dashed-and-
dotted lines in (b), and
(d) are the difference
absorption spectrum of
lumiretinochrome (L-Ret)
minus retinochrome (Ret)
at low temperature, and
that of metaretinochrome
(M-Ret) minus retinochrome
(Ret) at room temperature,
respectively.

Fig.8: The time-resolved
absorption spectra at
(a) t_d= -100 ps, (b) 0 ps,
(c) 60 ps and (d) 1 ns. Two
different frames of the multichannel photodiode are shown
in this figure. The dashed-
and-dotted line in (d) is the
difference absorption spectrum
of lumiretinochrome (L-Ret)
minus retinochrome (Ret)
at -190°C.

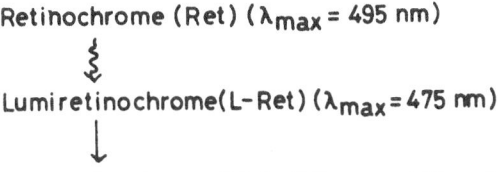

Retinochrome (Ret) (λ_{max} = 495 nm)

Lumiretinochrome(L-Ret) (λ_{max} = 475 nm)

Metaretinochrome(M-Ret)(λ_{max} = 470 nm)

Fig.9: Photobleaching process of retinochrome (Ret)

of the precursor of lumiretinochrome which may be the excited state of retinochrome or a new intermediate in the thermal process. We have not yet observed positive absorbance change due to a new absorption band relevant to either the excited state or the new intermediate in any wavelength region observed between 460 and 640 nm.

The formation time of the order of 100 ps is considered to be too long to be assigned to the excited singlet state of retinochrome since the lifetimes of other retinoid pigments such as rhodopsin and bacteriorhodopsin are of the order of subpicosecond or a few picoseconds [17,18]. Therefore we tentatively explain the time dependence in terms of the formation of lumiretinochrome from a newly-found intermediate. The new intermediate is called pre-lumiretinochrome (hereafter referred to as X-Ret).

The pre-lumiretinochrome may possibly be an intermediate similar to primerhodopsin, hypsorhodopsin, or bathorhodopsin in rhodopsin photocycle or I, J, or K intermediate in bacteriorhodopsin photocycle. Both primerhodopsin [18,19] and bathorodopsin [20] and also both J [21] and K [22] intermediates absorb longer wavelengths than the parent rhodopsin and bacteriorhodopsin, respectively. The absorption coefficients at the respective peaks of prime- and batho- and J and K intermediates are similar to or even larger than respective parent pigments. However, as already mentioned, there is no detectable absorbance change in the longer wavelength region than that of retinochrome at the delay time between 0 ps and 1 ns. Therefore the new intermediate possibly is correspondent with hypsorhodopsin or J intermediate. However further pico-second and femtosecond experiments are needed for the complete clarification of the very primary processes. The absorption spectroscopy at liquid helium temperature is also useful for the understanding and identification of the new intermediate.

(2) Absorption intensities at four wavelengths (440 nm, 500 nm, 620 nm, and 640 nm) keep constant between 30 ns and 9 μs (see Fig.5) and the time-resolved difference spectra at both t_d=200 ns and 7 ms at 23°C agree well with that of L-Ret at -190°C. These experimental results show that the L-Ret formed through X-Ret in the picosecond region ($\tau \simeq$ 100 ps), decays with much longer time constant than 7 ms.

(3) The absorbance at 460 nm decays with two time constants (τ fast and τ slow), 80 ms and 290 ms, and the time-resolved difference spectrum at t_d=1.6 s coincides relatively well with that of M-Ret minus Ret at 23°C. The two time constants are considered to be corresponding to two processes. L-Ret is converted into LM-intermediate, which is found in bacteriorhodopsin, with τ = 80±15 ms and M-Ret is formed from LM-Ret with τ = 290±30 ms. M-Ret and LM-Ret may have a relation to the two forms of M-Ret, i.e. tentatively the tight and loose forms of M-Ret [9].

The difference in absorption spectrum between LM-Ret and M-Ret is probably so small that the absorption spectrum of LM-Ret at low temperature has not been reported previously [10]. The measurement of the time-resolved absorption spectrum in a few hundred millisecond region is now in progress.

The schematic model of the photobleaching process of retinochrome constructed on the basis of the above discussions is shown in Fig.10. "K-like intermediate" (K-Ret) could not be observed in the time-region from picosecond to second in the spectra shown in Fig.7 and 8. This is a large difference between the photoisomerization process of retinal in rhodopsin (11-cis to all-trans) and that of Ret (all-trans to 11-cis). The absorption spectroscopy in the femtosecond region and the study with artificial pigments based on synthetic retinal analogues are useful to search for K-Ret and to make clear the difference between the photochemical process of rhodopsin and that of Ret.

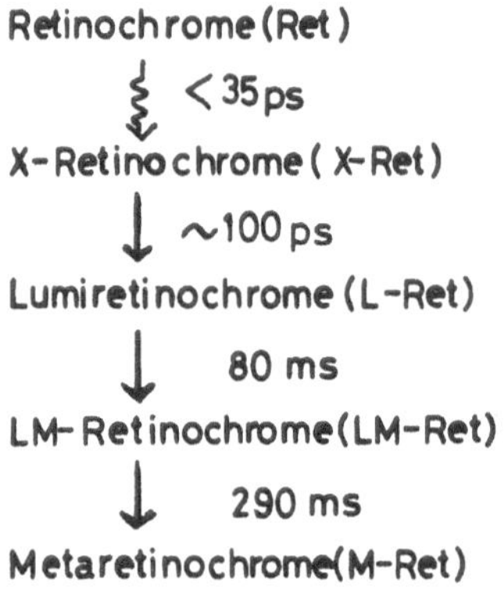

Retinochrome (Ret)

$\downdownarrows$ < 35 ps

X-Retinochrome (X-Ret)

$\downarrow$ ~100 ps

Lumiretinochrome (L-Ret)

$\downarrow$ 80 ms

LM-Retinochrome(LM-Ret)

$\downarrow$ 290 ms

Metaretinochrome(M-Ret)

Fig.10: Photobleaching process of retinochrome (Ret) excited by the green laser light (532 nm).

Acknowledgments
This work was partly supported by a Grant-in-Aid for Special Distinguished Research (56222005) from the Ministry of Education, Science and Culture of Japan, and also partly by the Toray Science and Technology Foundation, and the Kurata Science Foundation.

4 REFERENCES

1. T. Hara and R. Hara :Nature 206, 1331 (1965)
2. T. Hara and R. Hara :Nature 219, 450 (1968)
3. T. Hara and R. Hara :Nature 214, 573 (1967)
4. T. Hara and R. Hara :In Handbook of Sensory Physiology, vol VII/I, ed. by H. J. A. Dartnall, (Springer,Berlin,Heidelberg, New York 1972)
5. T. Hara and R. Hara :In Biochemistry and Physiology of Visual Pigments, ed. by H. Langer, (Springer,Berlin,Heidelberg, New York 1973)
6. T. Hara and R. Hara :Nature 242, 39 (1973)
7. T. Hara and R. Hara :J. Gen. Physiol. 67, 791 (1976)
8. T. Hara and R. Hara :Methods in Enzymology 81, 190 (1982)
9. R. Hara and T. Hara :Vision Res. 24, 1629 (1984)
10. R. Hara, T. Hara, F. Tokunaga and T. Yoshizawa :Photochem. Photobiol. 33, 883 (1981)
11. K. Ozaki, R. Hara and T. Hara :Exp. Eye Res. 34, 499 (1982)
12. K. Ozaki, R. Hara and T. Hara :Cell Tissue Res. 233, 335 (1983)
13. K. Hamdorf :In Handbook of Sensory Physiology, vol VII/6A, ed. by H. Autrum (Springer, Berlin 1979)
14. L. Sperling and R. Hubbard :J. Gen. Physiol. 65, 235 (1975)
15. T. Seki :J. Gen. Physiol. 84, 49 (1984)
16. J. Iwai, M. Ikeuchi, Y. Inoue and T. Kobayashi :In Protochlorophyllide reduction and greening, ed. by C. Sironvaland and M. Brouwers (Nijhoff/Dr. W. Junk, The Hague, 1984)
17. M. C. Nuss, W. Zinth, W. Kaiser, E. Kolling, and D. Oesterhelt : Chem. Phys. Letters, 117, 1 (1985)
18. T. Kobayashi :Photochem. Photobiol. 32, 207 (1980)

19. K. Peters, M. L. Applebury, and P. M. Rentzepis :Proc. Natl. Acad. Sci. USA 74, 3119 (1977)
    T. Kobayashi :FEBS Lett. 106, 313 (1980)
    B. Honig, T. Ebrey, R. H. Callender, U. Dinur, and M. Ottolenghi :Proc. Natl. Acad. Sci. USA 76, 2503 (1979)
    Y. Shichida, S. Matuoka, and T. Yoshizawa :Photobiochem. Photobiophys. 7, 221 (1984)
    S. Matuoka, Y. Shichida, and T. Yoshizawa :Biochem. Biophys. Acta 765, 38 (1984)
20. T. Yoshizawa :In Handbook of Sensory Physiology, ed. by Dartnall, (Springer-Verlag, Berlin 1972)
    T. Yoshizawa and Y. Kito :Nature 182, 1604 (1958)
21. M. L. Applebury, K. S. Peters, and P. M. Rentzepis :Biophys. J. 23, 375 (1978)
22. E. Kolling and D. Oesterhelt :Biophys. J. 49, 6510 (1986)

Part IV

Bacteriorhodopsin

Determination of Chromophore Structure and Photochemistry in Bacteriorhodopsin with Resonance Raman, NMR, and Chemical Analogues

R.A. Mathies[1], S.O. Smith[2], G.S. Harbison[3], J. Herzfeld[4], R.G. Griffin[2], and J. Lugtenburg[5]

[1]Chemistry Department, University of California, Berkeley, CA 94720, USA
[2]Francis Bitter National Magnet Lab., M.I.T., Cambridge, MA 02139, USA
[3]Department of Chemistry, State University of New York,
Stony Brook, NY 11794, USA
[4]Chemistry Department, Brandeis University, Waltham, MA 02254, USA
[5]Department of Chemistry, State University of Leiden,
P.O. Box 9502, NL-2300 RA Leiden, The Netherlands

Bacteriorhodopsin (BR) is an intrinsic membrane protein that transduces light-energy into a trans-membrane protonmotive force. To elucidate the mechanism of this important process, it is necessary to determine the structure of the retinal chromophore in BR and its intermediates and to characterize important chromophore-protein interactions. This paper will review recent physical chemical and bio-organic approaches that we have developed to study chromophore structure and environment in bacteriorhodopsin. A model for the molecular mechanism of proton-pumping based on inversion of the unprotonated Schiff base nitrogen is presented.

1. Resonance Raman Spectroscopy

Resonance Raman vibrational scattering provides a valuable *in situ* probe of chromophore structure in retinal-pigments [1,2]. In the resonance Raman experiment scattering from the chromophore is strongly and selectively enhanced so that only small amounts of material are required, background scattering from the protein is suppressed, and the high signal-to-noise ratio permits acquisition of time-resolved Raman spectra. Although a number of empirical vibrational assignments have been made [1-3], the full exploitation of these data has awaited a full understanding of the chromophore's normal modes.

1.1 Vibrational Analysis of BR_{568} and BR_{548}

We have recently completed a detailed study of the vibrational assignments of BR_{568} and BR_{548} using an extensive set of ^2H- and ^{13}C-isotopic derivatives in conjunction with normal mode calculations [4,5]. A summary of these assignments for BR_{568} is presented in Fig. 1. For the purposes of this paper we will concentrate on the 1100-1250 cm^{-1} fingerprint region. As we have shown in vibrational studies of all-*trans* retinal [10], the retinal isomers [11], and the all-*trans* retinal protonated Schiff base [12], the fingerprint normal modes are most easily discussed in terms of their C-C stretch character, which is often quite localized. Typically, the "$C_{12}-C_{13}$" and "C_8-C_9" modes are highest in frequency because they are pushed up by interaction with the $C_{13}-CH_3$ and C_9-CH_3 stretches, respectively. In BR_{568}, the C_8-C_9 mode is found at 1214 cm^{-1} while the $C_{12}-C_{13}$ stretch contributes to the 1255 cm^{-1} band. The highest frequency C-C stretch mixes strongly with the C-C-H

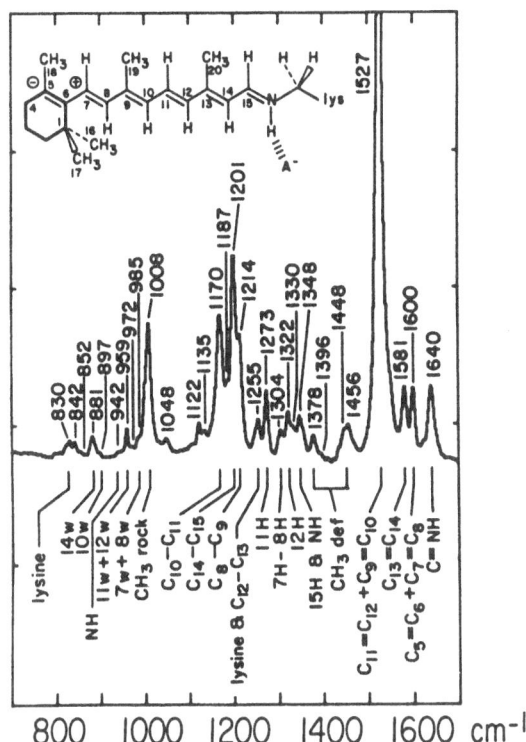

Figure 1. Resonance Raman spectrum of light-adapted bacterio-rhodopsin, BR_{568}. The dominant internal coordinate character in each normal mode has been indicated. The chromophore is shown in the, 6-s-*trans*, C_{15}=N *anti* structure, which has been demonstrated by recent ^{13}C-NMR, Raman and chemical analogue studies [6-9]

rocks, causing its C-C stretch character to be delocalized. In BR_{568} there is an additional complication caused by the near degeneracy of the "C_{12}-C_{13} stretch" with the lysine CH_2 rock. Thus, the 1255 cm^{-1} band is made up of two normal modes, the "C_{12}-C_{13} stretch" at ~1248 cm^{-1} and the CH_2 rock at 1255 cm^{-1} which are strongly mixed with one another. The majority of the C_{14} C_{15} stretch character is found in the 1201 cm^{-1} mode. Although this mode is nearly degenerate with the C_8-C_9 stretch at 1214 cm^{-1}, the ^{13}C-shifts indicate that these internal coordinates are relatively localized in their respective normal modes. The 1170 cm^{-1} mode is a fairly pure C_{10}-C_{11} stretch at a characteristic C-C stretch frequency for an unsubstituted linear polyene. It is important to note that the methyl substituents and the Schiff base moiety "break up" the C-C stretches into more localized modes. If these substituents were not present, a much more delocalized set of normal modes would result.

The vibrational assignments for BR_{548} are presented in Fig. 2 [5]. The major differences from the fingerprint assignments of BR_{568} are that the C_{14}-C_{15} mode has dropped ~34 cm^{-1} to 1167 cm^{-1} and the C_{12}-C_{13} mode has dropped 14 cm^{-1} to 1234 cm^{-1}. These changes are the expected consequences of the presence of the *cis* C_{13}=C_{14} and C_{15}=N bonds. Curry *et al.* [11] pointed out that a *cis* bend in a linear polene will result in a ~20 cm^{-1} frequency reduction of the adjacent C-C stretching modes. In BR_{548}, the C_{14}-C_{15} drops about double this value, consistent with its proximity to two such *cis* bends.

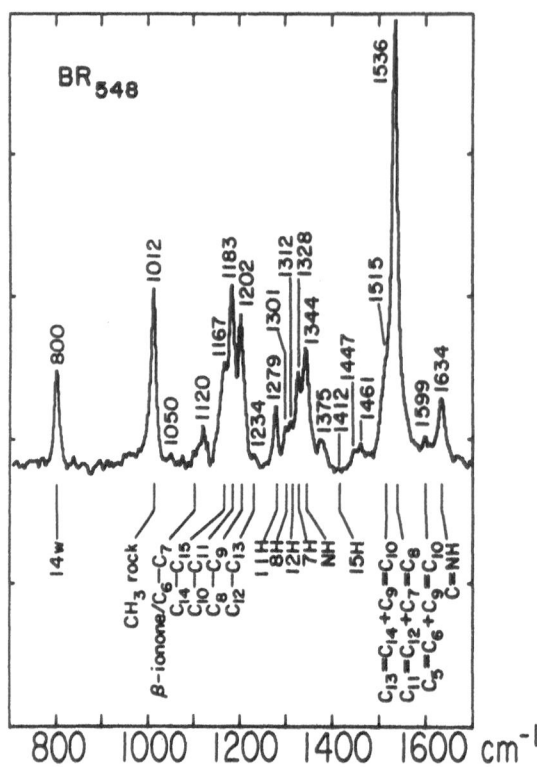

Figure 2. Resonance Raman spectrum of the 13-*cis* component of dark-adapted bacteriorhodopsin, BR_{548}. The dominant internal coordinate character in each mode has been indicated [5]

1.2 Structural Conclusions

These assignments can now be used to study chromophore structure in bacteriorhodopsin. First, the frequencies of the C-C stretching modes in the $1100-1250$ cm^{-1} fingerprint region of BR_{568} have been compared with those of the all-*trans* retinal protonated Schiff base [12]. The $C_{10}-C_{11}$, $C_{14}-C_{15}$, C_8-C_9 and $C_{12}-C_{13}$ modes in BR_{568} are all found to be elevated approximately 10 cm^{-1} from the all-*trans* retinal protonated Schiff base (PSB) values. This suggests that there is a general increase in the ground state electron delocalization upon protein binding. Second, the frequency of the $C_{12}D + C_{14}D$ in-plane rock can be used as a marker band for $C_{13}=C_{14}$ configuration [1]. Its characteristic frequency is $900-915$ cm^{-1} in 13-*trans* molecules and $935-948$ cm^{-1} in 13-*cis* molecules. The 916 cm^{-1} $C_{12}D + C_{14}D$ rock in BR_{568} demonstrates that BR_{568} contains a 13-*trans* chromophore while the corresponding mode in BR_{548} is found at 944 cm^{-1}, demonstrating a 13-*cis* chromophore. Third, the characteristic coupling of the N-H rock with the $C_{14}-C_{15}$ stretch can be used to determine C=N configuration [8]. In C=N *trans* molecules, the rock/stretch coupling is weak and N-deuteration has little effect on the $C_{14}-C_{15}$ mode frequency. However, in C=N *cis* chromophores these coordinates are strongly coupled so N-deuteration causes a large shift in the fingerprint modes having significant $C_{14}-C_{15}$ stretch character. Suspension in D_2O buffer results in a <1 cm^{-1} shift of the 1201 cm^{-1} $C_{14}-C_{15}$ mode in BR_{568} showing that the chromophore is C=N trans. However, there is a 41 cm^{-1} upshift of the

1167 cm^{-1} C_{14}-C_{15} mode in N-D BR$_{548}$. We conclude that the chromophore in the latter species is C=N *cis*, in agreement with solid state NMR studies [7].

The introduction of models for the BR→K transition that involve 14-s-*cis* structures [13,14] has made the *in situ* determination of chromophore <u>conformation</u> an important question. This can be probed with vibrational spectroscopy since the frequency of a C-C stretch is very sensitive to its conformation. As depicted in Fig. 3, s-*trans*→s-*cis* isomerization results in a ~100 cm^{-1} frequency reduction of the C-C stretching mode [15]. This arises from altered coupling of the stretch with the C-C-C bends. Based on this we would expect that the C_{14}-C_{15} stretching mode in K would lie ~100 cm^{-1} below the C_{14}-C_{15} mode in the 13-*cis* PSB. The observation of this mode at ~1188 cm^{-1} in K (and at ~1172 cm^{-1} in L), within 10 cm^{-1} of the C_{14}-C_{15} stretch in the 13-*cis* PSB, argues that K and L both contain C_{14}-C_{15} s-*trans* chromophores [15].

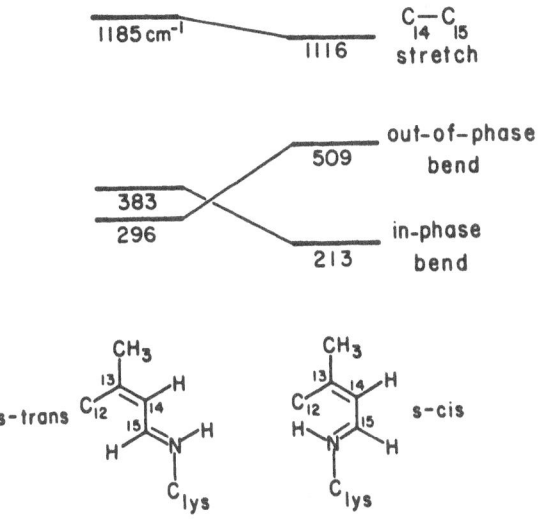

Figure 3. Calculated frequencies for C_{14}-C_{15} s-*trans* and s-*cis* Schiff base fragments

2. ^{13}C-Magic Angle Sample Spinning NMR

A second valuable probe of chromophore structure in membrane proteins is provided by solid-state NMR. In order to observe individual ^{13}C-NMR resonances in BR, it is necessary to regenerate the pigment with ^{13}C-labeled retinals. Solution NMR experiments on detergent solubilized BR have met with only modest success because the long rotational correlation time of the protein results in broad resonances with poor S/N ratios. Better resolution is obtained with hydrated pellets of intact purple membrane by spinning the sample rapidly (2-3 kHz) at the magic angle. It has recently been shown that high-quality ^{13}C-NMR spectra can be obtained of ^{13}C-labeled retinal in BR [16]. A complete series of experiments have now been performed using retinals labeled with ^{13}C from C$_5$ to C$_{15}$ [6]. The isotropic chemical shift of each resonance as well as the individual elements of the chemical shift tensor have been

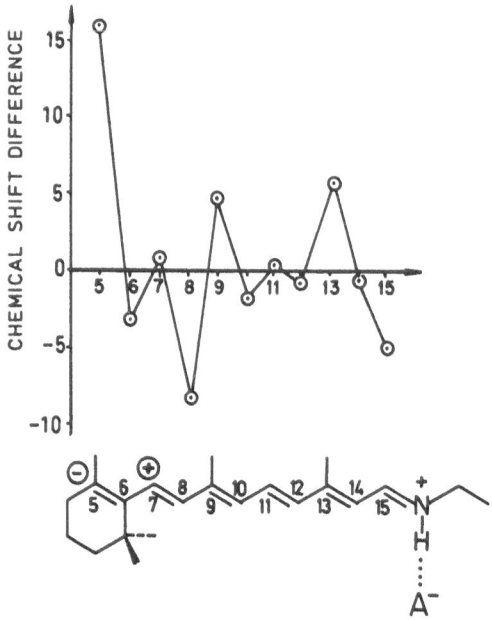

Figure 4. Plot of ^{13}C-NMR chemical shift differences between light-adapted BR and the all-*trans* retinal protonated Schiff base (6-s-*cis*, chloride salt) as a function of chain position [6]

determined. Figure 4 presents a plot of the "opsin shift" of the isotropic chemical shift in BR_{568} as a function of chain position. The most dramatic difference is the ~15 ppm deshielding of the C_5 resonance by the protein. This observation provides strong evidence for the perturbation of the chromophore by an electronegative protein residue near C_5 as originally put forward by NAKANISHI et al. [17]. The observation that the C_6 and C_7 resonances were only slightly perturbed suggested that the deshielding effect of the negatively charged residue near C_5 was cancelled out by a nearby positive counterion. A third result concerned the conformation about the C_6-C_7 single bond. In solution, the C_6-C_7 conformation is ~40° twisted 6-s-*cis*. However, in BR the shift tensor elements of the C_5 resonance provide clear evidence that <u>the protein-bound chromophore is 6-s-trans.</u> The implications of this observation for the mechanism of the opsin shift in BR are discussed below.

3. Chemical Analogue Experiments

Chemical analogue experiments can also provide useful information about retinal structure and environment [17,18]. Figure 5 presents the structures of the analogues we have studied along with the energy difference between the λ_{max} of the PSB in solution and that of the pigment which is called the opsin shift. The most interesting result is the observation that 7,8-dihydroretinal has an opsin shift of 3500 cm^{-1} [19,20]. This indicates that a large fraction of the overall opsin shift must be associated with chromophore-protein interactions near the Schiff base end of the chromophore. A similar conclusion has been reached by MURADIN-SZWEYKOWSKA et al. [21] based on the observation that retinal truncated at C_8 still has a large opsin shift (2700 cm^{-1}). This immediately suggests that more than half of the 5100 cm^{-1} opsin shift in BR can be attributed to an

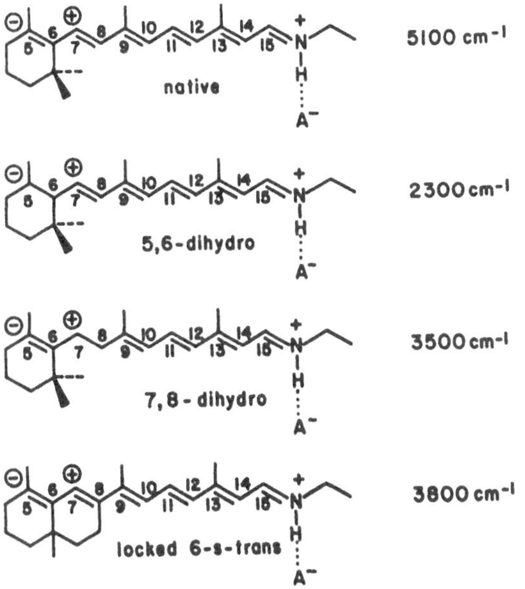

Figure 5. Retinal analogues that have been regenerated with bacteriorhodopsin and their corresponding opsin shifts [9,19]

altered Schiff base environment. This is in harmony with the ^{15}N-NMR [22] and resonance Raman data [23] on BR_{568} which indicate an unusually weak hydrogen-bond between the positive Schiff base nitrogen and its negative counterion.

The 6-s-*trans* conformation of the protein-bound chromophore must also be considered in attempts to determine the mechanism of the opsin shift. A retinal analogue with a locked 6-s-*trans* conformation has been synthesized [9]. It is found that twisted 6-s-*cis* to planar s-*trans* isomerization causes a 25 nm (1200 cm^{-1}) red shift. Thus, the 5100 cm^{-1} opsin shift of the native chromophore can be partitioned into ~1200 cm^{-1} due to the altered C_6-C_7 conformation and ~3500 cm^{-1} due to the weak Schiff base hydrogen-bond. This analysis is nicely supported by the 3800 cm^{-1} opsin shift observed for the s-*trans* locked retinal analogue in Fig. 5.

4. A Model for the Proton Pump

We present in Fig. 6 a model for the BR proton pump that is consistent with the results presented here. The primary photoisomerization from 13-*trans* to 13-*cis* translates the Schiff base moiety from A_1^- to HA_2 thereby storing energy [24] and lowering the pK_a of HA_2. In the K → L step, HA_2 deprotonates and in the L → M transition the Schiff base proton is donated to A_2^-. At this point a proton has been pumped and we must reset the pump mechanism. If the Schiff base picked up a proton from HA_2 this would constitute a "short circuit", which must be avoided. Thus we need a "reprotonation switch" which will change the orientation of the Schiff base nitrogen before reprotonation. One possibility, depicted in Fig. 6, is inversion about the Schiff base nitrogen. Inversion is the favored mechanism for the *anti* → *syn* conversion and the barriers are estimated to be from 12–28 kcal/mole [25]. A C=N *syn* geometry would position the lone pair on the nitrogen to pick up

141

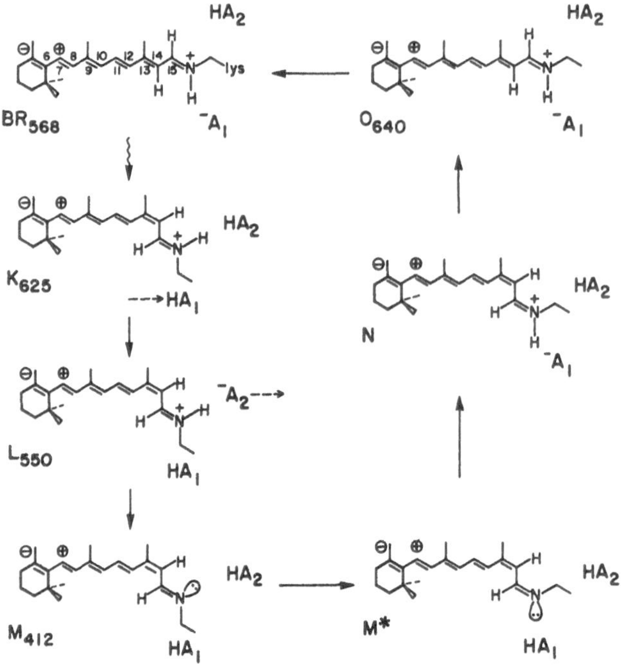

Figure 6. Model illustrating how inversion of the Schiff base nitrogen might act as a reprotonation switch in the BR proton pump. The direction of proton transfer is indicated by dashed arrows

a proton from HA_1, forming a dicis chromophore structure denoted N. A simple bicycle pedal mechanism [26], already observed in the $BR_{548} \rightarrow BR_{568}$ interconversion, would then lead to O_{640} and subsequently to BR_{568}. Experiments are now needed on the M → O part of the photocycle to probe for the presence of the structures proposed here.

5. Acknowledgment

We wish to thank P.L. Biesheuvel, J.M.L. Courtin, C. Heeremans, I. Hornung, M. Muradin-Szweykowska, J.A. Pardoen, R. van der Steen and C. Winkel for the synthesis of the retinal derivatives, and M. Braiman and A.B. Myers for assistance with the vibrational assignments. This research was supported by the NIH (GM-27057), the NSF (CHE-8116042), the Netherlands Foundation for Chemical Research (SON), and the Netherlands Organization for the Advancement of Pure Research (ZWO).

REFERENCES

1. S.O. Smith, J. Lugtenburg, R.A. Mathies: J. Membrane Biol. 85, 95 (1985).
2. M. Stockburger, T. Alshuth, D. Oesterhelt, W. Gärtner: Adv. Infrared Raman Spectrosc. 13, 483 (1986).
3. T. Alshuth, M. Stockburger: Ber. Bunsenges. Phys. Chem. 85, 484 (1981).
4. S.O. Smith, M.S. Braiman, A.B. Myers, J.A. Pardoen, J.M.L. Courtin, C. Winkel, J. Lugtenburg, R.A. Mathies: Vibrational analysis of the all-trans retinal chromophore in light-adapted bacteriorhodopsin. J. Am. Chem. Soc., in press.

5. S.O. Smith, J.A. Pardoen, J. Lugtenburg, R.A. Mathies: Vibrational analysis of the 13-*cis* retinal chromophore in dark-adapted bacteriorhodopsin. J. Phys. Chem., in press.
6. G.S. Harbison, S.O. Smith, J.A. Pardoen, J.M.L. Courtin, J. Lugtenburg, J. Herzfeld, R.A. Mathies, R.G. Griffin: Biochemistry 24, 6955 (1985).
7. G.S. Harbison, S.O. Smith, J.A. Pardoen, C. Winkel, J. Lugtenburg, J. Herzfeld, R.A. Mathies, R.G. Griffin: Proc. Natl. Acad. Sci. USA 81, 1706 (1984).
8. S.O. Smith, A.B. Myers, J.A. Pardoen, C. Winkel, P.P.J. Mulder, J. Lugtenburg, R.A. Mathies: Proc. Natl. Acad. Sci. USA 81, 2055 (1984).
9. R. van der Steen, P.L. Biesheuvel, R.A. Mathies, J. Lugtenburg: J. Am. Chem. Soc. 108, 6410 (1986).
10. B. Curry, A. Broek, J. Lugtenburg, R.A. Mathies: J. Am. Chem. Soc. 104, 5274 (1982).
11. B. Curry, I. Palings, A.D. Broek, J.A. Pardoen, J. Lugtenburg, R.A. Mathies: Adv. Infrared Raman Spectrosc. 12, 115 (1985).
12. S.O. Smith, A.B. Myers, R.A. Mathies, J.A. Pardoen, C. Winkel, E.M.M. van den Berg, J. Lugtenburg: Biophys. J. 47, 653 (1985).
13. R.S.H. Liu, D. Mead, A.E. Asato: J. Am. Chem. Soc. 107, 6609 (1985).
14. K. Schulten, P. Tavan: Nature (London) 272, 85 (1978).
15. S.O. Smith, I. Hornung, R. van der Steen, J.A. Pardoen, M.S. Braiman, J. Lugtenburg, R.A. Mathies: Proc. Natl. Acad. Sci. USA 83, 967 (1986).
16. G.S. Harbison, S.O. Smith, J.A. Pardoen, P.P.J. Mulder, J. Lugtenburg, J. Herzfeld, R. Mathies, R.G. Griffin: Biochemistry 23, 2662 (1984).
17. K. Nakanishi, V. Balogh-Nair, M. Arnaboldi, K. Tsujimoto, B. Honig: J. Am. Chem. Soc. 102, 7945 (1980).
18. K. Nakanishi: Pure Appl. Chem. 57, 769 (1985).
19. J. Lugtenburg, M. Muradin-Szweykowska, C. Heeremans, J.A. Pardoen, G.S. Harbison, J. Herzfeld, R.G. Griffin, S.O. Smith, R.A. Mathies: J. Am. Chem. Soc. 108, 3104 (1986).
20. J.L. Spudich, D.A. McCain, K. Nakanishi, M. Okabe, N. Shimizu, H. Rodman, B. Honig, R.A. Bogomolni: Biophys. J. 49, 479 (1986).
21. M. Muradin-Szweykowska, J.A. Pardoen, D. Dobbelstein, L.J.P. van Amsterdam, J. Lugtenburg: Eur. J. Biochem. 140, 173 (1984).
22. G.S. Harbison, J. Herzfeld, R.G. Griffin: Biochemistry 22, 1 (1983).
23. P. Hildebrandt, M. Stockburger: Biochemistry 23, 5539 (1984).
24. B. Honig, T.G. Ebrey, R.H. Callender, U. Dinur, M. Ottolenghi: Proc. Natl. Acad. Sci. USA 76, 2503 (1979).
25. M. Shanshal: Tetrahedron 28, 61 (1972); V. Bonacic-Koutecky, J. Michl: Theor. Chim. Acta 68, 45 (1985); H.-O. Kalinowski, H. Kessler: Top. Stereochem. 7, 295 (1973).
26. A. Warshel: Nature (London) 260, 679 (1976).

On the Nature of the Primary Photochemical Events in Rhodopsin and Bacteriorhodopsin

M. Ottolenghi[1] *and M. Sheves*[2]

[1]Department of Physical Chemistry, The Hebrew University of Jerusalem, Jerusalem 91904, Israel
[2]Department of Organic Chemistry, The Weizmann Institute of Science, Rehovot 76100, Israel

I. Introduction

All rhodopsins, whether responsible for the visual process (visual pigments-Rh), photosynthetic activity (bacteriorhodopsin-bR and halorhodopsin), phototaxis (sensory rhodopsin-sR), or photoisomerization (Retinochrome), share the same basic chromophore system: a retinyl polyene, bound to the opsin via a protonated Schiff base linkage with a lysine ϵ- amino group [1]. Moreover, most rhodopsins (e.g., Rh, bR and sR) exhibit photocycles with remarkably similar features [1]. Primarily (see Fig. 1): The occurrence of an early red shifted photointermediate, the deprotonation of the Schiff base during subsequent stages, the photoreversibility of most intermediates, the time scales associated with various stages of the photocycle, etc. It is the purpose of this overview to discuss in a comparative way the primary photochemical steps in Rh and bR, as monitored by optical, resonance-Raman, FTIR and NMR spectroscopy. Interpretation of such phenomena in terms of the primary molecular changes occurring in the retinal moiety and its protein environment is an essential pre-requisite for understanding the mechanism by which retinoid pigments convert, store and subsequently utilize, solar radiation. Such topics will be addressed in light of recent progress and still open questions.

II. The Evolution of BATHO and K_{610}: Experimental Phenomena.

The primary events in visual rhodopsins and in (light-adapted, all-trans) bacteriorhodopsin (bR^t_{570} abbreviated as bR) are associated with the generation of the red-shifted phototransients bathorhodopsin (BATHO) and K_{610}, respectively (for reviews, see Ref. [1]). A feature common to these two intermediates is their evolution kinetics, as monitored by picosecond absorption spectroscopy. Thus, it was recognized that both BATHO [2] and K_{610} [3] are generated from further red shifted species, denoted [4] as prebathorhodopsin (PBATHO) and J_{625}, respectively.

An analysis [4] of the early subpicosecond experiments of Ippen et al. [5] suggested that J_{625} grows-in, on an approximately 1 psec time scale, from a precursor denoted as I. Both $I \rightarrow J_{625}$ and $J_{625} \rightarrow K_{610}$ processes have been recently quantitatively re-examined by Polland et al. [6,7] and by Nuss et al. [8]. By applying low intensity excitation it was shown that, in variance with previous reports [3,9], neither of the two reactions is associated with an isotope effect upon replacement of the exchangeable protons in bR with deuterium. An analysis of the picosecond and femtosecond excitation data [7] indicated that J_{625} is red shifted by approximately 10 nm relative to K_{610}, while I is markedly blue shifted ($\lambda_{max} < 500$ nm). The time constants associated with the $I \rightarrow J_{625}$ and the $J_{625} \rightarrow K_{610}$ processes were established as 0.43+0.05 psec and 5 psec, respectively. Similar values (i.e.,

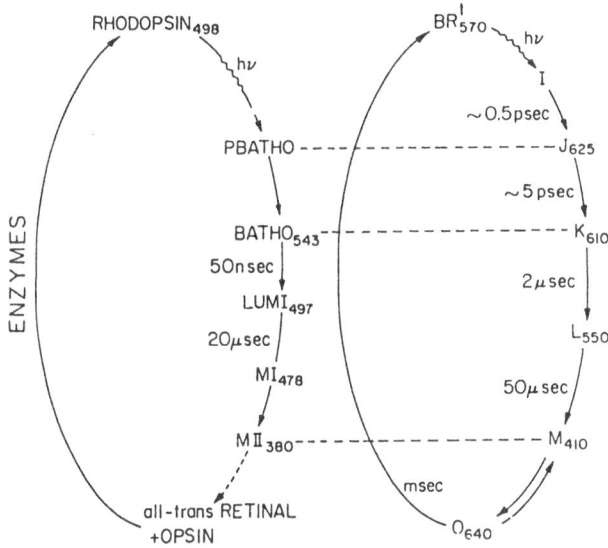

Fig. 1 The photocycles of rhodopsin (11-cis) and (all-trans) bacteriorhodopsin. Subscripts refer to wavelengths of maximum absorption. Time notations are approximate room temperature values. Horizontal dotted lines indicate analogous intermediates (see text).

0.7 psec [10] and 0.50+0.04 psec [11] for I → J_{625}, and 3.2+0.2 psec [11] for J_{625}→ K_{610}) were recently observed in other laboratories.

In the purple membrane of bacteriorhodopsin the three symmetrically arranged bR molecules are electronically coupled via exciton interactions [12]. The question arises as to the time scale in which the absorbed light energy, which is initially delocalized over the trimeric bR unit, is subsequently trapped by one of the three chromophores. Time–resolved linear dichroism experiments [13], as well as the comparison of femtosecond excitation data of trimeric bR with monomeric bR [8], indicate that the above energy localization process takes place within less than 50 femtoseconds.

A growing-in of PBATHO, parallel to the I → J_{625} step in bR, has not yet been observed in visual pigments. It appears, however, that the PBATHO → BATHO process is analogous to the J_{625} → K_{610} transition. Primarily, as in the case of the bR intermediates, PBATHO (or photorhodopsin, as termed by Yoshizawa and coworkers [14]) is red shifted by approximately 10 nm relative to bathorhodopsin. The rate constants for the PBATHO → BATHO process is 40 psec for bovine rhodopsin and approximately 200 psec in the cases of squid and octopus rhodopsin [14].

A complicating feature in the photocycles of several visual pigments has been associated with the observation, by continuous illumination at low temperatures, of an additional early intermediate absorbing at 430 nm (main band) and approximately 540 nm (secondary band) [15]. The question arose as to the origin and nature of this species (termed as hypsorhodopsin-HYPSO) which thermally decays into BATHO. Laser photolysis experiments carried out at room temperature show that a main path leading to HYPSO involves a biphotonic mechanism in which a second photon is absorbed by PBATHO or by an earlier precursor [14,16]. However, contribution of a direct dark (branching) reaction, from PBATHO to HYPSO, is suggested by the data of

Kobayashi for bovine rhodopsin [17], as well as by Yoshizawa et al. in the case of an artificial (9-cis-5,6- dihydro)rhodopsin pigment [14]. In any event, it appears that whenever observed, hypsorhodopsin originates from PBATHO rather than being a primary photoproduct formed in a parallel early event.

III. The Molecular Mechanism: The Structures of I, J_{625}, K_{610} and of PBATHO, BATHO and HYPSO

The observations described in the previous section strongly suggest an essentially identical model for the primary photochemical events, i.e., for the structure of the batho intermediates (BATHO and K_{610}) and their precursors, in both rhodopsin and bacteriorhodopsin. Early models (for a review see ref. 1c) based primarily on the isotope effects reported for the PBATHO → BATHO (2) and J_{625} → K_{610}[3] processes, invoked structures for BATHO and K_{610} involving proton translocation to the Schiff base or in the surrounding protein. No changes or only small distortions of the polyene chain were invoked. Numerous arguments rejecting this approach have been presented and extensively discussed [1c]. In the case of bR they are now further supported by the lack of an isotope effect in the steps leading to K_{610} [6-8].

The currently accepted model for the structures of BATHO and K_{610} is based on primary polyene photoisomerization processes: 11-cis(Rh) → all-trans (BATHO) and all-trans (bR) → 13-cis (K_{610}) respectively [18]. Evidence supporting isomerization at the early stages of BATHO and K_{610} is derived from several observations: (a) The photoequilibria attainable in visual pigments between bathorhodopsin, rhodopsin (11-cis), isorhodopsin (9-cis) and 7- cis rhodopsin [18-20]; (b) Resonance Raman data supporting (distorted) isomerized chromophores: all-trans in the case of BATHO [21] and 13-cis in the case of K_{610} [22]; (c) No BATHO product is observed upon blocking the $C_{11}=C_{12}$ isomerization in rhodopsin by preparing an artificial pigment with a C_{10}-C_{13} 7-membered ring [23]. Similarly, artificial bacteriorhodopsins in which the $C_{13}=C_{14}$ isomerization is blocked by five-membered [24] or epoxy [25] rings do not exhibit a photocycle. Normal photocycles are, however, observed upon blocking any other double bond in a similar way [25,26].

While there appears to be no doubt concerning the involvement of a specific double bond isomerization in the primary event of both photocycles, the question arises as to the isomerization of other C=C bonds or the C=N bond, as well as to the rotation about single C-C polyene bonds. The participation of the C=N bond in the primary event in bacteriorhodopsin has been excluded on the basis of NMR [27] and resonance Raman [28] studies. Similarly the involvement of any C=C isomerization other than $C_{13} = C_{14}$ in the photocycle of bR has been excluded on the basis of extensive studies with artificial pigments [25,26]. Of special interest is the observation that K_{610} is formed even when the synthetic chromophore is constituted by a single C=C bond (anologous to $C_{13}=C_{14}$). Similar comprehensive information is still unavailable in the case of the visual photocycle, but it appears that the C=N bond does not isomerize in the primary event of rhodopsins either [55,56].

Considerable attention has been devoted to theoretical models for the generation of BATHO and K_{610} involving rotations about several [29] or one specific [30-32] single bond(s). Mechanisms invoked are based on a concerted double bond isomerization - single bond rotation about adjacent C,C bonds: ($C_{13}=C_{14}$, C_{14}-C_{15}) in bR [30,31] and (C_{10}-C_{11}, $C_{11}=C_{12}$) in visual pigments [32]. The main argument favouring the concerted motion approach involves a minimal mass motion, i.e., minimal disturbance of the protein cavity. In the case of bR the mechanism is also claimed to account for the prevention of the back thermal reaction $K_{610} \rightarrow$ bR and for deprotonation of the Schiff base at the stage of the M_{410} intermediate [30]. Experimental evidence based on an artificial, C_9-C_{11} locked rhodopsin has been recently presented [33] indicating that the C_{10}-C_{11} rotation is not a prerequisite for the occurrence of a normal visual photocycle. The same applies to all C-C bonds in bR [25,26], except for C_{14}-C_{15} whose role in the photocycle is still controversial. Thus, the experimental confirmation of the detailed theoretical model of Schulten and Tavan [30] based on the concerted motion bR(all-trans) $-h\upsilon \rightarrow$ K_{610} (13,14 di-cis), is still subject to discussion [30,33,34] due to the argued interpretation of the vibrational spectra of bR, as derived from resonance Raman [33] and FTIR [34] experiments.

Early assignments of J_{625} and PBATHO as excited states have been discussed [1] and rejected [14] on the basis of: (a) the close resemblance of their absorption spectra with those of K_{610} and BATHO, respectively; (b) the inconsistency of the rate of the $J_{625} \rightarrow K_{610}$ transition with that of the fluorescence decay in bR [4,7]. It was concluded [4] that PBATHO and J_{625} are ground state species with polyene conformations identical to the (isomerized) structures of BATHO and K_{610}, respectively, but with a non-relaxed protein structure. This is in keeping with the failure to observe J_{625} and PBATHO in artificial pigments in which the photocycle is blocked by appropriate ring structures [7,14]. The exact nature of the environmental relaxations associated with the PBATHO $\rightarrow$ BATHO and $J_{625} \rightarrow K_{610}$ transitions is still unclear. In view of the apparent lack of an isotope effect in the latter process [7] it may be concluded that, at least in the case of bacteriorhodopsin, it probably cannot be associated with an intraprotein proton transfer [16]. An alternative suggestion [7] interprets the $J_{625} \rightarrow K_{610}$ process in terms of dissipation of excess thermal energy generated in the retinal moiety following excitation. Finally, it should be noted that the kinetic sequence in bR may be more complicated than that monitored by psec absorption spectroscopy. Thus (see G. Atkinson, in this volume) psec time-resolved resonance-Raman and fluorescence excitation, suggest that an additional (yet unidentified) intermediate may be present between J_{625} and the completely relaxed form of K_{610}.

A clear identification of the I species in the bR photocycle is still subject to the availability of further experimental evidence. A plausible assignment is that of identifying I with the fluorescent excited state of bR. In fact, the observed lifetime of the I$\rightarrow J_{625}$ transition at room temperature ($\tau_{I\rightarrow J}$ = 0.4-0.7 psec [7,10,11]), is qualitatively in keeping with the fluorescence lifetime as measured ($\tau < 2$ psec) by Sharkov et al. [35] or as estimated [$\tau_f = \phi_f \tau_r$ = 0.2-0.7 psec] from fluorescence quantum yields (ϕ_f = 1±0.8x10^{-4} [1c,36,7,37,38], using the expected radiative lifetime (τ_r = 5-10 nsec) for an allowed $^1B_u^+$ fluorescent state [1c,39]. In light of the marked dependence of ϕ_f on temperature [37,38], a more critical test of the coincidence between fluorescence and transient absorption kinetics could obviously involve studies over a relatively broad temperature range. Although several temperature-dependent psec absorption studies have been carried out in bR [3,40], the absence of experiments in which τ_f and $\tau_{I\rightarrow J}$ are measured simultaneously under the same excitation conditions, precludes any quantitative conclusions in this respect.

It has also been suggested [4,1c] that I may correspond to the "common excited state" (CES), along the reaction (isomerization) coordinate between the trans (bR) and 13-cis (J_{625}) configurations. The quantitative population of such a state, followed by partition to the two

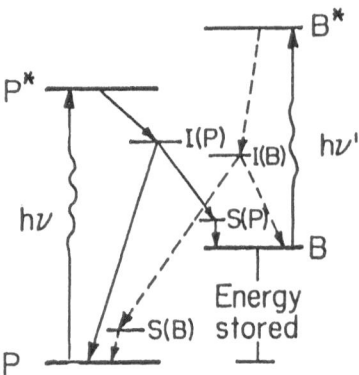

Fig. 2 Pathways involving primary photoreactions in rhodopsins according to the "common excited state" model [18]. The model has been modified [1c] so as to account for differences between the forward, ———→ (bR -hυ→ K_{610}) and backward ———→ (K_{610}-hυ→ bR) psec kinetics (see P.G. Kryukov, Yu. A. Lazarev, Yu. A. Matveetz, E.L. Terpugov, L.N. Chekulaeva, A.V. Sharkov: Studea Biophisica 83, 101 (1981); T. Iwasa, Y. Susuki, T. Nakayama, F. Tokunaga, M. Hirai: J. Phys. Soc. Japan 53, 2851 (1986), and discussion in Ref. 1c). P - Pigment (bR, all-trans or Rh, 11-cis); P* - Excited (Franck-Condon) state of pigment; B - Red shifted photoproduct (K_{610}, 13-cis, or BATHO, all-trans). B* - Excited (Franck- Condon) state of photoproduct; I(P) - "Common excited state" (I intermediate in bR?) with a (unrelaxed) protein configuration as in P; I(B) - "Common excited state" with protein configuration as in B. S - Transient states: J_{625} or PBATHO. Note that P*, B*, I(B) and I(P) (in rhodopsin) have not yet been detected experimentally.

corresponding ground states, was suggested as the basis of a model accounting for the primary event in bacteriorhodopsin and, analogously, in rhodopsin [18,39] (see Fig. 2). Since it is unclear whether the observed fluorescence in bR originates from CES [1b,38,39], it is not possible to use the correlation between τ_{I-J} and τ_f as a criterion for identifying I with CES. However, since from CES branching takes place to the ground states of bR (back reaction) and K_{610} (forward reaction), it may be expected [18,1c] that the original ground state of bR, depleted following light absorption, is (partially) repopulated at a rate which matches the growing-in of J_{625}. In other words, the observation of a I→bR process, kinetically matching the I→J_{625} transition, could serve as evidence for identifying I with CES. In spite of the recent progress in the femtosecond kinetics of bR [7,8,10,11], such an analysis is still unavailable.

The structure of hypsorhodopsin and the question as to its occurrence along the main (monophotonic) photobleaching pathway of rhodopsin under physiological conditions, are still unclear [14- 17]. The picture is further complicated by the observation [14,16] that the main hypsorhodopsin band around 440 nm is accompanied by a weak, long-wavelength, absorption at 540 nm. The latter may correspond to the same chromophore, or to a different species in fast equilibrium with hypsorhodopsin. It should also be noted that although, as in the case of BATHO, hypsorhodopsin appears to have a (twisted) all-trans retinylidene chromophore [14], it is unclear as to whether its Schiff base maintains its original state of protonation.

IV. Changes in Chromophore-protein Interactions in the Primary Event: Energy Storage in BATHO and K_{610}

In addition to the corresponding $C_{11}=C_{12}$ and $C_{13}=C_{14}$ isomerization processes, molecular models for the primary event in rhodopsin and bacteriorhodopsin must also account for two main observations: a) The bathochromic shift in BATHO and K_{610} with respect to their parent pigments (Rh and bR); b) The storage of a substantial fraction of the photon energy: ~35 kcal/mol in BATHO [41-43] and ~14 kcal/mol in K_{610} [44]. (We note that a BATHO intermediate has not been observed in the photocycle of Retinochrome [60] whose main task is apparently limited to retinal isomerization [61]. In variance with photosynthesis (bR) and vision (Rh), energy storage in the primary event may not be required in this case.) A model, qualitatively accounting for both observations was proposed [41], based on the separation (induced by photoisomerization) between the protonated Schiff base moiety and its (protein) counterion. Accordingly, the photon energy is stored primarily by electrostatic interactions[41,45,46,51] between the positively charged nitrogen and its protein counterion, and/or between such ion pairs and charges in the surrounding protein.

Evidence favouring a major change in the Schiff base environment at the stage of K_{610} has been presented in the case of bacteriorhodopsin. First, work with artifical bR has shown that most of the red shift associated with the bR $\rightarrow K_{610}$ transition is independent of retinal-opsin interactions, other than in the neighborhood of the Schiff base [25,47]. Second, the C=N stretch frequency in bR as monitored by resonance-Raman [48-50] or FTIR [50-52] spectroscopy, undergoes a substantial drop (20-30 cm^{-1}) upon conversion to K_{610}. However, a problem arises when considering analogous data in visual rhodopsins. Thus, both resonance Raman [53-55] and FTIR [56] data, indicate that the C=N stretch frequency in BATHO is identical to that of the mother pigment. Such observations, implying that the Schiff base environment in visual pigments does not change after isomerization, question the validity of any general model for the primary event which is common to visual rhodopsins and to bacteriorhodopsin. Specifically, they call for a molecular model for the visual photocycle which will account for both spectral shift and energy storage in BATHO, without requiring changes in the Schiff base environment - such as charge separation from its counterion.

More quantitative insights into some of the above problems have been obtained by a recent analysis of the factors which affect the C=N stretch frequency of protonated Schiff bases of retinal (PRSB) in solution [57]. Analysis of FTIR data, from a variety of retinal analogues in different solvents, has shown that the above frequency is highly sensitive to non-conjugated charges in the vicinity of the NH$^+$ moiety, but not to those located near the ring or along the polyene chain. The C=N frequency is also markedly affected by hydrogen bonding of the nitrogen to its counterion or to solvent molecules. Comparison of the C=N frequencies ($\upsilon_{C=N}$) of model systems with those of the biological pigments, indicates that in bacteriorhodopsin the frequency ($\upsilon_{C=N} \simeq 1639$ cm^{-1}) [50-52] is similar to that prevailing in a weakly H-bonded PRSB ($\upsilon_{C=N} = 1640\text{-}1649$ cm^{-1}). Also comparable are the respective drops induced by deuteration (14-17 cm^{-1}). This picture is in keeping with the conclusions derived from ^{15}N NMR experiments [62]. A different situation prevails in rhodopsin, characterized by $\upsilon_{C=N} = 1659$ cm^{-1} (deuteration shift ~30 cm^{-1}) [53-56], indicating a Schiff base environment similar to that of a strongly H-bonded PRSB ($\upsilon_{C=N} = 1652\text{-}1656$ cm^{-1}, deuteration shift ~20 cm^{-1}).

Accordingly, the primary event in the two systems may be interpreted as follows: a) bR -hυ→ K_{610}. Isomerization induces charge separation and/or weakening of H- bonds, in keeping with the optical red shift, the 20-30 cm^{-1} drop in $\upsilon_{C=N}$, and (electrostatic) energy storage. It should be noted that in respect to the above effects, charge separation is equivalent to neutralization of the counterion due to intraprotein proton transfer. It is still unclear if this possibility may be related to the protonation of a tyrosinate group at the stage of K_{610} [58,59]; b) Rh -hυ→ BATHO. One possibility is that charge separation takes place (as in bR), accounting for energy storage and for the (optical) red shift. The fact that $\upsilon_{C=N}$ does not change in BATHO, may be attributed to strong H-bonds (with protein moieties, or with H_2O), counterbalancing the effect caused by separation from the counterion. Alternatively, it is possible that the Schiff base environment does not change (no charge separation). In such a case, non-electrostatic energy storage such as conformational distortion of the polyene chromophore, must be invoked [45,46,55]. In fact, the increased C_8-C_9 and C_{14}-C_{15} stretching frequencies and the anomalous properties of the hydrogen out-of-plane wagging vibrations in BATHO, are indicative of chromophore-protein interactions in the center of the chain that might be involved in the energy-storage mechanism [55].

V. Conclusions Due to the extensive application of novel experimental techniques, considerable progress has been obtained in clarifying the mechanism of the primary photochemical events in rhodopsins. Basic features, such as cis-trans isomerization in the polyene, the state of protonation of the Schiff base, the ultrafast generation of the early ground-state photoproducts, etc., are now well established. However, a quantitative comprehensive mechanism is still unavailable. Some basic open questions are: a. The clear identification of all (Franck-Condon, relaxed and fluoresecent) states. b. The direct observation of the reverse process regenerating the original pigment, which is responsible for the reduction of the quantum yield to below unity. c. A model which will account for the unchanged $\upsilon_{C=N}$ frequency in the BATHO product of visual pigments as well as for the spectral shift and energy storage at this stage. Further work, hopefully assisted by the clarification of the structure of the pigments, will be required for solving these problems.

References

[1] For reviews, see: a). W. Stoeckenius, R.H. Lozier, R.A. Bogomolni: Biochim. Biophys. Acta 505, 215 (1979); b) M. Ottolenghi: Adv. Photochem. 12, 97 (1980); c) M. Ottolenghi: in: Biomembranes, Part I, Visual Pigments and Purple Membranes, II. Methods in Enzymology, ed. by L. Packer, p. 470 (Academic Press, N.Y., 1982).

[2] K. Peters, M.L. Applebury, P.M. Rentzepis: Proc. Nat. Acad. Sci. USA 74, 3119 (1977).

[3] M.L. Applebury, K.S. Peters, P.M. Rentzepis: Biophys. J. 23, 375 (1978).

[4] U. Dinur, B. Honig, M. Ottolenghi: in: Development in Biophysical Research, ed. by A. Borsellino, P. Omodeo, R. Strom, A. Vecli, E. Wanke, pp. 209-222 (Plenum, New York, London 1980); Photochem. Photobiol. 33, 523 (1981).

[5] E.P. Ippen, A. Shank, A. Lewis, M.A. Marcus: Science 200, 1279 (1978).

[6] H.J. Polland, W. Zinth, W. Kaiser: In Ultrafast Phenomena, ed. by D.H. Auston, K.B. Eisenthal (Springer, Berlin 1984) p. 456.

[7] H.J. Polland, M.A. Franz, W. Zinth, W. Kaiser, E. Koling, D. Oesterhelt: Biophys. J. 49, 651 (1986).

[8] M.C. Nuss, W. Zinth, W. Kaiser, E. Koling and D. Oesterhelt: Chem. Phys. Lett. 117, 1 (1986).

[9] M.C. Downer, M. Islam, C.V. Shank, A. Harootunian, A. Lewis: In Ultrafast Phenomena, ed. by D.H. Auston, K.B. Eisenthal (Springer, Berlin 1984) p. 500.

[10] A.V. Sharkov, A.V. Pakulev, Y.A. Martreetz: Biochim. Biophys. Acta 808, 94 (1985).

[11] J.L. Martin, J.W. Petrich, J. Breton, A. Orszag: This volume.

[12] B. Becher and T.G. Ebrey: Biochem. Biophys. Res. Commun. 69, 1 (1976).

[13] M.A. El Sayed, B. Karvali and J. Fukumoto: Proc. Natl. Acad. Sci. USA 78, 7512 (1981).

[14] Y. Schichida, S. Matuoka, T. Yoshizawa: Photochem. Photobiophys. 7, 221 (1984); T. Yoshizawa, Y. Shichida, S. Matuoka: Vision Res. 24, 1455 (1984).

[15] T. Yoshizawa, S. Horiuchi: In Biochemistry and Physiology of Visual Pigments, ed. by H. Langer (Springer-Verlag, Berlin 1973) p. 69.

[16] N. Sasaki, F. Tokunaga, T. Ogurusu, T. Yoshizawa: Photochem. Photobiophys. 7, 341 (1984).

[17] T. Kobayashi: Photochem. Photobiol. 32, 207 (1980); T. Kobayashi, S. Nagakura: In Methods in Enzymology, Vol. 81 f1 (Academic Press, N.Y., 1982) p. 51.

[18] T. Rosenfeld, B. Honig, M. Ottolenghi, J. Hurley, T.G. Ebrey: Pure Appl. Chem. 49 341 (1977); J. Hurley, T.G. Ebrey, B. Honig, M. Ottolenghi: Nature (London) 540 (1977).

[19] B. Aton, R. Callender, B. Honig: Nature (London) 273, 784 (1978).

[20] S. Kawamura, S. Miyatani, H. Matsumoto, T. Yoshizawa, R.S.H. Liu: Biochemistry 19, 1549 (1980).

[21] G. Erying, B. Curry, A. Broek, J. Lugtenburg, R. Mathies: Biochemistry 21, 384 (1982).

[22] M. Braiman and R. Mathies: Proc. Natl. Acad. Sci. USA 79, 403 (1983).

[23] H. Akita, S.P. Tanis, M. Adams, V. Balogh-Nair, K. Nakanishi: J. Am. Chem. Soc. 102, 6370 (1980); B. Mao, M. Tsuda, T.G. Ebrey, H. Akita, V. Balogh-Nair, K. Nakanishi: Biophys. J. 35, 543 (1981); T. Yoshizawa, Y. Shichida, S. Matuoka, N. Sasaki, K. Nakanishi, V. Balogh-Nair, H. Akita: Zool. Mag. 92 535, 1983.

[24] J.O. Fang, J. Carriker, V. Balogh-Nair, K. Nakanishi: J. Am. Chem. Soc. 105, 5162 (1983); C.H. Chang, R. Govindjee, T. Ebrey, K.A. Bagley, G. Dollinger, L. Eisenstein, J. Marque, H. Roder, J. Vittitow, J.O. Fang, K. Nakanishi: Biophys. J. 47, 509 (1985).

[25] M. Sheves, N. Friedman, A. Albeck, M. Ottolenghi: Biochemistry 24, 1260 (1985).

[26] A. Albeck, N. Friedman, M. Sheves, M. Ottolenghi: J. Am. Chem. Soc. 108, 4614 (1986).

[27] G.S. Harbison, S.O. Smith, J.A. Pardoen, C. Winkel, J. Lugtenburg, J. Harzfeld, R. Mathies, R.G. Griffin: Proc. Natl. Acad. Sci., USA 81, 1706 (1984).

[28] S.O. Smith, A.B. Myers, J.A. Pardoen, C. Winkel, P.P.J. Mulder, J. Lugtenburg, R. Mathies: Proc. Natl. Acad. Sci. USA 81, 2055 (1984).

[29] A.Warshel: Proc. Natl. Acad. USA 75, 25587 (1978).

[30] K. Schulten, P. Tavan: Nature (London), 272, 85 (1978); P. Tavan, K. Schulten: Biophys. J., 50, 81 (1986).

[31] R.S.H. Liu, D. Mead, A.E. Asato: J. Am. Chem. Soc. 107, 6609 (1985).

[32] R.S.H. Liu, A.E. Asato: Proc. Natl. Acad. Sci. USA 82, 259 (1985).

[33] S.O. Smith, A.M. Myers, J.A. Pardoen, C . Winkel, P.P.J. Mulder J. Lugtenburg, R. Mathies: Proc. Natl. Acad. USA 81, 2055 (1984); S.O. Smith, J. Lugtenburg, R. Mathies: J. Membr. Biol. 85, 109 (1985).

[34] K. Gerwert and F. Siebert: The EMBO J. 5, 805 (1986).

[35] A.V. Sharkov, Yu.A. Matveetz, S.V. Chekalin, A.V. Konyashchenko, O.M. Brekhov, B.Yu. Rootskov: Photochem. Photobiol. 38, 108 (1983).

[36] R.R. Alfano, W.Y.R. Govindjee, B. Becher, T.G. Ebrey: Biophys. J. 16, 541 (1976).

[37] S.L. Shapiro, A.J. Campillo, A. Lewis, J. Perrault, J.P. Spoonhower, R.K. Klayton, W. Stoeckenius: Biophys. J. 23, 383 (1978).

[38] T. Kouyama, K. Kinosita, A. Ikegami: Biophys. J. 47, 43 (1985).

[39] R.R. Birge: Ann. Rev. Biophys. Bioeng. 10, 315 (1981).

[40] T. Gillbro, V. Sundstrom: Photochem. Photobiol. 37, 455 (1983).

[41] B. Honig, T. Ebrey, R.H. Callendar, U. Dinur, M. Ottolenghi: Proc. Natl. Acad. Sci. USA 76, 2503 (1979).

[42] A. Cooper: Nature (London) 282, 531 (1979).

[43] F. Boucher, R.M. Le Blanc: Photochem. Photobiol. 41, 459 (1985).

[44] R.R. Birge, T.H. Cooper: Biophys. J. 42, 61 (1983).

[45] R.R. Birge, L.M. Hubbard: J. Am. Chem. Soc. 102, 2195 (1980); R.R. Birge, L.M. Hubbard: Biophys. J. 34, 517 (1981).

[46] A. Warshel and N. Barboy: J. Am. Chem. Soc. 104, 1469 (1982).

[47] M. Sheves, T. Baasov, N. Friedman, M. Ottolenghi, R. Feinmann-Weinberg, V. Rosenbach, B. Ehrenberg: J. Am. Chem. Soc. 106, 2435 (1984).

[48] J. Terner, C.L. Hsieh, A.R. Burns, M.A. El-Sayed: Proc. Natl. Acad. Sci. USA 76, 3046 (1979); J. Terner, A.M. El- Sayed, Acc. Chem. Res. 18, 331 (1985).

[49] A.B. Meyers, R.A. Harris, R.A. Mathies: J. Chem. Phys. 79, 603; D. Stern and R. Mathies: In: Time - Resolved Vibrational Spectroscopy, ed. by M. Stockburger, A. Laubreau, Vol. 4, 250 (Springer-Verlag Proceedings in Physics (1985).

[50] K.J. Rothschild, H. Marrero, M. Braiman, R. Mathies,   : Photochem. Photobiol. 40, 675 (1984).

[51] K.J. Rothschild, H. Marrero: Proc. Natl. Acad. Sci. USA 79, 4045 (1982); K.J. Rothschild, P. Roepe, J. Lugtenburg, J.A. Pardoen: Biochemistry 23, 6103 (1984).

[52] K. Bagley, G. Dollinger, L. Eisenstein, A.K. Singh, L. Zimenyi: Proc. Natl. Acad. USA 79, 4972 (1982).

[53] G. Eyring, R.A. Mathies: Proc. Natl. Acad. Sci. USA 76, 33 (1979).

[54] B. Aton, A.G. Doukas, D. Narva, R.H. Callender, U. Dinur, B. Honig: Biophys. J. 29, 79 (1980); D. Narva, R.H. Callender; Photochem. Photobiol. 32, 273 (1980).

[55] I. Palings, J.A. Pardoen, E. Van den Berg, C. Winkel, J. Lugtenburg, R. Mathies: Biochemistry, in press (1987).

[56] K.A. Bagley, V. Balogh-Nair, A.A. Croteau, G. Dollinger, T.G. Ebrey, M. Hong, K. Nakanishi, J. Vittitow: Biochemistry 24, 6055 (1985).

[57] M. Sheves, T. Baasov, N. Friedman: Biochemistry, in press (1987).

[58] K. Rothschild, P. Roeppe, P.L. Ahl, T.N. Earnest, R.A. Bogomolni, S.A. Das Gupta, C.M. Mulliken, J. Herzfeld: Proc. Natl. Acad. Sci. USA 83, 347 (1986).

[59] G. Dollinger, L. Eisenstein, S-L Lin, K. Nakanishi, J. Termini: Biochemistry 25, 6524 (1986).

[60] T. Kobayashi, K. Ogasawara, S. Koshihara, K. Ichimura, R. Hara: This volume.

[61] T. Hara, R. Hara: Methods in Enzymology 81, 190 (1982).

[62] G. Harbison, J. Herzfeld, R. Griffin: Biochemistry 22, 1 (1983).

Two Kinds of Bacteriorhodopsin Analogues Synthesized from Naphthylretinal

F. Tokunaga[1], M. Takao[1], T. Iwasa[1], and K. Tsujimoto[2]

[1]Department of Physics, Faculty of Science, Tohoku University,
Aobayama, Sendai 980, Japan
[2]Department of Material Science, University of Electrocommunications,
Chofu 182, Japan

Abstract: Two kinds of bacteriorhodopsin analogues were formed with
all-trans-naphthylretinal and bacterioopsin. One has an absorption
maximum at 503 nm (Np-bR503) and the other, at 442 nm (Np-bR442).
An equimolar mixture of naphthylretinal and bacterioopsin produced
Np-bR442, while successive addition of naphthylretinal gave Np-
bR503. The binding sites of naphthylretinal in Np-bR442 and in
Np-bR503 were the same as retinal in the native bacteriorhodopsin.
The CD spectrum of Np-bR503 shows characteristic exciton CD bands,
but that of Np-bR442, only a positive band. Np-bRs produced own
batho- and meta-intermediates different from each other. The pH
increase and temperature elevation change Np-bR503 to Np-bR442
reversibly. Solubilization of Np-bR503 in Triton X-100 also changes
Np-bR503 to Np-bR442. These observations indicate that Np-bR503
and Np-bR442 are in equilibrium.

1 Introduction

Bacteriorhodopsin (bR) is the sole protein component of the
purple membrane of Halobacterium halobium and transports protons
across the membrane upon absorbing light. The color of the protein
results from a retinal moiety bound to Lys-216 of the apoprotein via
a protonated Schiff-base linkage [1, 2]. Two different
configurations of the chromophore are found in bacteriorhodopsin;
all-trans and 13-cis [3, 4]. Only the pigment having all-trans-
retinal as its chromophore (trans-bR) has the ability to transport
protons [5]. Upon absorption of light, trans-bR undergoes a cyclic
photoreaction through K, L, and M intermediates, which have been
investigated by several authors [6 - 8] and us [9 - 11].

Many kinds of retinal analogues have been used for studying the
role of the chromophore in the functions [12]. TOKUNAGA et al. [13]
showed that retinal interacts with the protein at its polyene chain
from reconstitution experiments with C17 aldehyde. Also, TOWNER et
al. [14] concluded that the specific binding site for the β-ionone
ring does not exist on the basis of reconstitution studies using β-
ionone. We describe here the formation and the properties of two
kinds of bacteriorhodopsin analogues produced from 3-methyl-7(2-
naphthyl)-2,4,6,-octatrienal (Np-retinal, Fig. 1), which has a large
ring group. One has its absorption maximum at 503 nm (Np-bR503) and
the other at 442 nm (Np-bR442).

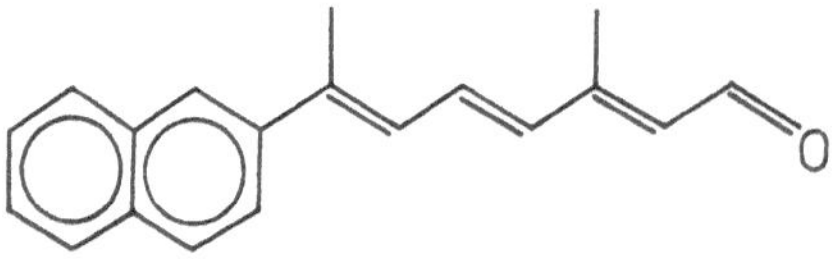

Naphthylretinal

Fig. 1   Structure of
Naphthylretinal.

154

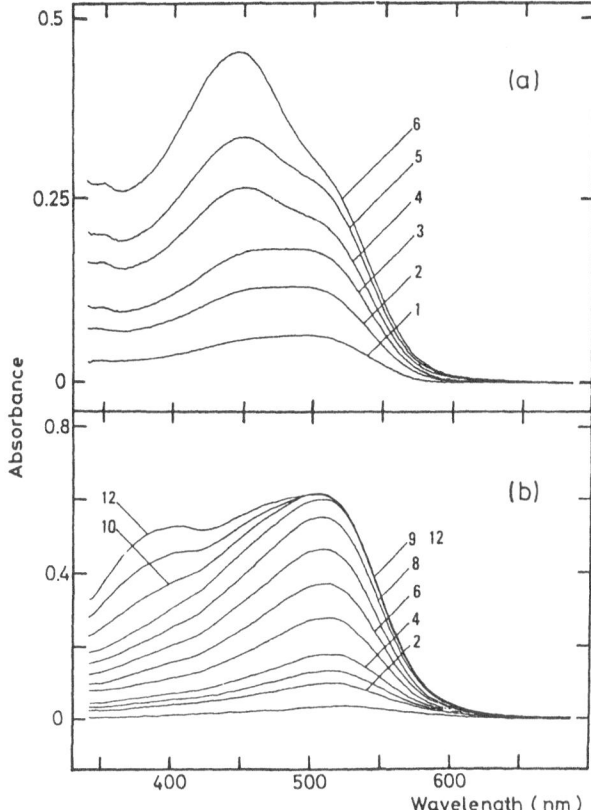

Fig. 2 Formation of two different pigments from Np-retinal and bacterio-opsin in 10 mM HEPES buffer pH 6.8. (a) The formation of the pigments after addition of Np-retinal in different retinal/protein ratios (1/12 to 1/1). Curves 1-6: the products from bacterioopsin (0.7 mL, 12 μM) after addition of Np-retinal (A370 = 60 OD) of 0.5, 1.0, 1.5, 3.0, 5.0, and 6.0 μL, respectively. (b) The formation of the pigments after successive addition of Np-retinal. Curves: the products after successive addition of 0.5 μL (curves 1-4) and 1 μL (curves 5-12) of Np-retinal (A370 = 75 OD) to bacterioopsin (0.7 mL, 13 μM). The absorption spectra were measured after each addition followed by two-hour incubation in the dark.

2   Formation of Np-bRs

Various amounts of Np-retinal were added to a given amount of bacterioopsin and the absorption spectra were measured after incubation for 2 h at 20 °C. The absorption maxima of the products were different depending on the amount of the added Np-retinal (Fig. 2a). When the amount of added Np-retinal was less than a half mole of bacterioopsin, the products showed broad absorption spectra between 450 and 530 nm. When the amount of added Np-retinal increased, the absorption maxima shifted toward 442 nm (curves 4-6 in Fig. 2a) due to formation of a different pigment. This pigment is designated Np-bR442.

A different phenomenon was observed when a small amount of Np-retinal was added successively to a larger amount of bacterioopsin at two-hour intervals (Fig. 2b): When the amount of Np-retinal was below one-third saturated, the absorption maximum was at 503 nm (curves 1-4 in Fig. 2b). Until the amount of added Np-retinal reached saturation, the absorbance increased at around 500 nm.

The above results indicate that two different types of pigments were formed from Np-retinal and bacterioopsin. The wavelength for the maximum absorbance of the product depended on the molar ratio of added Np-retinal to bacterioopsin on mixing. The addition of the small amount of Np-retinal (less than one-tenth of bacterioopsin) furnished the pigment with its absorption maximum at 503 nm (Np-bR503). When the equimolar amount of Np-retinal was added to bacterioopsin, the absorption spectrum showed an absorption maximum at 442 nm (Np-bR442) together with a very small shoulder around 505 nm probably due to the concomitant production of Np-bR503. Further overnight incubation in the dark caused a small increase in the absorbance at 503 nm. When a small amount of Np-retinal was added to bacterioopsin the absorbance increase at 400 nm was observed within 12 s after addition of Np-retinal. Since the free Np-retinal showed its absorption maximum around 380 nm (curves 10-12 in Fig. 2b), the absorbance increase at 400 nm is concluded to be an intermediate in the formation of Np-bR503. This intermediate gradually changed to Np-bR503 with an isosbestic point at 440 nm. Such a kind of intermediate is observed during the process of the formation of native bR from all-trans-retinal and bacterioopsin [15].

In the case of pigment formation using a larger amount of Np-retinal, the absorbance increase at 400 nm was not observed even at 26 s after addition of Np-retinal. The intermediate that appears in the formation step of Np-bR503 was not observed. Absorbance at 442 nm increased with concomitant decreases in absorbances over a range of wavelengths shorter than 410 nm.

The values of absorption maximum of the synthesized pigments did not change within a day in the dark at room temperature. This contrasts with all-trans- or 13-cis-retinal, where the absorption maximum of the regenerated pigment (570 nm for trans-bR or 550 nm for 13-cis-bR) [16] shifted to that of the dark-adapted form (560 nm) within several hours in the dark at room temperature [3].

Hydroxylamine is a reagent which attacks the retinal-Schiff base and forms retinal-oxime. In the case of bR and cattle rhodopsin, the retinal-Schiff base is inaccessible to hydroxylamine in the dark. Both of the pigments generated from Np-retinal and bacterioopsin slowly react with hydroxylamine in the dark. The decays of both pigments were a single exponential. But these spectral changes had different time courses. The difference in the reaction rate (half-time of Np-bR442 = 40 min; that of Np-bR503 = 200 min) was due to a difference in environment around the Schiff-base portion of the chromophore. The UV spectra of both products had their absorption maxima at 365 nm with shoulders at 350 and 385 nm, suggesting that the product is Np-retinal-oxime.

It was investigated whether or not the binding sites of both pigments are different from that in native bR. When sodium borohydride was added under illumination, the main band due to Np-bR503 or Np-bR442 disappeared and the sharp peak at 343 nm having two shoulders at 325 and 360 nm appeared. Then the sodium borohydride-treated protein was washed several times to remove residual sodium borohydride, and all-trans-retinal was added. If Np-retinal binds to a different site from Lys-216, absorbance due to trans-bR would be expected to increase. But no significant absorbance increase was observed over a range of 500-600 nm after a two-hour dark incubation.

When all-trans-retinal was added to the suspension of fully synthesized Np-bR442 or Np-bR503 without sodium borohydride treatment, the chromophore exchange took place; the absorbance of Np-bR decreased with a concomitant increase of trans-bR. The exchange rates were much slower than the regeneration rate with trans-retinal. These observations also suggest that the Np-retinal in Np-bRs and retinal in trans-bR occupy the same site of the bacterioopsin.

3 Circular Dichroic Spectra of Np-bRs

The CD spectrum of bacteriorhodopsin in the purple membrane consists of two oppositely directed bands due to an exciton interaction between the trimerically arranged chromophores. No CD signal was observed in bacterioopsin in the visible region. After

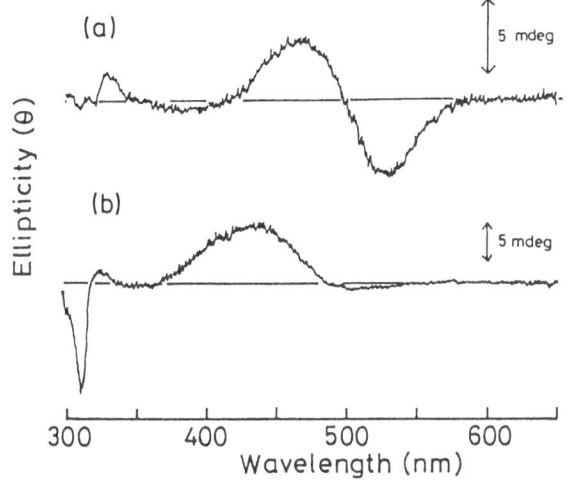

Fig. 3 Circular dichroic spectra of the following samples in 50 % (w/v) sucrose solution. (a) Np-bR503 (A503 = 0.86 OD). (b) Np-bR442 (A442 = 1.15 OD).

addition of Np-retinal to the bacterioopsin suspension, the CD bands appeared. But their spectral shapes were different. The CD spectrum of Np-bR503 had a shape typical of the exciton-coupled chromophores (Fig.3a), but only a positive peak was observed at 440 nm in the CD spectrum of Np-bR442 (Fig. 3b).

The difference between CD bands in Np-bR503 and Np-bR442 means that the two pigments have different chromophoric arrangements. Exciton coupling of the CD spectrum of Np-bR503 suggests that the chromophore alignment of Np-bR503 is similar to that of native bR.

4 Photochemical Reaction of Np-bRs

On irradiation of Np-bR442 with 420 nm light at 18 °C, the absorption maximum shifted towards longer wavelengths up to 480 nm (Fig. 4a). The light-adapted state formed from Np-bR442, designated as Np-bR480, was different from Np-bR503. The subsequent dark incubation caused the reverse reaction to the original pigment (Fig. 4b) not to Np-bR503. The amount of the final product of the dark-reaction, however, was less than that of the original. This indicates that a small amount of pigment degraded during the forward and/or back reactions. The difference spectrum of the dark reaction (curve 14-12 in Fig. 4b) is coincident with that of the early stage of the photoreaction (curve 8-1 in Fig. 4a) but not that of the later stage (curve 12-8 in Fig. 4a), suggesting that the degradation of the pigments took place during the photoreaction. When Np-bR503 was irradiated with 450 nm light at room temperatures, a drastic spectral change was not observed but the absorbances decreased a little around 400 nm and increased around 500 nm. The difference spectrum of this change was the same as that of the light adaptation of Np-bR442. Therefore, apparently, Np-bR503 is photoinsensitive at room temperatures.

The isomeric composition in the chromophore of the irradiated products of Np-bR442 was more than 60 % of the isomer which has cis configuration at the same double bond in the polyene chain as 13-cis retinal. This suggestes that the light adaptation of Np-bR442 may be the isomerization of the chromophoric Np-retinal from the all-trans to the 13-cis one. This phenomenon was different from the light adaptation in native bR, where chromophoric retinal isomerizes from the 13-cis to all-trans form. The bR analogue containing α-retinal as the chromophore showed a similar phenomenon: the chromophoric retinal isomerizes from the trans to 13-cis form on light isomerization [17].

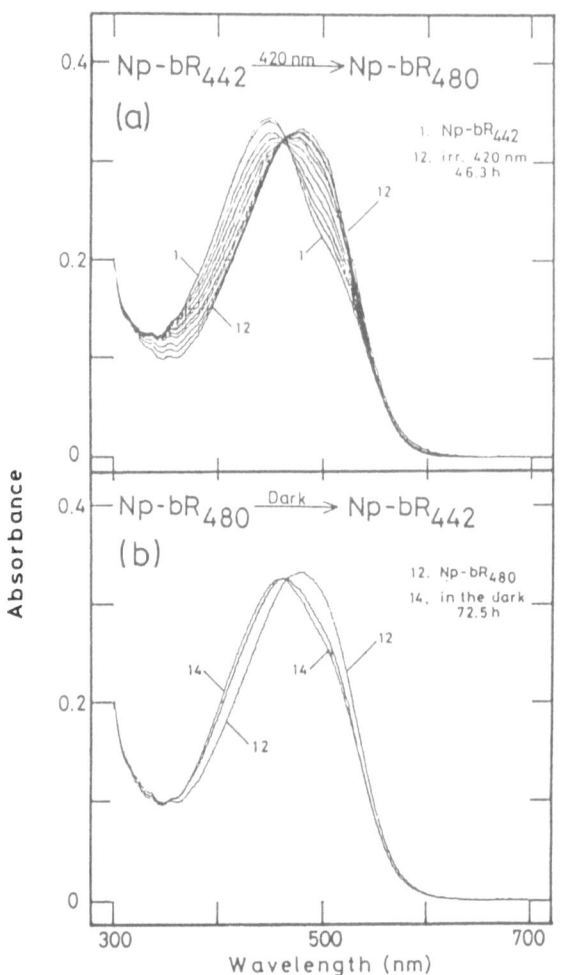

Fig. 4 Formation (a) and dark reversion (b) of Np-bR480. (a) The photoconversion from Np-bR442 (curve 1) to Np-bR480 at 18 °C with successive irradiation at 420 nm for 1, 5, 5, 10, 20, 40, 80, 160, 320, and 640 min (curves 2-11, respectively). Np-bR480 (curve 12) was obtained after 46.3 h irradiation. (b) The dark reversion of Np-bR480 to Np-bR442. In the dark at 18 °C, Np-bR480 (curve 12, redrawn from a) slowly reverts to Np-bR442 (curves 13 and 14). After 72.5 h in the dark, curve 14 was recorded.

When native bR absorbs photons, it is converted to the batho-intermediate K, which has the absorption maximum at a longer wavelength and is stable below -120 °C. When Np-bRs (Np-bR442 and Np-bR503) were irradiated with yellow light (420 nm) at liquid nitrogen temperature, the spectrum was shifted to longer wavelengths (Fig. 5b and d, curves 2, 3). This suggests the formation of batho-intermediates. This photosteady state is designated PSS-420. Upon irradiation of PSS-420 with light of wavelengths longer than 600 nm which is absorbed only by the photoproducts having the absorption maximum at the longer wavelength, the absorbance decreased at wavelengths longer than 540 nm and increased around 510 nm (curve 4 in Fig. 5d), suggesting that the batho-intermediate produced from Np-bR503 reverted to the original pigment (Np-bR503). The mixture in this state was designated PSS-600. Further irradiation of PSS-600 with light of wavelengths longer than 540 nm caused a decrease of absorbance around 530 nm and finally the absorption spectrum reverted to the original one. This change should be the reversion of the batho-intermediate produced from Np-bR442 to its original pigment. These results indicate that each spectral change corresponds to the reversion of each batho-intermediate to the parent Np-bR pigment.

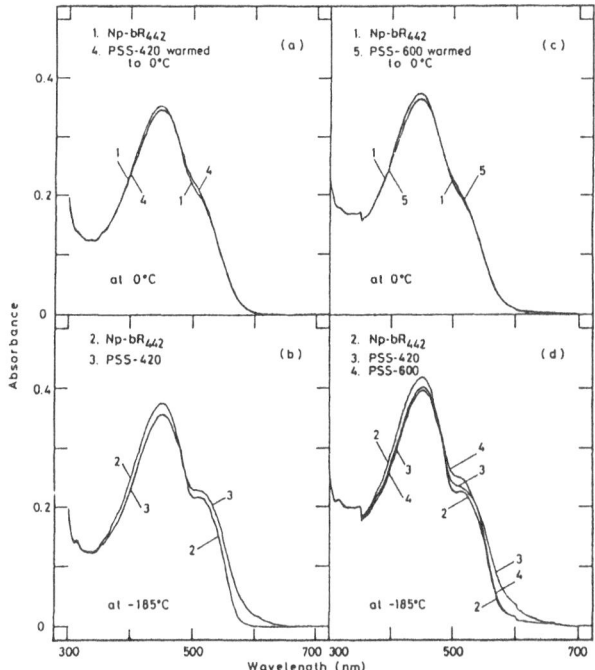

Fig. 5 The formation of batho-intermediates and their thermal reactions. (a) and (b): The formation of PSS-420 and the final products of its thermal reaction. The mixture of Np-bR503 and Np-bR442 at 0 °C (curve 1 in a) was cooled to -185 °C and the absorption spectrum was recorded (curve 2 in b). After irradiation at 420 nm for 32 min PSS-420 was formed (curve 3 in b), containing the batho-intermediates of both Np-bR503 and Np-bR442. It was then warmed to 0 °C and its absorption spectrum recorded (curve 4 in a). (c) and (d): The formation of PSS-600 and thermal reaction. The mixture of Np-bR503 and Np-bR442 (0 °C; curve 1 in c, -185 °C; curve 2 in d) was irradiated with light at 420 nm for 32 min (curve 3 in d) and then with wavelengths longer than 600 nm for 16 min (curve 4 in d). The product, PSS-600 (curve 4 in d), containing only the batho-intermediate of Np-bR442, was warmed to 0 °C and the absorption spectrum was measured (curve 5 in c).

The above interpretation was confirmed using fairly pure Np-bR503. Irradiation of Np-bR503 with 420 nm light at liquid nitrogen temperatures caused an increase of absorbances at wavelengths longer than 540 nm and a concomitant decrease around 500 nm. The reverse reaction with light of wavelengths longer than 600 nm showed an isosbestic point at 540 nm, which is the same as that of the formation. Therefore, it was confirmed that both Np-bRs form their own batho-intermediates (K442 and K503) and that the photoreactions between batho-intermediates and their parent pigments are photoreversible.

Meta-intermediate is regarded as a key intermediate for physiological functions. We confirmed that each meta-intermediate was also produced from Np-bR442 and Np-bR503 by the irradiation at about -50 °C.

In order to clarify the final product from batho-intermediates of Np-bRs, two different photosteady states (PSS) were formed at

-185 °C and warmed to 0 °C. One PSS contains batho-intermediates of both Np-bRs (PSS-420) and another, only that of Np-bR442 (PSS-600). After the Np-bRs (curve 1, Fig. 5a and c) were cooled to -185 °C (curve 2, Fig. 5b and c), PSS-420 (curve 3, Fig. 5b) and PSS-600 (curve 4, Fig. 5d) were formed. When PSS-420 or PSS-600 was warmed to 0 °C, the final absorption spectra (curve 4 in Fig. 5a and curve 5 in Fig. 5c) showed an absorbance increase around 500 nm and a decrease around 430 nm. The shape of the difference spectrum between curves 1 and 4 in Fig. 5a was similar to that between Np-bR480 and Np-bR442. The shape of the difference spectrum between curves 1 and 5 in Fig. 5c was also similar. These results suggest that meta- or batho-intermediate from Np-bR442 converts to Np-bR480 by warming and that from Np-bR503 reverts to Np-bR503. These results coincided with those of the irradiation of Np-bR442 and Np-bR503 at room temperatures (Fig. 4).

5 Proton Pumping Activity

Np-bR503 incorporated in DMPC vesicles has its absorption maximum around 470 nm. Scattering was corrected by subtracting the spectrum of vesicles treated with hydroxylamine. The shift of the absorption maximum of Np-bR503 in vesicles is probably related to changes in protein conformation resulting from incorporation into the vesicles. In the case of bR, incorporation into the vesicles caused a shift of the absorption maximum to a shorter wavelength (our unpublished observation).

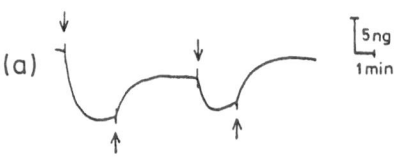

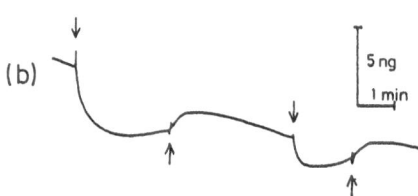

Fig. 6 Light-dependent proton pumping activity of vesicles containing the following pigments: (a) Np-bR442, and (b) Np-bR503. Proton pumping activity of vesicles was measured in 150 mM KCl solution at 25 °C with white light illumination. Light-on and light-off are represented by the arrows (↑) and (↓), respectively. The scale bars on the right represent the amount of protons (5 ng; vertical bar) and time (1 min; horizontal bar), respectively.

The vesicles with Np-bR503 or Np-bR442 showed proton pumping activity under illumination with white light (Fig. 6a, b). Several kinds of measurements were performed in order to exclude the possibility that the observed proton pumping activity was derived from bR that remained in the bacterioopsin preparation. The vesicles with the same amount of bacterioopsin used for reconstitution of Np-bRs showed only less than 6 % of the activity of Np-bRs. The vesicles formed with Np-bRs showed a larger activity on irradiation with 440 nm light than that observed with 560 nm light adjusted to have the same number of quanta. Thus, Np-bRs can transport protons by light, indicating that the naphthalene ring that is larger than the β-ionone ring in retinal did not abolish the proton pumping activity.

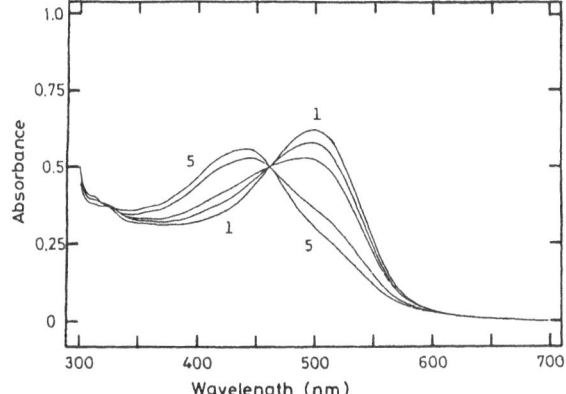

Fig. 7 Effect of pH on the absorption and CD spectrum of Np-bR503. Np-bR503 in distilled water was successively alkalized by KOH. (a) Absorption spectra (curves 1-5) of Np-bR503 were measured at pH 5.1, 7.0, 7.6, 9.0 or 9.9 respectively. (b) CD spectra of Np-bR503 at pH 4.9 and pH 10.0.

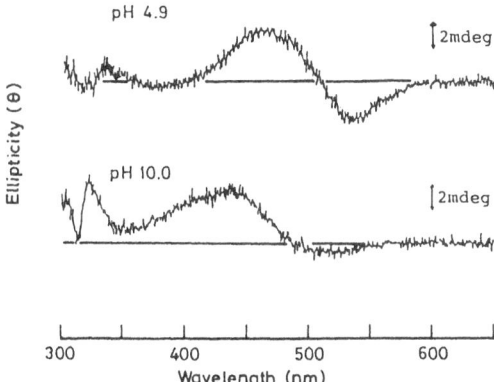

6   Effect of pH, Temperature and Solubilization

In order to clarify the relationship between Np-bR503 and Np-bR442, environmental effects on both pigments were investigated.

Upon raising the pH of Np-bR503 suspension from 4.3 to 10.2, the absorbance at 440 nm increased with an isosbestic point at 462 nm (Fig. 7a). When the pH of the suspension was subsequently lowered to 7.0, almost all the pigments returned to Np-bR503. The reverse process from pH 10.3 to 7.0 was much slower (nearly 20 h) than that of raising the pH (15 min).

The CD spectrum of Np-bR503 in a 0.1 M acetate buffer (pH 4.9) was essentially the same as that obtained at pH 6.8, whereas the characteristics of Np-bR442 appeared in its CD signal in 0.2 M borate buffer (pH 10.0) (Fig. 7b). The blue shift of the absorption maximum of bR was observed when bR was solubilized by Triton X-100 [19]. Addition of Triton X-100 to the Np-bR503 suspension resulted in the blue shift of the absorption maximum to 440 nm within 30 min. The CD spectrum of this solubilized state showed a single positive peak at its absorption maximum like Np-bR442. The negative CD signal at 315 nm appeared in the near uv region like Np-bR442. In contrast, Np-bR442 with Triton X-100 did not significantly change its absorption and CD spectra except for a decrease of the absorbance around 510 nm, probably due to the solubilization of contaminating Np-bR503. These results indicate that the protein state of Np-bR442 is quite similar to that of solubilized Np-bR503.

In order to fluctuate the interaction of proteins and lipids mildly, the temperature was elevated. Upon warming to 60 °C from 24 °C, the absorption maximum of Np-bR503 largely shifted to 450 nm with an isosbestic point at 462 nm, which indicates the conversion from Np-bR503 to another molecular species. Subsequent cooling back to 24 °C reproduced the initial Np-bR503 without any loss of absorbance. The absorption spectrum at 60 °C was able to be reconstituted by those of Np-bR442 and Np-bR503 at room temperature (55 % and 45 %, respectively). Therefore, the above spectral changes must be the conversion from Np-bR503 to Np-bR442. The CD spectrum obtained from Np-bR503 kept at 65 °C showed the characteristics of both Np-bRs; the negative band around 530 nm was weakened, the positive one at 440 nm was intensified, and the sharp negative peak at 315 nm appeared. From these observations, it was concluded that the interconversion between Np-bR503 and Np-bR442 took place depending on the temperature and that the latter pigment favored the conditions at high temperatures.

7 Discussion

Np-bR503 and Np-bR442 have the all-trans form of Np-retinal as the chromophore and are in equilibrium with each other. The formation of the pigments depends on the mixing ratio of retinal and bacterioopsin. The large head of Np-retinal may perturb the smooth fitting of chromophore and bacterioopsin. A small amount of Np-retinal and a long enough time interval after addition allow the formation of a stable state (Np-bR503) in which the chromophore-protein interaction may be similar to that of trans-bR as suggested by its CD spectrum. A large amount of Np-retinal, however, may not allow such smooth fitting and result in another stable state between chromophore and bacterioopsin.

Both kinds of Np-bRs contain almost exclusively the all-trans isomer of Np-retinal. The difference between Np-bR503 and Np-bR442 is attributable to the extent of the interactions between the chromophore and the protein, as shown by the difference in sensitivity to hydroxylamine (data not shown) and in CD spectra: (1) The exciton CD band, which is characteristic of the trimer structure, was not observed in Np-bR442. (2) The negative CD band at 310 nm, which probably results from steric hindrance of the chromophore [18], was larger in Np-bR442 than in Np-bR503, though Np-retinal formed a Schiff-base linkage in both pigments.

Iodation of the purple membrane induces a blue shift of the absorption spectrum of bR [20, 21]. This blue shift is strongly coupled with disorder of the purple membrane structure [22]. From the fact that bR solubilized with Triton X-100 shows no biphasic CD spectrum, it has been considered that bR molecules in Triton X-100 micelles exist as a monomer. However, our preliminary experiments by small angle x-ray scattering show that bR should exist as a trimer in Triton X-100 micelles [23]. It is possible that the trimer of bR shows no biphasic CD spectrum. Therefore, the following picture could be possible: The large naphthyl group may induce some changes in the structure of the apoprotein and then Np-bR fluctuates in the trimer.

References

1. H.-D. Lemke, D. Oesterhelt: FEBS Lett. 128, 255 (1981)
2. H. Bayley, K.-S. Huang, R. Radhakrishnam, A.H. Ross, Y. Takagi, H.G. Khorana: Proc. Natl. Acad. Sci. U.S.A. 78, 2225 (1981a)
3. A. Maeda, T. Iwasa, T. Yoshizawa: J. Biochem. (Tokyo) 82, 1599 (1977)

4. M.J. Pettei, A.P. Yudd, K. Nakanishi, R. Henselman, W. Stoeckenius: Biochemistry 16, 1955 (1977)
5. M. Yoshida, K. Ohno, Y. Takeuchi, Y. Kagawa: Biochem. Biophys. Res. Commun. 75, 1111 (1977)
6. W. Stoeckenius, R.H. Lozier: J. Supramol. Struct. 2, 769 (1974)
7. R.H. Lozier, R.A. Bogomolni, W. Stoeckenius: Biophys. J. 15, 955 (1975)
8. N. Dencher, M. Wilms: Biophys. Struct. Mech. 1, 259 (1975)
9. F. Tokunaga, T. Iwasa, T. Yoshizawa: FEBS Lett. 72, 33 (1976)
10. T. Iwasa, F. Tokunaga, T. Yoshizawa: FEBS Lett. 101, 121 (1979)
11. T. Iwasa, F. Tokunaga, T. Yoshizawa: Biophys. Struct. Mech. 6, 253 (1980)
12. W. Stoeckenius, R.H. Lozier, R.A. Bogomolni: Biochim. Biophys. Acta 505, 215 (1979)
13. F. Tokunaga, T.G. Ebrey, R. Crouch: Photochem. Photobiol. 33, 495 (1981)
14. P. Towner, W. Gaertner, B. Walckhoff, D. Oesterhelt, H. Hopf: Eur. J. Biochem. 117, 353 (1981)
15. T. Schreckenbach, B. Walckhoff, D. Oesterhelt: Eur. J. Biochem. 76, 499 (1977)
16. T. Iwasa, F. Tokunaga, T. Yoshizawa: Photochem. Photobiol. 33, 539 (1981)
17. P. Towner, W. Gaertner, B. Walckhoff, D. Oesterhelt, H. Hopf: FEBS Lett. 117, 363 (1980)
18. B. Becher, J.Y. Cassim: Biophys. J. 16, 1183 (1976)
19. T. Iwasa, F. Tokunaga, T. Yoshizawa: Can. J. Chem. 63, 1891 (1985)
20. P. Scherrer, L. Packer, S. Seltzer: Arch. Biochem. Biophys. 202, 589 (1981)
21. T. Iwasa, K. Takeda, F. Tokunaga, P.S. Scherrer, L. Packer: Biosci. Rep. 2, 949 (1982)
22. M. Kataoka, K. Takeda, Y. Morimoto, N. Sato, N. Tanaka, F. Tokunaga: Photobiochem. Photobiophys. in press
23. M. Kataoka, F. Tokunaga, N. Sato, M. Nakasako, T. Ueki, Y. Hiragi, Y. Izumi, H. Tagawa, Y. Muroga, Y. Amemiya: PF Act. Rep. 1984/85, VI-111 (1985)

A New Intermediate in the Photocycle of Bacteriorhodopsin

R. Diller and M. Stockburger

Max-Planck-Institut für Biophysikalische Chemie,
D-3400 Göttingen, Fed. Rep. of Germany

Introduction

The retinal-binding protein Bacteriorhodopsin in the purple membrane of Halobacteria acts as a "light-driven proton pump". This function is controlled by a chromophoric group which contains a retinal molecule. This is bound to a lysine residue of the protein via a Schiff Base (SB) linkage. On illumination the chromophore runs through various intermediate states and under normal conditions is reconstituted within a few milliseconds. This photochemically induced and completely reversible reaction (photocycle) initiates the proton pumping mechanism /1/.

On the basis of optical transient- and low-temperature spectroscopy the cyclic reaction scheme depicted in Fig. 1 was first suggested by Lozier et al. /2/.

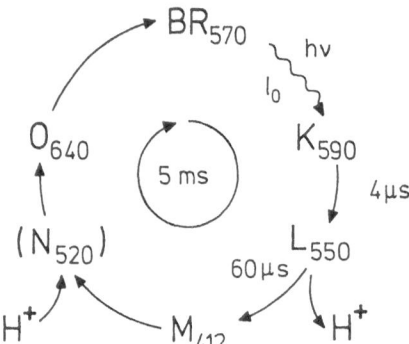

Fig. 1. Photochemical cycle of bacteriorhodopsin as suggested in ref /2/ The parent, BR-570 and the intermediates are designated by capitals BR,K,L,M,N,O. The lower labels give the wavelengths of maximum absorption. The thermal decay constants refer to room temperature and neutral pH

In this scheme the parent chromophore BR-570 is photochemically converted into the primary intermediate K-590 which subsequently runs through a linear chain of thermally controlled reaction steps before the original species is reconstituted.

It was demonstrated that resonance Raman (RR) spectroscopy is a powerful method to study the various intermediates of the photocycle /3,4/. Indeed, the rich manifold of vibrational features in the RR spectra provides us with important structural information. Thus it could be concluded from the spectra that in the primary photoreaction step an isomerization about the 13-14 double bond of the retinal chain takes place and that during the cyclic reaction a proton is released from the chromophore and later is taken up from its environment to reconstitute the original species. On the other hand, the vibrational bands in the RR spectra can also be used to study the temporal evolution of the intermediates. In the present contribution the RR spectra of a new intermediate shall be presented and the role it plays in the cyclic reaction of BR-570 shall be discussed.

Time-Resolved Raman Experiments

Time-resolved or kinetic RR studies on the photocycle can be performed in a double beam pump-probe experiment. In the most simple design an aqueous sample of purple membrane is flown across two parallel CW laser beams which are aligned perpendicularly to the flow direction and focussed into the sample. In the pump beam the sample is photolyzed and the mixture is probed by monitoring the RR radiation which is excited in the probe beam. The degree of photolysis in the pump laser is given by the product of the photochemical rate constant, l_0, and the residence time, Δt, of the sample in the beam. One has to choose the experimental conditions in such a way that sufficient photolysis of the parent BR-570 occurs in the pump beam but photolysis of any species in the probe beam can be largely neglected /5/.

The delay time (δ) between photolysis and probe event can be varied by the lateral distance between the two beams or by the flow velocity. In this way an RR spectrum of a certain intermediate can be obtained by choosing a value of δ which fits the appearance time of this species in the cycle. Its time-evolution then can be studied by following its diagnostic RR bands as a function of δ /5/.

In our work flow experiments were performed with the help of a spinning cell as described in Fig. 2. When the cell is rotated a cylindrically shaped liquid layer is formed at the external cell wall which is crossed by the two beams. It is important to note that the rotational period (T) of the cell has to be large compared with the period of the photocycle so that the parent species BR-570 is completely reconstituted when the sample again enters the pump beam. Typical parameters of our flow experiments are displayed in Fig. 2.

Pump Beam   Probe Beam   Pump B. Probe B.

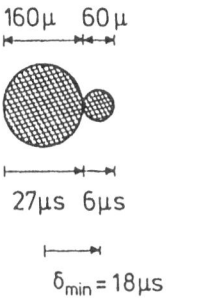

$$v \quad \Delta s$$
$$\delta = \Delta s/v$$
$$D$$
$$T$$

160μ   60μ

27μs 6μs

$\delta_{min} = 18$μs

D=40mm, T=20ms, v=6ms^{-1}
$\delta/\Delta s = 170$μs/mm

Fig. 2. Lay-out of a pump-probe flow experiment using a spinning cell. Typical parameters are: D = 40 mm, T = 20 ms, v = 6 ms^{-1}, $\delta/\Delta s$ = 170 μs/mm. On the right-handside of the diagram the diameters of the focal areas and the respective residence times are given

The L-to-M Transition

During the transition from L-550 to M-412 a proton is released from the chromophore and it is thought that this step plays an important role in the proton pump mechanism. We have therefore studied this reaction step in more detail.

In order to record the RR spectra of L-550 one is confronted with two principle problems The first is that BR-570 can only be partially converted into the intermediate (to about 30 to 40 percent). The second is that the absorption spectra of the two species strongly overlap, so that selective RR enhancement of L-550 is not possible. The spectra which are probed by photolysis thus contain a large contribution of BR-570. In order to obtain the pure RR spectrum of L-550 we have developed a procedure in which the spectra of the parent and of the mixture are recorded quasi-simultaneously. This allows to subtract the large contribution of BR-570 in a fairly accurate way, giving pure spectra of L-550 /5/.

The basic problem in a kinetic RR experiment is to determine the concentration of the various species in a photolyzed sample on the basis of diagnostic vibrational RR bands. It turned out that the strong C=C stretching bands are most useful for this purpose since they are characteristic of each species and also are the strongest bands in the spectra. In order to deconvolute the fairly narrow spectral range of the C=C stretching region band-fitting techniques could be successfully applied /5/.

In a previous work /5/ we have studied the time-behaviour of the L-to-M transition on the basis of the strong C=C stretches. Thus for L-550 one finds a double-peak at 1539 / 1551 cm^{-1} while the deprotonated species M-412 is represented by a single and sharp peak at 1565 cm^{-1}. The analysis therefore could be performed by band-fitting. It turned out that the "L" signal does not decay completely into M-412 as would be expected from the linear reaction scheme in Fig. 1. Instead a residual "L" component of about 25 percent of the maximum value was found. While the major part of "L" decays with a lifetime of 62 μs into M-412, the lifetime of the residual "L" component is a few milliseconds and thus comparable to that of M-412. In our previous work /5/ it had been assumed that the molecular structure of the residual "L" component would be the same as that of the fast-decaying fraction, since, according to the band-fitting analysis, the frequency and shape of the C=C stretching bands did not seem to change significantly on the time-scale of the two species. In order to cheque this assumption the total RR spectra were recorded at delay times of 20 μs for the fast- and of 500 μs for the slow component. The two spectra are displayed in Fig. 3.

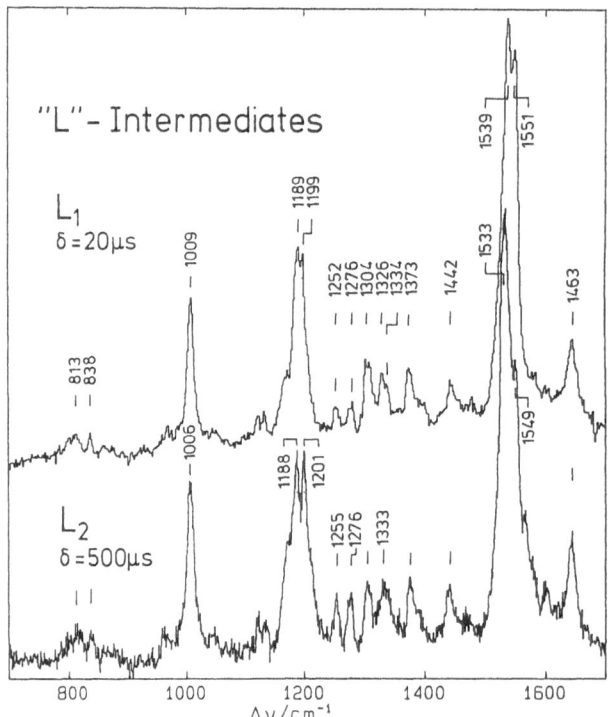

Fig. 3. Resonance Raman spectra of "L" intermediates at delay times of 20 μs and 500 μs. Pump laser at 647 nm, probe laser at 514 nm.

The RR Spectra of the two "L" Components

As expected, the two spectra in Fig. 3 are very similar. Indeed, each band in one spectrum has its related band in the other. But it is also evident that the two spectra are not identical. Thus the two bands at 1189 and 1199 cm^{-1} in the finger-print region of the 20 μs spectrum are slightly shifted apart at 500 μs. Moreover, the two weak bands at 1252 and 1272 cm^{-1} correspond to bands of significantly higher intensity at 1255 and 1276 cm^{-1}. Since the fingerprint region mainly reflects the geometry of the retinal chain, this geometry must be somewhat different in the two "L" components. However, this difference cannot be attributed to isomerization about a double bond which would lead to much more pronounced spectral changes /3, 4/. We therefore conclude that the "crude" geometry of the retinal chain in the two "L" components is the same but the spectral changes rather reflect minor conformational distortions.

The most distinct changes in the spectra of Fig. 3, however, occur in the C=C stretching region where the low-frequency peak of the "L"-doublet is shifted from 1539 to 1533 cm^{-1} and the relative intensity of the two peaks is changed. It is well established that the frequency of the "in-phase" C=C stretching mode of the retinal chain is extremely sensitive to π-electron delocalization /4/. This effect mainly depends on electrostatic interaction between the positively charged SB group and negative charges in its environment. The observed spectral changes therefore indicate that this interaction must be different in the two "L"-components. But on the other hand it is also evident from the spectra that in both forms the SB group in the "L" chromophores is protonated.

From the spectra in Fig. 3 we finally conclude that at delay times of 20 μs and 500 μs two species of slightly different geometry and electronic structure are observed. In the following they are labelled by "L_1" and "L_2", respectively.

Time-Dependence of "L_1", "L_2" and M-412

From our previous work /5/ it can be concluded that in the time-interval between $\delta_0 = 20$ μs and $\delta_1 \approx 500$ μs the time-dependence of the two "L" forms can be approximately displayed as

$$L_1 + L_2 = A e^{-(\delta - \delta_0)/\tau} + B , \qquad (1)$$

where the time-constant, τ, of the fast decay has a value of 62 μs. The same value was also found for the appearance-time of M-412. Since in this time-domain only the intermediates "L_1", "L_2" and M-412 are observed it follows from (1) that

$$M = 1 - (L_1 + L_2) \qquad (2)$$

if one sets A+B=1.

Various types of reaction mechanisms are conceivable which all would fit the time-dependence displayed in (1) and (2). In the following we discuss three of them in the light of our experimental data.

Equilibrium Reaction between "L" and M-412

In our previous work the existence of a second "L"-form had not been recognized /5/. The long-lived "L"-component therefore was explained by assuming that an intermediate equilibrium is established between "L" and M in the form

$$L \underset{k_2}{\overset{k_1}{\rightleftharpoons}} M \quad . \tag{3}$$

For the initial conditions $L(\delta_0)=1$ and $M(\delta_0)=0$ one finds

$$L = k_1 \tau e^{-(\delta-\delta_0)/\tau} + k_2 \tau \quad , \tag{4}$$

where $\tau = (k_1 + k_2)^{-1}$.

The identification of the long-lived form "L_2" makes the equilibrium reaction (3) less likely. In spite of this the time-dependence (4) would be observed if both, "L_1" and "L_2", would equilibrate with M according to (3) with identical rate constants.

In order to check if such an equilibrium reaction takes place we designed a special 3-colour laser experiment. After the normal pump beam a second one was used in the violet by which the intermediate M-412 could be selectively pumped back to the parent species BR-570. The relative concentration of "L" and M-412 was monitored by a third beam before and after the sample is perturbed by the violet beam. If "L" and M would react with each other according to (3) they would relax after perturbation to the original equilibrium ratio. Our experiments, however, revealed that such an effect does not occur. The reaction (3), therefore, can be excluded.

Branching Reaction at the Stage of "L"

For this reaction it is assumed that a single "L" species, namely L_1, is originally formed from its precursor, K-590, in the photochemical cycle. L_1 then undergoes a branching reaction of the form

$$L_1 \overset{k_1}{\underset{k_2}{\diagdown}} \begin{matrix} M \\ \\ L_2 \end{matrix} \tag{5}$$

from which one obtains

$$L_1 = e^{-(\delta-\delta_0)/\tau}, \qquad\qquad L_2 = k_2\tau(1 - e^{-(\delta-\delta_0)/\tau}), \tag{6}$$

$$L_1 + L_2 = k_1\tau\, e^{-(\delta-\delta_0)/\tau}+k_2\tau, \qquad M = 1 - (L_1+L_2), \tag{7}$$

where $\tau = (k_1 + k_2)^{-1}$.

It is evident from (6) and (7) that branching can only be distinguished from other reactions which give the same time-behavior of (L_1+L_2) if the time-dependence of the two "L" species can be measured separately.

"L_1" and "L_2" Originate from two Different Protein States

So far it was tacitly assumed that the two "L"-intermediates originate from protein molecules which all have the same uniform chemical structure. However, it is conceivable that the two species refer to chemically different forms or substates of the protein. Although we are unable to specify the difference, we have in mind only minor structural modifications. In the following the chromophores of the principle protein form shall be denoted as usual by BR, K, L, ..., while those of the substate are assigned as BR', K', L', ... If the structural difference of the two protein states is localized at a site which is not in the immediate neighbourhood of the chromophoric group it may happen that its structure would not be affected in the parent state. In this case BR and BR' might not be distinguished by their RR spectra and the same situation may hold for the primary photointermediates K and K'.

In the present picture it is now assumed that K and K' relax to L_1 and L_2', respectively, and the subscripts "1" and "2" shall indicate that the two chromophores L and L' have different structures, which, as we have seen above, can be distinguished by the RR spectra. In the time-interval between 20 μs and 500 μs the reactions of the two "L"-species can be described in the form

$$L_1 \xrightarrow{1/\tau} M + H^+ \quad , \tag{8}$$

$$L_2' = B \quad ,$$

which means that L_1 decays completely to M (with a time-constant τ of 62 μs) while L_2' remains constant. The time-dependence of $L_1 + L_2'$ again is given by (1) and therefore would fit the experimental data of our previous work /5/.

At a first glance the model presented here looks somewhat arbitrary. In particular one asks why only at the stage of "L" the chromophores should become distinguishable in structure and undergo completely different reactions. For an answer we have to consider the specific structural difference between L_1 and L_2' as expressed in the spectra of Fig. 3. - It must be emphasized that a distinction between L_2' and L_2 is only defined on the basis of the two models expressed in the reactions (8) and (5). Both model species, however, refer to the same 500 μs spectrum of Fig. 3. - From such spectra it was concluded that L_1 and L_2' differ by a slight conformational distortion of the retinal chain, but more importantly in changes of the electrostatic interaction of the protonated SB group with negative charges. In the parent species the structure of the chromophore is mainly determined by interaction of the SB group with a carboxylate group as a negative counterion /6/. This interaction is stabilized by water molecules /7/. It was argued that in the primary photochemical event which involves all-<u>trans</u> to 13-<u>cis</u> isomerization of the retinal chain the interaction with the counterion is weakened and in the subsequent relaxation step from K to L the SB group is attached to a different counterion /8/. If the structural difference between the two protein states would be localized in the vicinity of the new counterion this could explain both, the different structure and reactivity of the two L-chromophores.

Time-Dependence of $L_2(L_2')$

In principle it is possible to distinguish between the reactions (5) and (8) if one succeeds to disentangle the spectra of L_1 and L_2 (L_2') in the time-interval between 20 and 500 μs. The crucial question would be if L_2, according to (5) evolves from L_1 or if L_2' is formed synchronously with L_1 from a different protein substate.

We have measured the RR spectra of the L-intermediates in the C=C stretching region for different delay times (Fig. 4). It is seen that the spectra change significantly between 20 and 300 μs but beyond this limit remain fairly unchanged, reflecting the long-lived components L_2 (L_2'). From the spectra in Fig. 4, however, it cannot be definitely concluded if a species L_2' already partially is involved in the 20-μs spectrum, or if, according to (5) L_2 is formed from L_1.

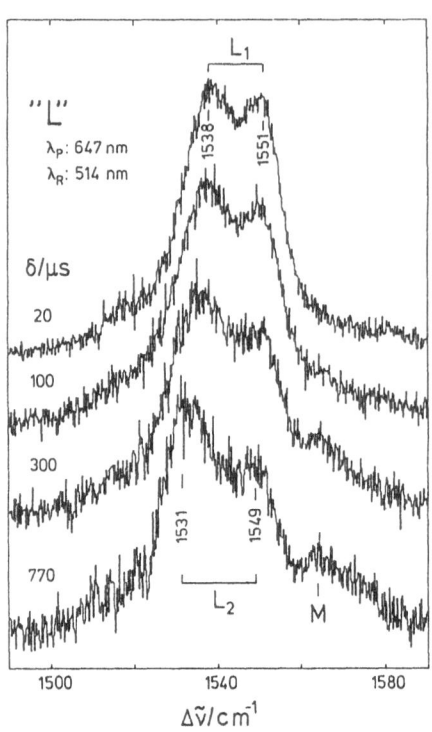

Fig. 4. Resonance Raman spectra
of the L-intermediates in the
region of the strong C=C
stretching bands at four different delay times. Pump beam
at 647 nm, Raman excitation at
514 nm

Role of $L_2(L_2')$ in the Photocycle

The question is now which role the partial reactions (5) and (8) of L_1 and $L_2(L_2')$ would play in the complete cyclic reaction of BR-570. From our previous experiments it can be concluded that $L_2(L_2')$ under normal conditions (room temperature, pH 7) has a lifetime of about 5 ms which is close to that of M-412. During the decay of $L_2(L_2')$ no indication of an M-like intermediate was found which implies that in the reconstitution mechanism of BR-570 via $L_2(L_2')$ no deprotonated species is involved.

Until now the role of 0-640 in the cyclic process is not yet definitely clarified. It was reported that all kinetic and mechanistic data which are available are inconsistent with a sequential reaction path of the form M-412 $\longrightarrow$ 0-640 $\longrightarrow$ BR-570 as used in the cyclic scheme of Fig. 1 /9,10/. In the following we therefore discuss the possibility if 0-640 is an intermediate in the reconstitution reaction of BR-570 via $L_2(L_2')$ so that

$$L_2(L_2') \xrightarrow{k_3} 0\text{-}640 \xrightarrow{k_4} BR\text{-}570 . \tag{9}$$

If this were so one would expect that the maximum concentration of 0-640 which is established during the photocycle is correlated with that of $L_2(L_2')$. It was found by Alshuth that $L_2(L_2')$ with respect to M-412 as an internal standard increases by nearly a factor of 2 when H_2O is exchanged by D_2O /11/ or when the external pH is decreased from 7.4 to 4.6 /5/. In both cases the maximum value of 0-640 with respect to M-412 also increases by about the same factor /10,12/.

It should be noted, however, that the correlation between $L_2(L_2')$ and 0-640 we invoke here is only valid if the rate constants k_3 and k_4 for the formation and decay of 0-640 in reaction (9) are independent of the above cited modifications or at least their ratio. It is not unreasonable to assume that this is indeed the case, since the principle reaction step between $L_2(L'_2)$ and 0-640 is 13-<u>cis</u> to all-<u>trans</u> isomerization of the retinal chain /13/ which should not depend on pH between 7.4 and 4.6 nor on H_2O/D_2O exchange. The same argument holds for the step from 0-640 to BR-570 which mainly involves a rearrangement of the chromophore. In this context it is important to note that it could be confirmed by Mäntele /10/ that the ratio k_3/k_4 is independent of H_2O/D_2O exchange.

From the two reactions (5) and (8) and considering (9) one obtains two different cyclic reaction schemes which are displayed in Fig. 5. Let us first consider the scheme on the left side which involves a branching at the stage of L. There is an argument against such a reaction. Thus it was found that the rate constant k_1 for the L_1-to-M transition is decreased by a factor of 4 in D_2O /11/. Such a kinetic isotope effect would indeed be expected for the proton transfer involved in this transition. For the transition from L_1 to L_2, on the other hand, no significant isotope effect should occur. This would imply that in D_2O the ratio k_1/k_2 should decrease by a factor of 4 while the intermediate concentration L_2/M should increase by the same factor. However, only an increase by a factor of about 1.7 was observed /11/.

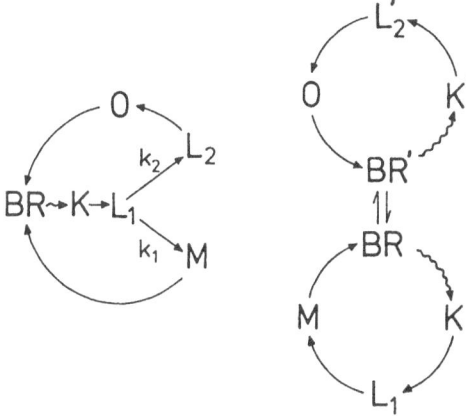

Fig. 5. The reaction scheme on the left side is based on a branching at the stage of L. The scheme on the right is based on two different protein states in which the chromophores at the stage of L undergo different reactions.

The reaction scheme on the right-hand side of Fig. 5 is based on the two parent chromophores BR and BR' which refer to different substates of the protein. As was noted above it may happen that the local structure of the chromophore is not influenced by the difference in protein structure so that BR and BR' cannot be distinguished spectroscopically.

This would explain why in the RR spectra no clear evidence for two different parent species BR and BR' was found. The same situation may hold for K and K'. Only at the stage of L the different protein structure modifies the chromophore in a way, as discussed in the previous section, that two different forms L_1 and L'_2 can be distinguished spectroscopically and kinetically.

In this scheme it is proposed that an equilibrium is established between BR and BR'. The time-constant for equilibration is assumed to be much longer than that of the cyclic reaction so that BR and BR' independently run through this process. The changes in concentration of L'_2 or 0-640 with respect to M-412 which were observed for environmental modifications (pH or H_2O/D_2O exchange) then would be due to changes in the equilibrium between BR and BR'.

Conclusions

In the present study a new "L"-intermediate in the photocycle of bacteriorhodopsin could be identified by time-resolved resonance Raman spectroscopy. This may have important consequences for our understanding of the photocycle which controls the biological function. On the basis of the presently available data two different cyclic reaction schemes were proposed in which the new "L" intermediate could be incorporated. A distinction between these schemes would require more sophisticated time-resolved RR studies. On the other hand a combination of chemical modification- and spectroscopic studies should help to answer the question if two chemically different states of the protein are already involved in the unphotolyzed parent species, as is proposed in one of the schemes.

References

1. D. Oesterhelt, W. Stoeckenius:
   Proc. Natl. Acad. Sci., USA 70, 2853 (1974)
   W. Stoeckenius, R.A. Bogomolni:
   Ann. Rev. Biochem. 51, 587 (1982)

2. R.H. Lozier, R.A. Bogomolni, W. Stoeckenius:
   Biophys.J. 15, 955 (1975)

3. S.O. Smith, J. Lugtenburg, R.A. Mathies:
   J. Membrane Biol. 85, 95 (1985)

4. M. Stockburger, T. Alshuth, D. Oesterhelt, W. Gärtner:
   In Advances of Spectroscopy, ed. by R.J.H. Clark and R.E. Hester, Vol. 13,
   Spectroscopy of Biological Systems, J. Wiley a. Sons, (1986) p. 483

5. T. Alshuth, M. Stockburger:
   Photochem. Photobiol. 43, 55 (1986)

6. U. Fischer, D. Oesterhelt:
   Biophys. J. 28, 211 (1979)
   M. Engelhard, K. Gerwert, B. Hess, W. Kreutz, F. Siebert:
   Biochemistry 24, 400 (1985)

7. P. Hildebrandt, M. Stockburger:
   Biochemistry 6, 5539 (1984)

8. P. Tavan, K. Schulten:
   Biophys. J. 50, 81 (1986)

9. W.V. Sherman, R.R. Eicke, S.R. Stafford, F.M. Wasacz:
   Photochem. Photobiol. 30, 727

10. W. Mäntele:
    Ph.D. Thesis, University of Freiburg, Freiburg, F.R.G. (1982)

11. T. Alshuth:
    Ph.D. Thesis, University of Göttingen, Göttingen, F.R.G. (1985)

12. Q. Li, R. Govindjee, T.G. Ebrey:
    Proc. Natl. Acad. Sci. USA, 81, 7079 (1984)

13. S.O. Smith, J.A. Pardoen, P.P.J. Mulder, B. Curry, J. Lugtenburg, R. Mathies:
    Biochemistry 22, 6141 (1983)

Structure Change of Bacteriorhodopsin and the Mechanism of Proton Pump

A. Ikegami, T. Kouyama, K. Kinosita, Jr., H. Urabe, and J. Otomo

The Institute of Physical and Chemical Research, Wako-shi,
Saitama 351-01, Japan

1. Introduction

Bacteriorhodopsin(bR) is a transmembrane protein in the purple membrane of Halobacterium halobium. It pumps protons across the membrane using light energy absorbed by retinal in it. Although the mechanism of the proton pump has been the subject of considerable investigation, the relations between the mechanism and bR protein structure are obscure, yet.

The electron density map of purple membrane studied with electron diffraction analysis [1] shows that each bR molecule is composed of the seven α-helices that cross the membrane. As a next step to understand pump mechanism in relation to the bR structure, we determined the three-dimensional location and orientation of the retinal chromophore in the purple membrane by the fluorescence energy transfer and polarized resonance Raman spectroscopy. By combining the results and the segments of amino acid sequence predicted as being seven transmembrane α-helices, we estimated the possible arrangements of the bR polypeptide chain, then amino acids, in the three-dimensional electron density map. Furthermore, we found the charge movements in bR during light-dark adaptation by electric dichroism and pH measurements. The results indicate that pK values of several amino acids change with the cis-trans isomerization of the retinal. Based on these studies and on other experimental results, we propose a novel model of proton pumping.

2. Three-Dimensional Disposition of the Retinal Chromophore in the Purple Membrane

A) Location and Orientation in the Membrane Plane

First, we estimated in-plane location and orientation of the retinal chromophore by the measurements of energy transfer from the fluorescent derivatives of retinal to the native retinal chromophore in partially reduced purple membrane[2]. In the analysis we applied a fairly unique feature of fluorescence energy transfer; it appears from an excited singlet of an energy donor (D) to an acceptor (A) quite far away from it by the dipole-dipole coupling between them. The rate of the fluorescence energy transfer calculated by Forster depends strongly on the distance between D and A, and depends on the directions of the emission dipole moment of D and the absorption dipole moment of A.

The retinal chromophores were partially converted to fluorescent derivatives by the reduction according to Peters et al. [3] without affecting the crystal structure of the membrane. As the emission spectrum of the fluorescent derivative overlaps with the absorption spectrum of the native retinal chromophore, the fluorescence energy transfer from the derivatives to the native chromophores is expected. The observed fluorescence intensity of the derivative, the donor, increases with the increase of the degree of reduction P, because the probability of finding acceptors, the native chromophores, at any acceptor site decreases with the increase of the conversion. Because of the crystallographic arrangement of bacteriorhodopsin in the purple membrane, fluorescence quantum yield of the derivative can be calculated as a function of three coordinates of the retinal chromophore in a unit cell and P, if the conversion occurs in a random way. The random conversion of the chromophore was confirmed from the initial fluorescence decay curves. By the least square comparison between the calculated and the experimentally obtained quantum yields, we estimated the most probable position and orientation of the retinal chromophore.

By the symmetry of the hexagonal lattice, there are 12 equally probable dispositions of the chromophore in the membrane. Among these twelve dispositions, we proposed that the position and orientation shown by the thick arrow in Fig. 1 were the most likely, because it located inside of the bacteriorhodopsin molecule, and did not overlap with the electron density map of the seven α-helices. Recently, our proposed disposition was supported by neutron diffraction studies; Jubb et al. [4] proposed almost the identical disposition, and Seiff et al. [5] proposed almost the same position as ours, but a somewhat different direction. Since the estimated position of retinal in a unit cell was confirmed by the diffraction methods, relative accuracy of the orientation estimated by fluorescence energy transfer should be increased; the rate of fluorescence energy transfer is sensitive to both position and orientation of the donor and acceptor.

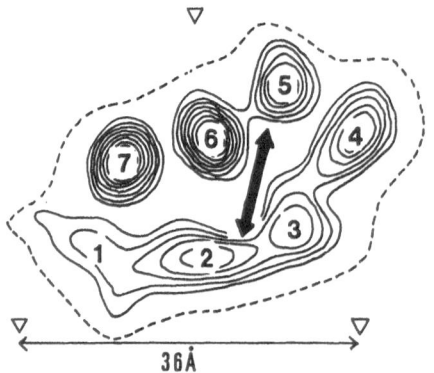

Fig. 1. In-plane location of the retinal determined by the fluorescence energy transfer [2].

B) Transmembrane Location

We applied the energy transfer technique to the determination of transmembrane location of the retinal chromophore in the purple membrane. Five different systems of donors and acceptors were used for this purpose. In the system 1 to 3, the fluorescent derivative of the retinal chromophore was used as a donor. In system 1 [6], fluorescent membranes were dispersed in the aqueous solution of the energy acceptor, cobalt-ethylenediaminetetraacetate (Co-EDTA), but in system 2 or 3 the membranes were embedded in solid Co-EDTA or tris(2,2-bipyridyl)ruthenium(II) complex. The cobalt complex was negatively charged whereas the ruthenium complex was positively charged. In system 4, energy transfer from the luminescent ruthenium complex interspersed between sheets to the native chromophore was observed. In system 5 [5], we observed the rate of fluorescence energy transfer from the fluorescent derivative to the native chromophore in membranes stacked in parallel.

In the theoretical analysis of the rate of energy transfer, we made the following assumptions.
a) The Forster mechanism and rate equation.
b) In the solid or stacked systems (2-5) random distribution of the donors or acceptors outside of the purple membrane, and random stacking of membrane sheets, irrespective of the asymmetric charge distribution of the membrane sheets.
c) Rapid-diffusion approximation for the dispersed system 1, because the validity of this approximation was theoretically estimated.

The results for these five systems were all consistent with a location of the retinal chromophore at a depth of 1.0 ± 0.3 nm from a surface of the purple membrane with a thickness of 4.5 nm.

The remaining problem, the sidedness of the chromophore in the purple membrane, was determined by the rate of energy transfer from the oriented purple membrane films with the fluorescent derivative of retinal. When purple membranes were spread on the cover glass coated with polylysine, the upper surfaces of most membranes adsorbed were the cytoplasmic surface at acidic pH, whereas both cytoplasmic and extracellular surfaces appeared in almost even ratio at neutral pH. We observed the fluorescence lifetime of the retinal derivatives in the oriented membrane covered with large acceptors like cytochrome C which can not penetrate into the space between the membrane and the cover glass. In all cases, the lifetime of the retinal derivative was longer at neutral pH than at acidic pH. Thus the results indicate that the retinal chromophore locates at 1.0 ± 0.3 nm below the cytoplasmic surface of the purple membrane.

C) Orientation of the Molecular Plane of Retinal

To detect the molecular direction of the retinal chromophore in the purple membrane, we measured polarized resonance Raman scattering due to C=C stretching vibrational modes[7]. The purple membranes were stacked on a conductive transparent glass by electrophoresis according to Varo [8]. The orientation of the membranes within the stacked film was checked by the small angle X-ray scattering.

Resonance Raman scattering of the oriented purple membrane was measured for three scattering geometries relative to the incident radiation(Ar laser 488nm). We observed the depolarization ratios of two strong lines at 1530 and 1570 cm^{-1} which indicate C=C stretching vibrational bands of bR and M intermediate, respectively. By the analysis of the observed depolarization ratios we estimated that the molecular plane of the retinal chromophore is almost perpendicular to the membrane surface.

3) Conformational Prediction of Bacteriorhodopsin Molecule

To predict the amino acid sequences of the seven α-helical segments in the density map of bR, we calculated the free energies for all segments composed of 22 consecutive amino acid sequences of bR with three conformations, α-helix(α), antiparallel β-sheet of hairpin type (β) and random coil (γ), in both hydrophobic ("membrane-like") and hydrophilic ("water-like") environments. We chose segments of 22 consecutive amino acids because the lengths of α and β conformations composed of 22 amino acids are almost the same as the thickness of the hydrophobic region of membrane. From the calculated free energy values of six conformational states of 22 amino acid sequences, we predicted amino acid sequences of seven α-helical segments, which are labeled A – G, starting from the amino terminus[7]. Similar predictions of the amino acid sequences were reported by several investigators[9-14], and all results strongly resemble each other, except for F sequence. For example, the amino acid numbers of the central position of G helices predicted are all confined within a very narrow limit, 212 ± 2.

The transmembrane positions of the α-carbon of ionizable groups averaged over the reported predictions are shown in Fig. 2. In the figure the numbers of the central amino acids of seven segments A – G, averaged over the predictions, are denoted on the center line of the membrane. Transmembrane location of the retinal shown in the figure suggests that Asp 96 is the most probable counter ion of the Schiff base proton, though there remain the possibilities of some deprotonated tyrosine. Furthermore, it suggests that the side chain of Lys 216 is extended to the cytoplasmic surface without large flexibility because Lys 216 locates near the center of the membrane. Therefore, the possible locations of G helix in the membrane plane should be restricted either helix 2 or 5 in Fig. 1, on account of short distances from the in-plane location of the retinal ends. For a similar reason, the probable location of the C helix which hold Asp 96 may be helix 6 or 3.

In-plane location of the helix B was estimated by the method of fluorescence energy transfer combined with fluorescence depolarization[7]. The fluorescent retinal derivative and a fluorescent probe NBD (7-chloro-4-nitrobenzo-2-oxa-1,3-diazole) bound to Lys-41 were used as an energy donor and acceptor, respectively. The fluores-

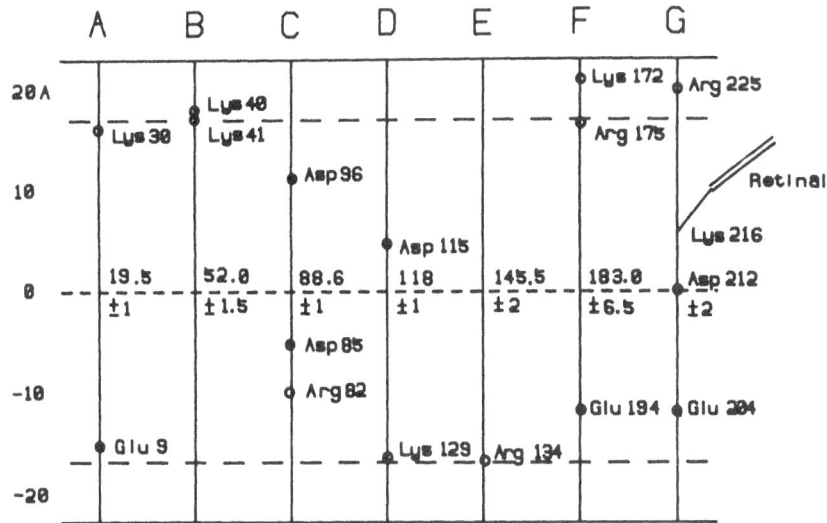

Fig. 2. The transmembrane positions of the α-carbon of ionizable groups averaged over 7 reported predictions (see text).

cence intensity and anisotropy of the donor and acceptor were observed for samples with various labeling ratios of NBD to the protein. The crystallographic analysis of these results suggested that Lys 41 of the helix B locates near the helix 7. Recently, Trewhella et al.[15] reported that the most probable locations of A and B helices are either helices 1,7 or helices 7,6 by the neutron diffraction method.

In the light of these results, we tentatively assigned the seven segments(A B ......G) to the seven helices 1-7-6-2-3-4-5 or 1-7-6-5-4-3-2 in the electron density map. We prefer the former assignment rather than the latter because of the electrostatic and hydrophobic interactions. Since the arrangement of the five sequences C D E F G in one model makes nearly the mirror image for the retinal plane of the other model, the mutual distances between the retinal and these helices are almost the same between two models. If we assume either of two models, we can imagine the trans membrane distribution of ionizable groups around the chromophore as shown in Fig. 3.

4) Charge displacements between bR^L and bR^D

Bacteriorhodopsin molecules in the dark and thermodynamic equilibrium state take a dark adapted form (bR^D) with absorption maximum near 560 nm. After exposure to yellow light, they become a light-adapted form (bR^L) with absorption maximum near 570 nm.

Recently, we estimated the difference in the surface charge distributions between bR^L and bR^D with the electric dichroism and pH measurements from initial pH 5 to pH 10 [16]. Purple membrane sheets in solution are oriented in a week electric field by their permanent dipole moments, which is due to the uneven distribution of membrane

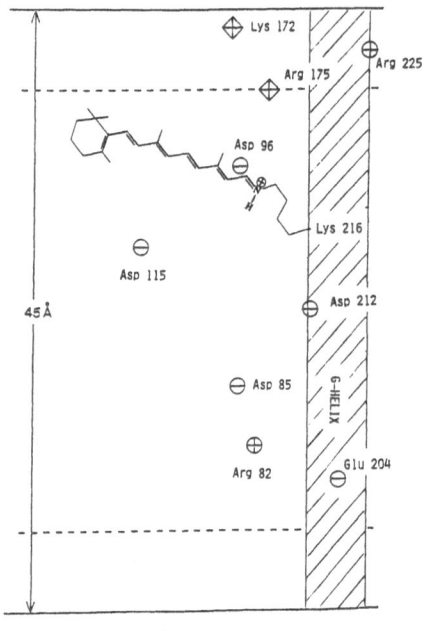

Fig. 3. The trans-membrane location of the α-carbon of ionizable residues around the chromophore predicted from a proposed structure model (see text)

charges between two surfaces. From the degree of orientation observed by the electric dichroism, the difference in the permanent dipole moment between light and dark adapted form was estimated. To simplify the following discussion, dipole moments are shown as the charge difference between two surfaces by the assumption that every charged group is located in either of the two surfaces. Then, the change in permanent dipole moment per bR, $\Delta \mu$, during the light adaptation is

$$\Delta \mu = (\mu_L - \mu_D) = [(\rho^e_L - \rho^c_L) - (\rho^e_D - \rho^c_D)] \cdot 1 ,$$

where ρ^e and ρ^c denote the number of charges per bR at external and cytoplasmic surfaces and 1 denotes the distance between the center and two surfaces of the membrane. The results are shown in Fig. 4.

The difference in the net charge between the light- and dark-adapted forms was obtained from the light-induced pH change in purple membrane solution. The number of protons taken up by bR during light adaptation shown in Fig. 4, $\rho_L - \rho_D$, was estimated from the pH titration measurements at each initial pH. The value is described by

$$\rho_L - \rho_D = (\rho^e_L + \rho^c_L) - (\rho^e_D + \rho^c_D) .$$

The changes in surface charges calculated from these relations, $\Delta \rho^e = \rho^e_L - \rho^e_D$, or $\Delta \rho^C = \rho^c_L - \rho^c_D$ should be due to the change in dissociation states of amino acids of bR between bR^L and bR^D. Specifi-

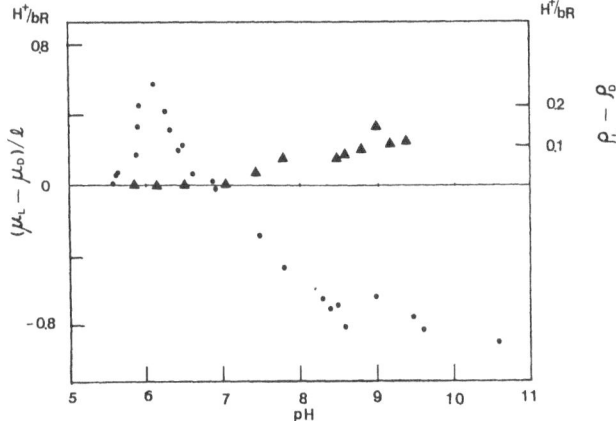

Fig. 4. pH dependence of the difference in permanent dipole moments ($\mu_L - \mu_D$) and surface charge ($\rho_L - \rho_D$) between the light- and dark-adapted form.

cally, we suppose that the isomerization of retinal with the light-dark adaptation might shift the pK values of some amino acids residues around it. Since the changes in pK values of more than four amino acid residues are necessary to explain the pH dependences of the estimated surface charges, we took Arg 175, Asp 96, Asp 85, and Arg 82 around the retinal in Fig. 3, and estimated their pK values to fit the observed changes. The results indicate that the pK values of Arg 175 and Asp 85 were increased, but those of Asp 96 and Arg 82 were decreased by the light adaptation.

5) A Model of Proton Pumping

From the conformational points of view described above, we propose a model of proton pumping, and discuss the mechanism.

In the dark and thermodynamic equilibrium state of bR, bR^D, two con-formations, bR^t with all-trans isomer of retinal and bR^c with 13-cis isomer, exist at almost even ratio. As the all-trans configuration is more stable than cis configuration for free retinal molecules in solution, interaction between the protein and cis retinal stabilizes the bR^c con-formation. That is, the conformation of bR polypeptide chain in bR^c (we denote bR-C) should be more stable than that in bR^t (bR-T).

Several experimental evidences indicate small structural difference between bR^L and bR^D, that is between bR-T and bR-C, but no electron diffraction analysis indicates the displacement of any α-helix of bR in the membrane plane. Furthermore, we could not detect any rotation of the absorption moment of the chromophore between bR^L and bR^D with linear dichroism measurements[16]. This means that most of the polyene chain is tightly fixed by the protein structure. Thus, both ends of the chromophore should not be affected during the dark-light isomerization.

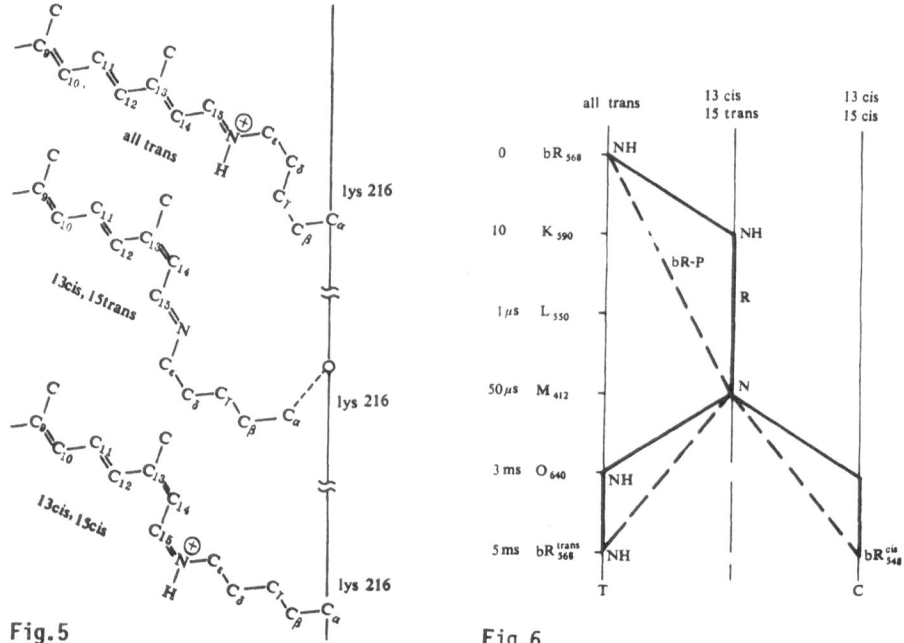

Fig. 5

Fig. 6

Fig. 5. (Left) Possible configurations of the retinal Schiff base in bR^t, bR^c and intermediate M.

Fig. 6. (Right) Schematic model for the conformation change of bR protein (dotted line) and the isomerization of retinal (full line).

Recent Raman scattering studies show that the retinal chromophore takes 13-cis and 15-cis conformation for bR^c, but 13-cis and 15-trans conformation for K, L, and M intermediates[17]. In our proposed structure model, the retinal of bR^t or bR^c can take stable conformations even when the both ends of the chromophore are fixed during the light-dark adaptation (Fig. 5). In the case of M and other intermediate states, however, we can not connect 13-cis and 15-trans retinal and Lys 216 without disturbing the stable conformation. Rather large distortion in the retinal configuration or the small displacement of the G-helix to the extracellular surface direction should be necessary to pass the M intermediate state in the photochemical reaction. The protein structure of the M intermediate must be more unstable than bR-T or bR-C.

By the absorption of a flash light, bR^t converts into the primary photoproduct K by the isomerization of retinal within very short time of about 3 ps. The primary photo product K, then undergoes a series of thermal reaction K -> L -> M changing protein conformations and reaching the metastable intermediate M. During or before the first thermal reaction, K -> L, changes in apparent pK values of several amino acids near the retinal are expected, in accord with the isomerization of retinal. Since the location of Schiff base nitrogen of K, L and M intermediates is the same as bR^c, we assume tentatively that the pK values of these intermediates are equal to those of bR^c estimated with the electric dichroism measurements. Then, protons redistribute between the ionizable groups during K -> L and L -> M reactions. The displacement of the G-helix supposed to occur during the thermal reaction L -> M, probably accelerates the deprotonation of Schiff base. The direction of the proton migration according to these processes is the same as that of proton pumping.

The deprotonated 13-cis, 15-trans retinal in metastable intermediate M converts spontaneously either to the all trans or to the 13-cis, 15-cis isomer by thermal agitation. The activation energy of cis-trans conversions should be reduced by the deprotonation and by the distorted conformation of M intermediate. After the isomerization of M intermediate, the metastable conformation of M intermediate changes into either bR-T or bR-C in the milli-second time range. The pK changes followed by proton migration take place also, and the bR molecule will return to either bR^t or bR^c conformations.

Overall scheme of the conformation change and the isomerization is shown in Fig. 6. The details will be published elsewhere.

References

1) Henderson R., and Unwin P.N.T. (1975). Three-dimensional model of purple membrane obtained by electron microscopy. Nature 257, 28-32.
2) Kouyama T., Kimura K., Kinosita K., Jr., and Ikegami A. (1981). Location and orientation of the chromophore in bacteriorhodopsin: Analysis by fluorescence energy transfer. J. Mol. Biol. 153, 337-359.
3) Peters J., Peters R., and Stoeckenius W. (1976). A photosensitive product of sodium borohydride reduction of bacteriorhodopsin. FEBS Lett. 61, 128-134.
4) Jubb J.S., Worceste D.L., Crespi H.J.L., Zaccai G. (1984). Retinal location in purple membrane of Halobacterium halobium: A neutron-diffraction study of membranes labeled in vivo with deuterated retinal. EMBO J. 3, 1455-1461.
5) Seiff F., Wallat I., Ermann P., and Heyn P. (1985). Neutron-diffraction studies on the location of retinal in bacteriorhodopsin. Proc. Natl. Acad. Sci. USA 82, 3227-3231.
6) Kouyama T., Kinosita K., Jr., and Ikegami A. (1983). Fluorescence energy transfer studies of transmembrane location of retinal in purple membrane. J. Mol. Biol. 165, 91-107.
7) Ikegami A., Kouyama T., Kinosita K., Jr., Otomo J., Urabe H., Fukuda K., and Kataoka R. Conformational analysis of bacteriorhodopsin. in Retinal proteins (Ovchinnikov Y.A. ed.) VNU Science Press, Utrecht, in press.
8) Varo G. (1983). Dried oriented purple membrane samples. Acta Biologica Acad. Sci. Hung. 32, 301-310.
9) Engelman D.M., Henderson R., McLachlan A.D., and Wallace B.A. (1980). Path of the polypeptide in bacteriorhodopsin. Proc. Natl. Acad. Sci. USA 77, 2023-2027.

10) Agard D. A. and Stroud R. M. (1982). Linking regions between helices in bacteriorhodopsin revealed. <u>Biophys. J.</u> <u>37</u>, 589-602.

11) Kimura K., Mason T. L. and Khorana H. G. (1982). Immunological probes for bacteriorhodopsin. <u>J. Boil. Chem.</u> <u>257</u>, 2859-2867.

12) Ovchinnikov Y. A. (1982). Rhodopsin and bacteriorhodopsin:structure and function relationships. <u>FEBS Lett.</u> <u>148</u>, 179-191.

13) Trewhella J., Anderson S., Fox R., Gogol E. and Engelman D., and Zaccai G. (1983). Assignment of segments of the bacteriorhodopsin sequence to positions in the structural map. <u>Biophys. J.</u> <u>42</u>, 233-241.

14) Katre N. V., Finer-Moore J., Stroud R. M. and Hayward S. B. (1984). Location of an extrinsic label in the primary and tertiary structure of bacteriorhodopsin. <u>Biophys. J.</u> <u>46</u>, 195-204.

15) Trewhella J., Popot J-. L., Zaccai G., Engelman D. M. (1986) Localization of two chymotryptic fragments in the structure of renatured bacteriorhodopsin by neutron diffraction. <u>EMBO J.</u> <u>5</u>, 3045-3049.

16) Otomo J., Ohno K., Takeuchi Y. and Ikegami A. (1986). Surface charge movements of purple membrane during light-dark adaptation. <u>Biophys. J.</u> <u>50</u>, 205-211.

17) Smith S. O.,Myers A. B., Pardoen J. A., Winkel C., Mulder P. P. J., Lugtenberug J., and Mathies R. (1984). Determination of retinal Schiff base configuration in bacteriorhodopsin. <u>Proc. Natl. Acad. Sci. USA</u> <u>81</u>, 2055-2059.

Activity of Bacteriorhodopsin in the Presence of a Large pH Gradient

T. Kouyama, A.N-. Kouyama, and A. Ikegami

The Institute of Physical and Chemical Resarch, Wako-shi, Saitama 351-01, Japan

1. Introduction

Bacteriorhodopsin (bR) is a membrane protein found in Halobacterium halobium. Due to the retinal chromophore bound to the apoprotein via a protonated Schiff base linkage, bR shows a strong absorption band in the visible region. Light excitation of the chromophore initiates a photo-chemical cycle of bR, through which protons are actively translocated from the cytoplasmic side to the outside, thereby generating a pH gradient across the plasma membrane [1]. It has been shown that forma-tion of a yellow intermediate M_{412} is accompanied with deprotonation of the Schiff base [2], and it has been widely accepted that its deprotona-tion (and subsequent reprotonation) is coupled with a unidirectional flow of protons in bR.

It has been believed that the proton pump activity of bR is in-fluenced by a pH gradient and an electric potential across the membrane. The effect of membrane potential on the photochemical reaction has been studied by several workers. It has been reported that the efficiency of formation of M_{412} and its decay kinetics are dependent on membrane potential [3,4]. On the other hand, little is known about the effect of pH gradient, mainly due to difficulty in generating and maintaining a large pH gradient. Very recently, we have developed the procedure to get a bR-containing vesicle that generates a large pH gradient under light illumination [5]. Within our knowledge, H. halobium envelope vesicle prepared from a strain JW3 by the freeze-thaw method and then treated with a weak acid exhibited the largest light-induced pH change; under an optimal condition, the pH change in the external medium was as large as 4 pH units (Fig. 1). By using this system, we were able to study the effects of the internal pH, external pH and pH gradient on the activity of bR.

In the first part of this communication we report the procedure to get the envelope vesicle exhibiting a large light-induced pH change; in the second part, the factors determining the extent of proton release are discussed; in the third part, the activity of the light-driven proton pump in the presence of a large pH gradient is reported; in the last part, the effect of pH gradient on the photochemical cycle of bR, especially on the process of proton uptake, is represented.

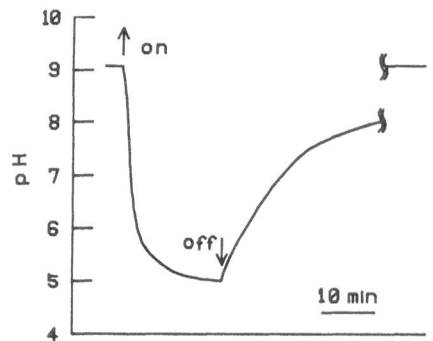

Fig. 1. Light-induced pH change in a suspension of H. halobium envelope vesicles (OD570~1). The solvent condition: 3M KCl, 0.4M MgCl$_2$, at 20°C.

2. Preparation of H. halobium Envelope Vesicles Exhibiting a Large Light-Induced pH Change

For investigation of the effect of pH gradient on the activity of bR, it is desirable to get a vesicle in which all bR molecules are oriented in the same direction. In the present preparation of the envelope vesicles from a strain JW3 [6], we have employed the freeze-thaw method by which cells were mildly broken. Briefly, cells were grown on complex medium consisting of "basal salt" solution (250g of NaCl, 20g of MgCl$_2$·2H$_2$O, 2g of KCl in 1 litter of water) supplemented with amino acids. The cells harvested from a stationary state culture were suspended in basal salt solution containing DNase I. The suspension was frozen at -70°C and thawed at room temperature, resulting in the breakage of cells [7]. The envelope vesicles thus obtained were precipitated by centrifugation at 17000 rpm: a reddish pellet which was sometimes found on a purple zone was discarded. The procedure of freeze-thaw and centrifugation was repeated three times with basal salt solution and twice with 4M NaCl or KCl solution. Cell wall proteins were not removed in the present preparation; in the presence of them, the light-induced pH changes were more reproducible than those observed in the absence of them. In the purified envelope vesicles, bR accounted for a large part (>70%) of membrane proteins.

Although the purified envelope vesicles contained many membrane proteins that interfered the action of bR, some of them were irreversibly modified in an acidic solution. As a result of modification of these proteins, the light-induced pH change came to reflect the activity of bR more directly. That is, complicated pH responses to light were usually observed for untreated samples: just below neutral pH, for instance, light illumination caused quick acidification followed by slow alkalization of the external medium. The latter component completely disappeared after the envelope vesicles had been incubated in an acidic solution (pH 3.5) at 40°C for about 1 hour. It is conceivable that some membrane protein like halorhodopsin [8] was destroyed by acid treatment. It is also suggested that a different kind of membrane protein showed different resistance to acid: i.e., treatment with a little stronger acid resulted in a considerable reduction in the proton permeability of the membrane. Fortunately, bR was one of membrane proteins surviving even at pH 1.

3. Light-Induced pH Changes of the Envelope Vesicle Suspension

The buffering action of the interior of the envelope vesicle was not
strong in the alkaline region, as long as pH buffer molecules were not
added. Figure 2 shows the result of pH titration of the envelope vesicle
suspension in 3M KCl. It is likely that a large part of the buffering
action came from cell wall proteins. Purple membrane suspension containing the same amount of bR showed much smaller buffering action (about
one fifth). Since bR was the main constituent in the membrane of the envelope vesicle, it was estimated that, in the alkaline region, extraction of only a few protons per one bR molecule from the vesicle interior
would result in a pH increase more than 1 pH unit. In fact, the pH
change in the suspension in 3M KCl was almost saturated within a few
seconds after turning on light and no more than 5 protons per bR were
released into the external medium.

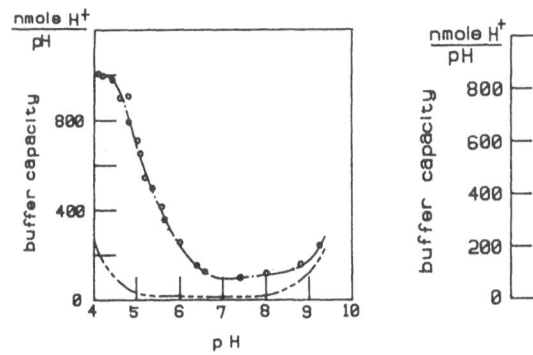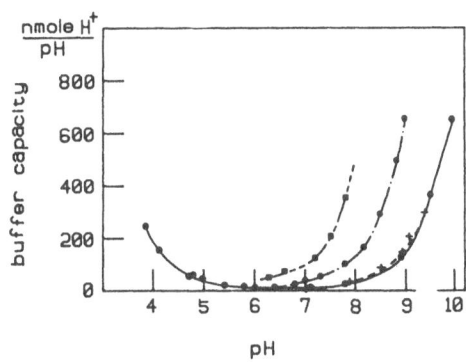

Fig. 2. The left panel: Buffering capacity of the envelope vesicle
suspension containing 18 nmole bR. The dashed line shows the contribution of the medium (3M KCl) to the buffering action.
The right panel: Buffering capacity of salt solution containing 80 mM
divalent cation; (+) Ca^{++}, (○) Mg^{++}, (■) Mn^{++}.

To observe a much larger extent of the light-induced proton release,
the buffering capacity of the vesicle interior must somehow be increased.
The simplest way to do this can be learned from nature. It should be
noted that the culture solution of halobacteria usually contains a high
concentration (80 mM) of Mg^{++}, and that it shows a strong buffering action in the alkaline region: when 3M KCl solution containing 80 mM Mg^{++}
was titrated with NaOH, the buffering capacity increased sharply above
pH 9.4 where precipitation of magnesium hydroxides appeared. Thus it
is not surprising to observe that the extent of the light-induced proton
release was greatly enhanced by Mg^{++}. In Fig. 3, the number of protons
released during the first 100 sec illumination (at 40°C) was plotted as
a function of Mg^{++} concentration. In the presence of low concentration
of Mg^{++} ion, the light-induced pH change was described with two phases;
a fast phase in which a few protons per bR were burst into the medium,
and a slow phase in which the pH of the medium continued to decline. Although the initial rate of proton release was not affected by Mg^{++}, the

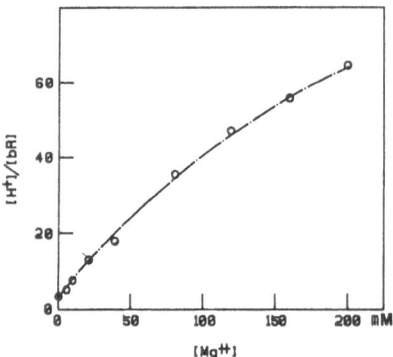

Fig. 3 The number of protons released during the first 100 sec illumination vs. the concentration of Mg^{++}. The data shown were obtained 1 or 2 hours later after addition of Mg^{++} ions to the suspension in 3M KCl and 2mM HEPES (pH 7.6).

rate of pH change in the slow phase increased almost linearly with increasing concentration of Mg^{++}.

Transition metal ions (Mn^{++}, Ni^{++} and Zn^{++}) were also effective to enhance the light-induced pH change. On the other hand, neither Ca^{++} nor Sr^{++} nor Ba^{++} affected the light-induced pH change in the entire pH region investigated. That is, only the divalent cations that formed the hydroxides at a relatively low pH (<10) were effective. From these results, it is suggested that the upper limit of the internal pH for a high activity of the proton pump is 10 more or less.

In order to estimate the absolute activity of the proton pump from the light-induced pH change, one has to understand what is the rate-determining factor in the steady-state translocation of protons. To satisfy the electroneutrality, there should exist a flow of counterions and/or co-ions electrically coupled with proton translocation. In the present preparation of envelope vesicles, monovalent cations played a minor role for the maintenance of the electroneutrality; the light-induced proton release took place well even in the complete absence of monovalent cations (e.g., in 1M $MgCl_2$). When 1M $MgCl_2$ was replaced with 1 M $MgSO4$, the apparent activity of the proton pump was somewhat lowered, suggesting a relatively high permeability to chloride ion. At the same time, the proton release in the absence of monovalent ions suggested the possibility of the inwards flow of magnesium (or metal) ions. This possibility is consistent with a linear dependence of the apparent activity of the proton pump on the concentration of magnesium ion (Fig. 3). In addition, under an adequate solvent condition, the number of net protons released during a prolonged illumination exceeded the number of divalent cations initially existing inside the vesicles.

Figure 4 schematically represents the movements of proton and magnesium or metal ion in the light. Light excitation of bR generates the outwards flow of proton. This flow is experimentally equivalent to the inwards flow of OH^- ion, and the OH^- ions accumulated inside the vesicle

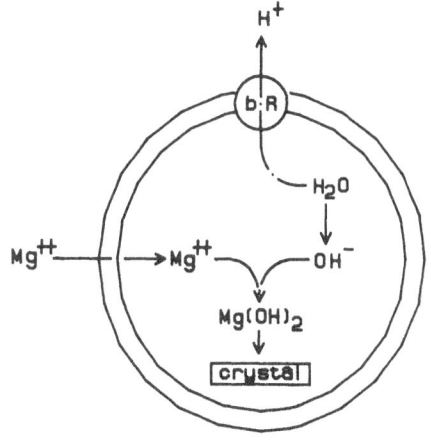

Fig. 4. Schematic representation
of the proton translocation and
the associated flow of divalent
cations.

react with magnesium (or metal) ions, producing their hydroxides.   Due
to the low solubility, solid crystals of the hydroxides are formed.
Lowering in the internal concentration of free divalent cation generates
the concentration gradient of divalent cation across the membrane,
which, besides electrical potential difference, accelerates the inwards
flow of divalent cation. The flows of divalent cation and proton con-
tinue until a large part of the vesicle interior is occupied by solid
crystals of the hydroxides, provided that the pH gradient is not
large.

4.   The Effect of pH Gradient on the Proton Pump Activity

The asymmetric buffering action of magnesium or metal ion made it possible to regulate the internal pH of the envelope vesicle in the light:
within a few seconds after turning on light, the internal pH was expected to increase to the value at which the buffering action of these
ions became noticeable. On the other hand, it was easy to keep the external pH at a substantially constant value during the first 1 or 2
minutes' illumination. Therefore the proton pump activity in the presence
of a large pH gradient can be examined by monitoring pH changes induced
by a long duration light pulse. In Fig. 5, the number of protons
released during the first 100 sec illumination is plotted against the
initial pH. In this experiment, a small amount of pH buffer molecules
(2mM or 4mM HEPES and MES) was added so that the pH change in the external medium was restricted within 0.5 pH unit.

   In the absence of Mg^{++}, more protons were pumped out at lower pH. This
pH dependence is exactly the same as the previous result [9]. It is
likely that the bump of the apparent activity at the acidic pH was due
to the buffering action of carboxyl groups distributed on the internal
membrane surface. In the presence of magnesium or metal ion, on the
other, more protons were pumped out at higher pH.   That is, the enhancement of the light-induced pH change by Mg^{++} was notable only when the
external pH was higher than 5.5. When Mg^{++} was replaced with Mn^{++}, a
similar pH dependence was observed, except that the effect of Mn^{++} was
notable above pH 4.5. The shift of the pH dependence seemed to cor-

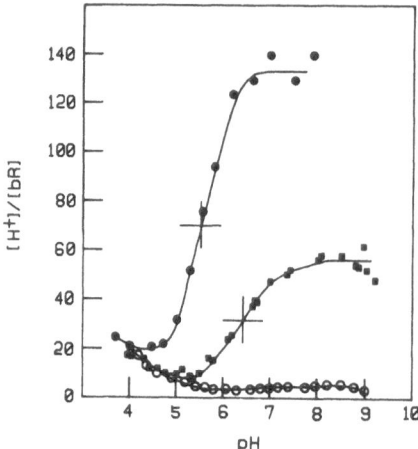

Fig. 5. The number of protons released during the first 100 sec illumination was plotted as a function of the initial pH of the medium.
The different symbols represent the data obtained in the absence ($\bigcirc$)
and in the presence of 80 mM Mg^{++} ($\blacksquare$) and 80 mM Mn^{++} ($\bullet$), respectively.

respond to the shift of the internal pH maintained in the light; $Mn(OH)_2$
precipitates at a lower pH (by 1.2 pH unit) than $Mg(OH)_2$. This result
suggests that the activity of the light-driven proton pump does not
depend much on the external pH, and that the extent of proton release is
reduced by a large pH gradient. From the observation of the more notable
effect of Mn^{++} than Mg^{++}, it is suggested that there exists some important residue with a pK value of about 9.5 whose dissociation state is
sensitive to the internal pH. Since the extent of proton release was almost constant in the alkaline pH region, it is likely that either the
proton motive force or proton leakage is non-linearly regulated by the
pH gradient.

5. Photochemical Cycle of bR in the Presence of $\triangle pH$

Since little is known about the effect of pH gradient on the proton
leakage, it is difficult to definitely answer whether the proton motive
force is directly regulated by the pH gradient. Another approach to this
problem may be to investigate the photochemical cycle of bR in the
presence of a large pH gradient. That is, understanding of the effects
of the internal pH, external pH and pH gradient itself on the
photochemical cycle will help to clarify the influence of pH gradient on
the proton motive force and therefore the mechanism of the light-driven
proton pump.

In the present preparation of the envelope vesicle, a large pH
gradient can be generated and maintained only under light illumination.
Thus a special spectrophotometer was constructed in which monitoring
beam and actinic beam are mechanically separated with synchronous chop-

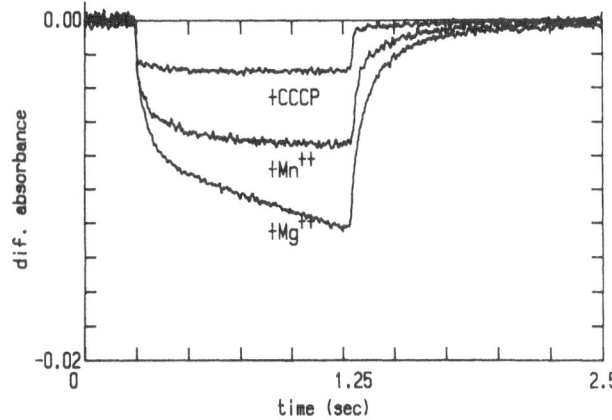

Fig. 6. Light-induced absorption changes (at 570 nm) of the envelope vesicle suspension at 40°C. The three curves were obtained in the presence of 20 mM CCCP, 80 mM Mg^{++}, 80 mM Mn^{++}, respectively. The medium contained 3M KCl and 1 mM HEPES (pH 7.2).

pers (100–400 Hz) so that absorption and fluorescence spectra in the entire visible region can be recorded in the presence of intense actinic light. The light-induced pH change was monitored with a fluorescent pH indicator (pyranine).

Figure 6 show typical examples of the light-induced absorption changes in the envelope vesicle suspension observed at the initial pH of 7.2. When 20 μM CCCP (a proton ionophore) was added so as to dissipate pH gradient, the absorption kinetics were not different much from those of the purple membrane suspension under the same solvent condition. The light-induced difference absorption spectrum was characterized with a positive peak around 660nm besides the one at 410 nm, and the lifetime of M_{412} was only slightly longer than that observed for purple membrane suspension. When a large pH gradient was allowed to develop, however, different absorption kinetics were observed. The most clear difference is disappearance of the positive peak at the long wavelength. It is known that, in purple membrane suspension, O_{640} intermediate can be seen at pH lower than 8. The present result indicates that the internal pH is the main factor determining the quantum yield of its formation (or its lifetime). As a result of disappearance of O_{640}, the light-induced difference absorption spectrum was very similar to that observed for purple membrane in a salt solution saturated with ether, in which only M_{412} is a detectable long-lived intermediate.

Another effect of developing pH gradient was seen in the lifetime of M_{412}. In the presence of 80 mM Mg^{++}, for instance, the light-induced absorption change at 410 nm increased with time and leveled off about a few seconds after turning on light; as long as the external pH was not high (<8), similar absorption kinetics (but with opposite sign) were observed at 570 nm. Since the amount of M_{412} accumulated in the photo-stationary state was almost proportional to its lifetime (multiplied by the quantum yield of its formation), the above result suggests the possibility that the lifetime of M_{412} became longer with developing pH gradient. In fact, the absorption recovery observed after turning off light became slower with increasing time duration of light illumination. Since the external pH was kept at a substantially constant value in the above experiment, either an increase in the internal pH or the pH gradient itself was responsible for the increase in the lifetime of M_{412}. When the absorption kinetics observed in the presence of different species of divalent cations (Mg^{++} and Mn^{++}) were compared, it was suggested that the internal pH had the strongest influence on the lifetime of M_{412}: that is, when 80 mM Mg^{++} was replaced with 80 mM Mn^{++} so that the internal pH did not increase above ~8.5, the amount of M_{412} accumulated in the light became smaller.

The present result suggests that the lifetime of M_{412} is dependent strongly on the dissociation state of some residue which is located near the cytoplasmic membrane surface. Since the proton pump activity becomes negligible as the internal pH becomes higher than 10, it seems likely that the residue under consideration participates in the proton transfer process from the vesicle interior to the deprotonated Schiff base. If this proton transfer process was blocked, bR might be enforced to undergo a photocycle which is ineffective for proton pump; e.g., the deprotonated Schiff base might get a proton from the extracellular side. With this respect, it was found that, as the external pH was lowered, a smaller amount of M_{412} was accumulated at the photo-stationary state. Thus the possibility was suggested that, at a fixed internal pH, the decay of M_{412} might be accelerated with developing pH gradient or with increasing proton concentration at the extracellular side.

When the external pH was kept at a high value (>8), the absorption kinetics became a little complicated, mainly due to the appearance of a long-lived intermediate. The lifetime of this intermediate was an order of second at room temperature, and the associated difference absorption spectrum had a negative peak at 580 nm and no large positive peak in the entire wavelength region [10]. This intermediate may correspond to 'L-like intermediate' reported by MAEDA et al. [11], 'P intermediate' recently reported by DRACHEV et al. [12], or 'N intermediate' originally reported by LOZIER and NIEDELBERGER [13]. Due to the existence of the very slow relaxation process of bR, even purple membrane suspension at alkaline pH exhibited the absorption kinetics that depended on the duration of excitation light pulse and so on. The absorption changes induced by a long-duration light pulse were shown in Fig. 7. Under the experimental condition used, the lifetime of M_{412} increased monotonously with increasing pH. This is consistent with the observation that the lifetime became longer with increasing internal pH. On the other hand, experiments with short light pulses (10 μs) provided a slightly different result; the decay kinetics at 410 nm was characterized with two components; the decay constant of the fast component was almost independent of pH, whereas the other component decayed more slowly at higher pH. In addition, it was found that the relative amplitude of the slow component became larger with increasing repetition rate of light pulses. This result is consistent with the previous report that the relative amplitude of the slow component became larger with increasing intensity of light pulses [14]. It is very possible that the slow component originates from the long-lived intermediate (or the bR state becoming predominant under continuous illumination).

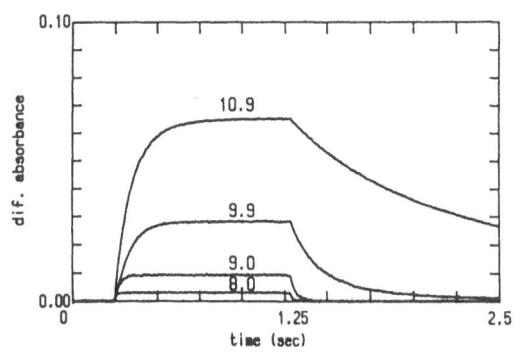

Fig. 7. Light-induced absorption changes (at 410 nm) of purple membrane suspension in 3M KCl at 40 ℃. The four curves shown were obtained at pH 8.0, 9.0, 9.9 and 10.9, respectively.

The proton pump activity measured under continuous illumination was not dependent much on the external pH (Fig. 5). With this respect, a slightly different result was reported by LI et al.[15]; by monitoring pH changes induced by short light pulses, they found a drop in the pump activity above pH 8. The discrepancy seems to be explained when the dependence of the photochemical reaction of bR on the duration of light pulses is taken into account.

It has been reported that membrane potential affects the lifetime of M_{412} [3,4]. At the present stage, we can not exclude the possibility that, like pH gradient, membrane potential altered the ionization state of some residues that would have the strongest influence on the lifetime of M_{412}.

6. Conclusion

(1) Bacteriorhodopsin (bR) does not work when the internal pH becomes higher than 10: there is some important residue with a pK value about 9.5 whose dissociation state is sensitive to the internal pH.
(2) The activity of bR does not depend on the external pH: It works in a wide range of the external pH values, at least between 4.5 and 9.5.
(3) In the presence of the pH gradient larger than 3 pH unit, the activity of bacteriorhodopsin becomes low, or (ii) proton permeability of the cytoplasmic membrane becomes remarkable.
(4) Some photochemical reaction of bR is sensitive to the internal pH; e.g., the lifetime of M_{412} (the slow component), the efficiency of formation of O_{640}. On the other hand, appearance of a very long-lived intermediate is determined by the external pH.

Acknowledgments:

We thank Drs. K. Kinosita and H. Nihei for fruitful discussions and Dr. J. Weber for kindly providing us a strain JW3. T. K. is grateful to Drs. M. K. Mathew and W. Stoeckenius for previous collaboration from which valuable information on a long-lived intermediate of bR was obtained. Research was supported by special coordination funds for promotion of science and technology and a research grant for 'Solar Energy-Photosynthesis' given by the Agency of Science and Technology of Japan.

References:

1. W. Stoeckenius, R. H. Lozier, R. A. Bogomolni: Biophys. Biochim. Acta, 505, 215 (1979)
2. A. Lewis, J. Spoonhower, R. A. Bogomolni, R. H. Lozier, W. Stoeckenius: Proc. Natl. Acad. Sci. USA, 71, 4462 (1974)
3. A. T. Quintanilha: FEBS Lett. 177, 8 (1980)
4. S. L. Helgerson, M. K. Mathew, D. B. Bivin, P. K. Wolber, E. Heinz, W. Stoeckenius: Biophys. J. 48, 709 (1985)
5. T. Kouyama, A. N-. Kouyama, A. Ikegami: Biophys. J. (in press)
6. H. J. Weber, R. A. Bogomolni: Methods in Enzymology, 88, 379 (1982)
7. A. E. Blaurock, W. Stoeckenius, D. Oesterhelt. & Scherphof G. L.: J. Cell Biol. 71, 1 (1976)
8. Y. Mukohata, Y. Kaji: Arch. Biochem. Biophys. 208, 615 (1981)
9. M. Eisenbach, H. Garty, E. P. Bakker, G. Klemperer, H. Rottenberg, S. R. Caplan: Biochemistry 17, 4691 (1978)
10. T. Kouyama, A. N-. Kouyama, A. Ikegami, M. K. Mathew, W. Stoeckenius: unpublished data
11. A. Maeda, T. Ogura, T. Kitagawa: Biochemistry, 25, 2798 (1986)
12. L. A. Drachev, A. D. Kaulen, V. P. Skulachev, V. V. Zorina: FEBS Lett. 209, 316 (1986)
13. R. H. Lozier, W. Niederberger: Fed. Proc. 36, 1805 (1977)
14. K. Ohno, Y. Takeuchi, M. Yoshida: Photochem. Photobiol. 33, 573 (1981)
15. Q. Li, R. Govindjee, T. G. Ebrey: Photochem. Photobiol. 44, 515 (1986)

Picosecond and Nanosecond Spectroscopies of the Photochemical Cycles of Acidified Bacteriorhodopsin

H. Ohtani[1*], *T. Kobayashi*[1], *and A. Ikegami*[2]

[1]Department of Physics, The University of Tokyo, Bunkyo, Tokyo 113, Japan
[2]Institute of Physical and Chemical Research, Wako-shi,
 Saitama 351-01, Japan

1. Introduction

The purple membrane of Halobacterium halobium [1] is composed of lipids and a single protein [1]. The protein is called bacteriorhodopsin, which has a protonated retinal Schiff base linked with lysin residue of an apoprotein. The light- and dark-adapted bacteriorhodopsins have absorption maxima at 568 and 558 nm, respectively. The former is a single species with all-trans retinal chromophore, hereafter denoted by bR_{568}. The latter is a mixture of the two proteins with all-trans and 13-cis retinals (50:50). The function of the light-adapted bacterio-rhodopsin, bR_{568}, is a light-driven proton pump from inside to outside of the plasma membrane of a cell.

The photochemical cycle of bR_{568} is initiated by absorption of a photon and followed by sequential thermal reactions. The following intermediates have been found in the photocycle of bR_{568} [2–5]. The significant chemical reactions in the

$$bR_{568} \xrightarrow{h\nu} J_{625} \longrightarrow K_{610} \longrightarrow KL_{596} \longrightarrow L_{543} \longrightarrow M_{412} \longrightarrow O_{640} \longrightarrow bR_{568}$$

photocycle of bR_{568} are (i) isomerization of retinal and (ii) dissociation of Schiff base proton. Intermediates K, L, and M have a 13-cis chromophore [6,7]. The Schiff bases in K and L are protonated and that in M is deprotonated [6,8].

The blue-colored membrane is formed by acidification or deionization of purple membrane [1,9,10]. The blue form sample denoted by bR_{605} has an absorption maximum at 605 nm. bR_{605} is a mixture of bacteriorhodopsins with all-trans and 13-cis retinals (60:40) [9,11]. The photochemical behavior of the blue sample differs from that of purple membrane. Earlier works show that the photocycle depends on the excitation wavelength and that the kinetics is complicated [9].

---

[*]Present Address: Hamamatsu Photonics K.K., Tsukuba Research Laboratory,
Toyosato, Ibaraki 300-26 Japan.

In this work, we studied the photocycle of the acidified bacteriorhodopsin (bR_{605}) with the aids of picosecond and nanosecond time-resolved absorption spectroscopies [12], and compared it with that of bR_{568} at neutral pH. We found that bR_{605} is composed of two acidified bacteriorhodopsins with all-<u>trans</u> and 13-<u>cis</u> retinals (abbriviated hereafter as $t-bR_{605}$ and $c-bR_{605}$, respectively) and a small amount of the neutral form bR_{568} and clarified all photocycles of these three bacteriorhodopsins in low pH suspension.

2. Experimental

2-1. Materials

Purple membrane fragments of <u>Halobacterium halobium</u> (R_1M_1) were purified according to conventional methods [13], and deionized on a cation exchange column (acidic form of Dowex 50W) [10]. This suspension obtained here was transparent and stable for long time enough for absorption measurements. The absorption maximum (605 nm) of the sample agreed well with that of the purple membrane in polyacrylamide gel at pH 2.0 [9].

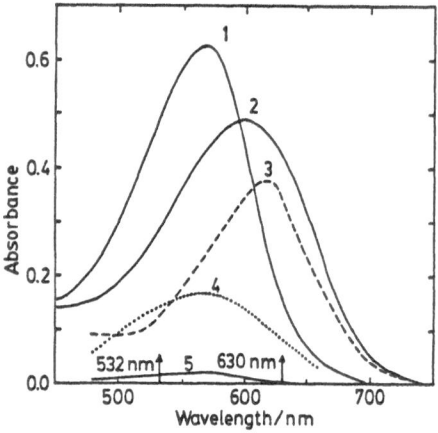

Fig 1. Absorption spectra of light-adapted purple membrane (curve 1) and acidified purple membrane (curve 2) containing $t-bR_{605}$ (curve 3), $c-bR_{605}$ (curve 4), and bR_{568} (curve 5).

The sample solution is a mixture of $t-bR_{605}$, $c-bR_{605}$, and a small amount of bR_{568}. Figure 1 shows the absorption spectra of the light-adapted purple membrane and the acidified membrane with an ion-exchange column. The composition and fractions of photon number absorbed by the components out of the total number of absorbed photons are given in Table 1. The fractions of the three components excited by lasers with different wavelength were obtained from their absorption cross sections.

Table 1. Absorption maxima (λ_{max}), molar extinction coefficients at λ_{max} (ε_{max}), and molar fractions (F_i) of t-bR$_{605}$, c-bR$_{605}$, and bR$_{568}$ in the sample and the number fractions (f_i) of photons absorbed by the three components out of the total absorbed photons. The number fraction of photons absorbed by component "i" is calculated by $F_i(\lambda) = f_i \, \varepsilon^i(\lambda)/(f_T \, \varepsilon^T(\lambda)+f_C \, \varepsilon^C(\lambda)+f_P \, \varepsilon^P(\lambda))$, where t-bR$_{605}$, c-bR$_{605}$, and bR$_{568}$ are denoted by T, C, and P, respectively.

	t-bR$_{605}$	c-bR$_{605}$	bR$_{568}$
λ_{max}/nm	610-620	550-570	568
ε_{max}/10^4M^{-1}cm^{-1}	6.3-6.6	4.1-4.2	6.3
f_i	0.57	0.40	0.03
F_i(630 nm)	0.77$\pm$0.07	0.22$\pm$0.07	0.01
F_i(532 nm)	0.45$\pm$0.02	0.50$\pm$0.02	0.05
F_i(266 nm)	0.57	0.40	0.03

2-2. Apparatuses for Picosecond and Nanosecond Laser Photolyses

The light source for picosecond spectroscopy was a passively mode-locked Nd:YAG laser. The excitation light source was the first Stokes of the Raman scattered light by acetone (630 nm, 20-ps FWHM, 76-90 µJ) generated by focusing the second harmonic of the laser (532 nm) [14]. A picosecond continuum was generated by focusing the 1064-nm light into a D$_2$O cell, and was divided into two beams, one of which was used for a probe and the other for a reference. They were separately detected by two coupled systems of a 512-channel photodiode array and a grating polychromator (f=20 cm, 600 grooves/mm). The output signals of the detectors were transfered to a minicomputer-microcomputer analyzing system [14].

Excitation light source for nanosecond spectroscopy was either the fourth (266 nm, 5-ns FWHM, 0.5 mJ) or the second (532 nm, 5-ns FWHM, 0.5 mJ) harmonic of a Q-switched Nd:YAG laser or the first Stokes Raman scattered light of 532-nm light by acetone (630 nm, 5-ns FWHM, 90 µJ). A probe light source was a xenon lamp. Output signals of a photomultiplier coupled with a monochromator (f=17 cm, 1350 grooves/ mm grating) were digitized with a transient recorder and averaged with a microcomputer [14].

The sample cells of 2- and 3-mm light path lengths were used for the picosecond and nanosecond time-resolved spectroscopy, respectively. All measurements were performed at 21$\pm$1°C.

3. Results and Discussion

3-1. Photocycle of t-bR$_{605}$

Of the three components in the acidified sample (bR$_{605}$), mainly t-bR$_{605}$ is excited
by the 630-nm light. Figure 2 shows the transient difference absorption spectra
of bR$_{605}$ just (a) and 217 ps (b) after excitation of the bR$_{605}$ sample. The negative absorbance change in the 550-680 nm region shown in Fig. 2a is due to the
bleaching of t-bR$_{605}$ in the ground state. The positive absorbance change in the
500-520 nm region attributed to the $S_n \leftarrow S_1$ transition of t-bR$_{605}$. The absorption
spectrum of bR$_{605}$ in S_1 state is located in the shorter wavelengh region than that
in the ground state. Similar result on bR$_{568}$ has been reported [15].

The time constants for the bleaching recovery at 595±25nm and formation of the
acidic form of K (K$_{acid}$) at 665±25 nm were obtained to be 10 ps by the least-
square best fitting. The lower and upper limits of the time constants determined
were 2 and 30 ps, respectively. The singlet lifetime of t-bR$_{605}$ is estimated to
be 7-17 ps from the the reported singlet lifetime of bR$_{568}$ (430±50 fs) [16] and
the fluorescence quantum yields in pH 2 and 7 (($2.5\pm0.1)\times10^{-4}$ and $(4.5\pm0.2)\times10^{-3}$,
respectively) [17] by assuming the same fluorescence yield of c-bR$_{605}$ as bR$_{548}$
(($0.7-1.2)\times10^{-4}$) [17]. The singlet lifetime of t-bR$_{605}$ measured in the present
study (2-30 ps) is consistent with this estimate. The singlet lifetime of t-bR$_{605}$
is longer than that of bR$_{568}$.

Figure 2b shows the formation of K$_{acid}$. The formation yield of K$_{acid}$ was determined to be 0.09±0.01. This value is smaller than that of K at neutral pH suspension (0.30±0.03) [18].

The primary event in the photocycle of bacteriorhodopsin is all-trans → 13-cis
photoisomerization of retinal. On analogy with photoisomerization of stilbene a
phantom excited state [19] is expected to be formed by twisting 13-14 carbon-carbon bond. The lowest excited singlet state (S_1) of bR$_{605}$ could relax to the phantom excited singlet state (S_p). The reaction scheme in the picosecond range is
given as follows.

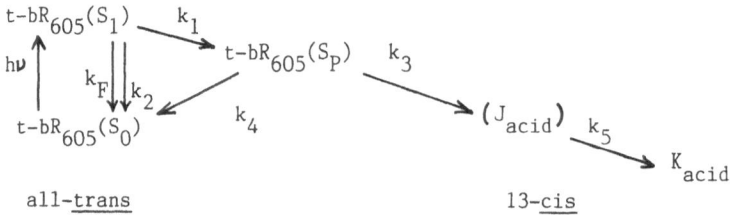

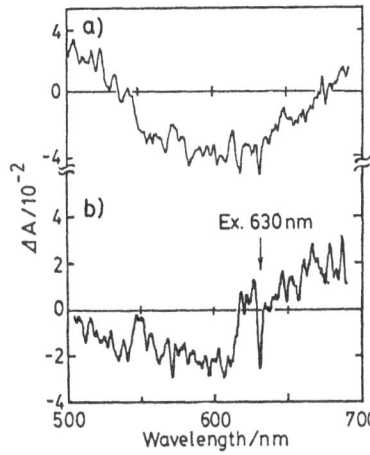

Fig. 2. Picosecond difference absorption spectra (a) just and (b) 217 ps after excitation of bR_{605} at 630 nm.

The increase in the fluorescence yield and the decrease in the quantum yield for photocycle with the decrease in pH are attributed to the reduction of k_1, and possibly of $k_3/(k_3+k_4)$. The kinetic pH effect may be caused by a protonation of an amino-acid residue near the chromophore which is dissociated in neutral pH suspension. Intermediate J_{acid} could not be detected in the photolysis of $t-bR_{605}$. The lifetime of J (3–5 ps) [15,20] may not be affected by the change in pH as well as K (KL) which has a similar structure to J. J_{acid} can not be accumulated since the lifetime is not much longer than to the formation time.

Figure 3 exhibits the transient difference spectra observed 150 ns and 7 µs after 630-nm excitation. The transient spectra at 150 ns (curve 1) and 217 ps (Fig. 2b) quite the same in main features. K_{acid} is converted to the acidic form of L (L_{acid}) as shown by curve 2 in Fig. 3. The time constant determined (1.5 ± 0.4 µs) is very close to that for the KL → L conversion at neutral pH (1.3 ± 0.3 µs). The time constants for these processes are not affected by pH change. L_{acid}-minus-

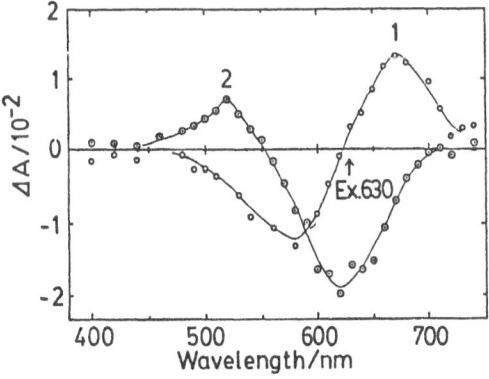

Fig. 3. Nanosecond difference absorption spectra 150 ns (curve 1) and 7 µs (curve 2) after excitation of bR_{605} at 630 nm.

K_{acid} [$\Delta A(7 \mu s)- \Delta A(150 ns)$] and L-minus-K difference spectra coincide with each other if one is shifted by 870 cm^{-1}. The spectra of K_{acid} and L_{acid} are slightly red-shifted from those of K and L, respectively.

On the other hand the following step after K is highly sensitive to the pH change. The behavior of L_{acid} was quite different from L in neutral pH suspension. L_{acid} has a longer lifetime than 700 μs. The formation efficiency of M from L_{acid} is too small to be estimated. It has been proposed that the deprotonation in Schiff base is induced by the dissociation of tyrosin neaby [21]. Our result is consistent with that the tyrosin dissociation is blocked in low pH suspension [22]. Presumably relevant proton-acceptor sites have already been protonated in L_{acid}. We could not clarify whether or not L_{acid} corresponds to a long-lived L reported in ref. [23].

3-2. Photocycle of $c-bR_{605}$

Both $t-bR_{605}$ and $c-bR_{605}$ are efficiently excited by 266-nm light. On the other hand the amount of excited bR_{568} in sample is negligibly small (see Fig. 1 and Table 1). Figure 4 shows the transient difference absorption spectra of the acidified sample excited by 266-nm light. An absorbance change maximum (630-640 nm) and an isosbestic point (610 nm) at 100 ns delay time are shifted to shorter wavelength from 630-nm excitation (curve 1 in Fig. 3). The shift could be due to the formation of an intermediate (C_{acid}) in the photocycle of $c-bR_{605}$. The difference spectrum between $\Delta A(7 \mu s)$ and $\Delta A(100 ns)$ is similar to that obtained by 630-nm excitation. Observed temporal changes in the 10 ns-7 us time region are attributed to the $K_{acid} \rightarrow L_{acid}$ conversion with a time constant of 1.2+0.2 μs. The concentration of C_{acid} does not change in this time region.

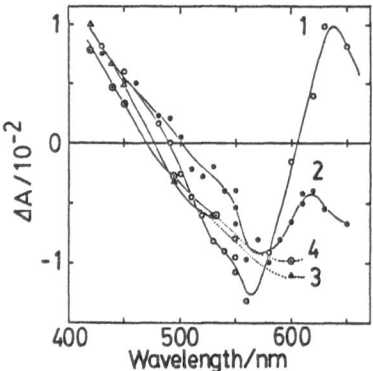

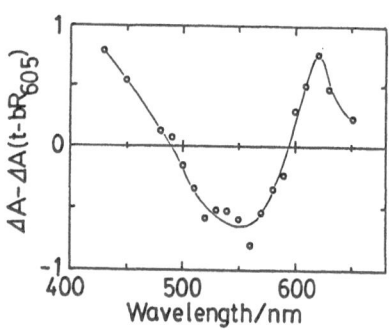

Fig. 4. Nanosecond difference absorption spectra 100 ns (curve 1), 7 μs (curve 2), 200 μs (curve 3), and 1.8 ms (curve 4) after excitation of bR_{605} at 266 nm.

Fig. 5. The difference absorption spectrum $\Delta A - \Delta A(t-bR_{605})$.

198

A difference absorption spectrum shown in Fig. 5 was obtained after correction for the absorbance change due to the photochemical process of t-bR$_{605}$. The spectrum in the 580-650 nm wavelength region is close to that in the photolysis of bR$_{548}$ at neutral pH. C$_{acid}$ corresponds to ^{610}C [24,25] and Batho-bR^{13} [26] intermediates in the photocycle of bR$_{548}$. The formation yield of C$_{acid}$ was estimated to be nearly equal to that of ^{610}C [25]. C$_{acid}$ does not decay as long as 200 μs after excitation (absorbance change maximum at 620 nm of curve 2 in Fig. 4). The spectroscopic and kinetic properties of the 13-<u>cis</u> pigment are less sensitive to pH than all-<u>trans</u> pigment.

An intense absorption was observed at wavelengths shorter than 450 nm (see Figs. 4 and 5). The M-like transient species (T) was not observed under 532- or 630-nm pulse irradiation. Time constants for both the decay of the absorbance change at 420 nm and the bleaching recovery at 600 nm were measured to be 6±1 ms. T may be a similar species which was formed from dark-adapted bacteriorhodopsin by a multiphoton process with visible light [25].

3-3. Photocycle of bR$_{568}$ contained in the blue sample

All components of bR$_{605}$ (t-bR$_{605}$, c-bR$_{605}$, and a small amount of bR$_{568}$) are excited by the 532-nm light (see Fig. 1 and Table 1). Figure 6 shows the transient difference absorption spectra 100 ns, 7 μs, and 90 μs after 532-nm excitation. The transient absorption spectrum at 100 ns is a superposition of the K$_{acid}$-minus-t-bR$_{605}$, KL-minus-bR$_{568}$, and C$_{acid}$-minus-c-bR$_{605}$ difference spectra. The difference in the absorbance change spectrum between ΔA(7 μs) and ΔA(100 ns) is given by a superposition of the L$_{acid}$-minus-K$_{acid}$ and L-minus-KL difference spectra. Both the time constants (1.3±0.2 μs) for the decay of absorption change in the 610-700

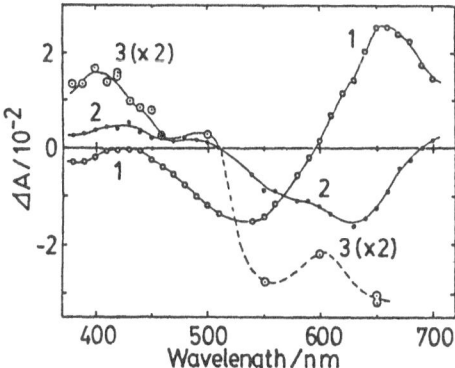

Fig. 6. Nanosecond difference absorption spectra 100 ns (curve 1), 7 μs (curve 2), and 90 μs (curve 3) after excitation of bR$_{605}$ at 532 nm.

nm region and the rise in the 460–500 nm region were close to those of the 630-nm excitation experiments (1.5±0.4 μs). Therefore the lifetime of the neutral form KL in the low pH suspension was found to be equal to that of K_{acid}.

The positive absorbance change in the 380–420 nm region shown by curve 3 in Fig. 6 is due to the formation of M, which is absent when the excitation wavelength was chosen at 630 nm. M could be formed by 532-nm excitation of the neutral form bR_{568} in the blue sample. The conversion time from L to M in low pH suspension was 32±10 μs, which is about 3 times shorter than in neutral pH (102±12 μs). The measured conversion efficiency from L to M (0.41) is lower than that at neutral pH. The low efficiency and the short conversion time constant (32±10 μs) are due to the opening of a new decay channel from L to $t-bR_{605}$ and/or $c-bR_{605}$.

4. Summary

The effect of acidic pH on the photocycle of bacteriorhodopsin are summarized as follows: (i) the decay rate of the lowest excited singlet state and (ii) the formation yield of K (K_{acid}) decreased. (iii) The formation of M is blocked.

In the early stages of the photochemical cycles for $t-bR_{605}$, $c-bR_{605}$, and bR_{568} are independently initiated by light. The observed transient difference spectra can be explained in terms of the superposition of changes caused in the three photochemical cycles shown in Fig. 7. L_{acid} and $c-bR_{605}$ have similar absorption spectra. If L_{acid} is actually $c-bR_{605}$, then $L_{acid} \rightarrow t-bR_{605}$ conversion process could correspond to the dark–adaptation of the acidified bR. The recovery time of $t-bR_{605}$ following 588-nm excitation of bR_{605} is 15±5 ms [9]. Thus the the fast dark–adaptation of bR_{605} can be explained.

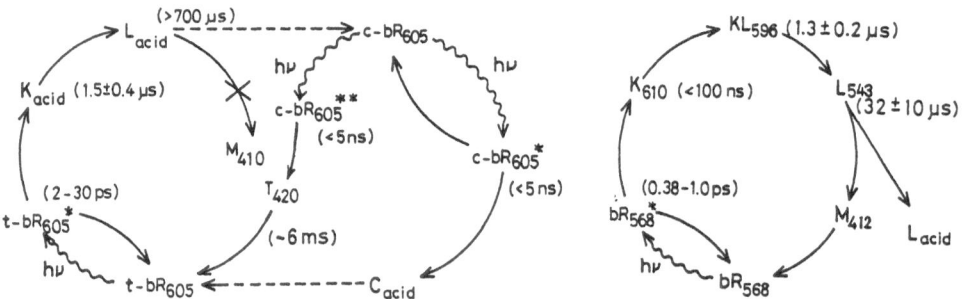

Fig. 7. Photochemical cycles of $t-bR_{605}$, $c-bR_{605}$, and bR_{568} in low pH suspension at 21±1 °C.

Acknowledgments

This work was partly supported by the following grants: a Grant-in-Aid for Special Distinguished Research (60115004) from the Ministry of Education, Science, and Culture, grants from Toray Science and Technology Foundation and Kurata Science Foundation to T.K., and Special Coordination Funds for the Promotion of Science and Technology from the Agency of Science and Technology of Japan to A.I. The authors thank Dr. Hisao Uchiki and Mr. Masayuki Yoshizawa, the University of Tokyo, for their assistance in data analysis.

References

1. D. Oesterhelt, W. Stoeckenius: Nature New Biol., 233, 149 (1971); W. Stoeckenius, R.H. Lozier, R.A. Bogomolni: Ann. Rev. Phys. Chem., 29, 31 (1979).

2. D. Oesterhelt, B. Hess: Eur. J. Biochem., 37, 316 (1973).

3. R.H. Lozier, R.A. Bogomolni, W. Stoeckenius: Biophys. J., 15, 955 (1975).

4. M.L. Applebury, K.S. Peters, P.M. Rentzepis: Biophys. J., 23, 375 (1978).

5. Y. Shichida, S. Matuoka, Y. Hidaka, T. Yoshizawa: Biochim. Biophys. Acta, 723, 240 (1983).

6. M. Tsuda, M. Glaccum, B. Nelson, T.G. Ebrey: Nature, 287, 351 (1980).

7. M. Braiman, R. Mathies: Biochemistry, 19, 5421 (1980); C.-L. Hsieh, M.A. El-Sayed, M. Nicol, M. Nagumo, J.-H. Lee: Photochem. Photobiol., 38, 38 (1983); S.O. Smith, A.B. Myers, J.A. Pardoen, C. Winkel, P.P.J. Murder, J. Lugtenburg, R. Mathies: Proc. Natl. Acad. Sci. USA, 81, 2055 (1984).

8. A. Lewis, J. Spoonhower, R.A. Bogomolni, R.H. Lozier, W. Stoeckenius: Proc. Natl. Acad. Sci. USA, 71, 4462 (1974); J. Turner, C.-L. Hsieh, A.R. Burns, M.A. El-Sayed: Proc. Natl. Acad. Sci. USA, 76, 3046 (1979); K.J. Rothschild, P. Roepe, J. Lugtenburg, J.A. Pardoen: Biochemistry, 23, 6103 (1984).

9. P.C. Mowery, R.H. Lozier, Q. Chae, Y.-W. Tseng, M. Taylor, W. Stoeckenius: Biochemistry, 18, 4100 (1979).

10. Y. Kimura, A. Ikegami, W. Stoeckenius: Photochem. Photobiol., 40, 641 (1984).

11. A. Maeda, T. Iwasa, T. Yoshizawa: Biochemistry, 19, 3825 (1980).

12. T. Kobayashi, H. Ohtani, J. Iwai, A. Ikegami, H. Uchiki: FEBS Lett., 162, 197 (1983); H. Ohtani, T. Kobayashi, J. Iwai, A. Ikegami: Biochemistry, 25, 3356 (1986).

13. D. Oesterhelt, W. Stoeckenius: Methods Enzymol., 31, 667 (1974).

14. J. Iwai, M. Ikeuchi, Y. Inoue, T. Kobayashi: in Protochlorophyllide Reduction and Greening (C. Sironval, M. Brouers, eds.) pp 99-112, Martinus Nijhoff / Dr. W. Junk Publishers, Hague (1984).

15. Yu.A. Matveetz, S.V. Chekalin, A.V. Sharkov: J. Opt. Soc. Am., B2, 634 (1985).

16. M.C. Nuss, W. Zinth, W. Kaiser, E. Kölling, D. Oesterhelt: Chem. Phys. Lett., 117, 1 (1985).

17. T. Kouyama, K. Kinosita, Jr., A. Ikegami: Biophys. J., 47, 43 (1985).

18. B. Becher, T.G. Ebrey: Biophys. J., 17, 185 (1977).

19. G.S. Hammond, J. Saltiel, A.A. Lamola, N.J. Turro, J.S. Bradshaw, D.O. Cowan, R.C. Counsell, V. Vogt, C. Dalton: J. Am. Chem. Soc., 86, 3197 (1964).

20. H.-J. Polland, M.A. Franz, W. Zinth, W. Kaiser, E. Kölling, D. Oesterhelt: Biophys. J., 49, 651 (1986).

21. O. Kalisky, M. Ottolenghi, B. Honig, R. Korenstein: Biochemistry, 20, 649 (1981).

22. P. Dupuis, T.C. Corcoran, M.A. El-Sayed: Proc. Natl. Acad. Sci. USA, 82, 3662 (1985).

23. T. Alshuth, M. Stockburger: Photochem. Photobiol., 43, 55 (1986).

24. N.A. Dencher, C.N. Rafferty, W. Sperling: Ber. Kernforsch., Nr.1374, 1 (1976).

25. O. Kalisky, C.R. Goldschmidt, M. Ottolenghi: Biophys. J., 19, 185 (1977).

26. T. Iwasa, F. Tokunaga, T. Yoshizawa: Photochem. Photobiol., 33, 539 (1981).

Structure of Bacteriorhodopsin and Halorhodopsin in Relation to the Pumping Function

A. Maeda

Department of Biophysics, Faculty of Science, Kyoto University, Kitashirakawa-Oiwake-cho, Sakyo-ku, Kyoto 606, Japan

1 Proton Pumping of Bacteriorhodopsin

Bacteriorhodopsin (BR) is a protein present in differentiated patches in the cellular membrane of a highly halophilic microorganism, Halobacterium halobium [1-3]. BR spans the cellular membrane and transports protons from the inside of the cell to its outer milieu upon receiving light energy into its retinylidene chromophore, which is linked to the lysine residue through the protonated Schiff base. The light-dependent proton pumping of BR is driven by two successive processes. The first is the release of proton from the membrane into the outer milieu to which the N-terminus of BR faces. The following is the uptake of proton from the opposite side of the membrane. As a whole, a concentration gradient for proton is established across the membrane. The present article summarizes the studies on the mechanism of energy conversion in BR.

BR gives a nice system for investigating the mechanism on the pumping process of proton in view of the presence of intermediates which are characterized well by the visible absorption and resonance Raman spectroscopic methods [1-6] (Fig.1). Each intermediate would reflect one of the consecutive steps for the pumping. The retinylidene chromophore of BR takes two different configuration at equilibrium in the dark [8,9]. The one shows the maximum at 548 nm with the chromophore retinal of 13,15-cis form [10] and the other at 568 nm with the all-trans chromophore. Bathochromic intermediates in the pico to nano second region are produced from both species of BR. In contrast, the blue shifted intermediates in the microsecond region, L(550) and M(412), arise only from the all-trans BR, an active species for the proton pumping. A similar situation is found with halorhodopsin (HR), another retinoid protein which transports chloride

BR WITH ALL-TRANS RETINAL
(ACTIVE)   BR(568) → → K(610) → → L(550) → → M(412) → → BR(568)

BR WITH 13-CIS RETINAL
(INACTIVE)   BR(548) → → K(608) —   ---   ---   → BR(568)

HR IN THE PRESENCE OF CHLORIDE
(ACTIVE)   HR(578) → → HR(630) → → HR(520) → → HR(640) — → HR(578)

HR IN THE ABSENCE OF CHLORIDE
(INACTIVE)   HR(567) → → HR(640) —   ---   ---   → HR(567)

RHODOPSIN (CATTLE)
    RH(498)   BATHO(543) METAI(478) METAII(380)

Fig. 1. Photochemical intermediates arising from all-trans and 13-cis BR, along with those of HR in the presence and absence of chloride. Only those concerned with discussion are listed. Intermediates from bovine rhodopsin [7], which are assumed to correspond to those intermediates of the pump proteins, are also listed. Number in parenthesis is the approximate wavelength for the maximum absorption. The same notation is used in the text.

in the opposite direction [3,11]. Its blue-shifted intermediate, HR(520), that corresponds to L(550) in BR, is produced only when chloride, a substrate anion of HR, is present [12,13]. In the photoreaction of inactive 13,15-cis BR [14] or HR in the absence of chloride [15], the bathochromic product returns back without circumvention of the blue shifted intermediates. These facts implicate an important role of the blue shifted intermediates in the pumping process of these retinoid proteins.

2 Release of Proton

The Fourier transform IR studies by ENGELHARD et al. [16] have suggested that the first step for the release of proton is triggered by the interaction of the protonated Schiff base in the photo-isomerized chromophore with carboxylate A_3^- (Fig.2b), which was present as protonated carboxylic acid before the isomerization of the chromophore. The dissociation of proton occurs with an accompanying formation of the blue-shifted intermediate, L(550). These events conversely result in the loss of the interaction of the protonated Schiff base with its previous partner in BR(568) shown as A_1^- in Fig.2a. L(550) having the 13-cis chromophore is similar to BR(548) with respect to visible absorption spectrum and C=C and C-C stretching frequencies but has some special features in terms of the absence or the shift of 1348-cm^{-1} line due to the N-H bending vibration and an appearance of the deuteration-sensitive 1446-cm^{-1} line [17]. These changes would reflect a different mode of the interaction between the chromophore and the protein but elucidation for these are now in progress at our hands.

In order to understand the role of the protein residues interacting with the protonated Schiff base, the pH dependence of the photoreaction at alkaline pH was

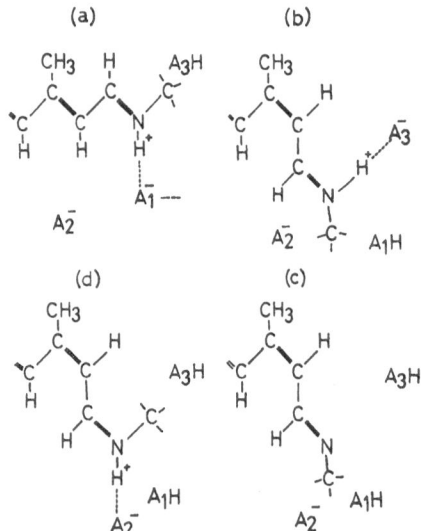

Fig. 2. Changes in the interaction of the protonated Schiff base during the photocycle of BR. Only the part from C_{12} of retinal to εC of lysine was shown for simplicity. (a); unphotolyzed BR(568) with all-trans chromophore. (b); L(550) with 13-cis chromophore. (c); M(412) with unprotonated 13-cis chromophore. (d); a postulated intermediate in the way from M(412) to the regenerated BR(568), having 13,15-cis chromophore. The residues indicated by A_1, A_2 and A_3 are those in anionic states. A_1^- is the counterion to the unphotolyzed BR(568) when the chromophore takes all-trans configuration (a). A_2^- is the counterion to the unphotolyzed BR(548) with the 13-cis chromophore. It was also supposed to be a counterion in a postulated intermediate (d). Anionic A_3^- is only present in L(550) (b). See the text for details.

studied by resonance Raman spectroscopy. The long-lived photoproduct of BR which accumulates in the spinning cell at alkaline pH under intense CW laser [18] does not exhibit the 1348 cm^{-1} line of N-H bending vibration but exhibits the 1188 cm^{-1} line that is also characteristic of L(550). The 1446 cm^{-1} line instead of the 1456 cm^{-1} line is another feature common to L(550) and the alkaline photoproduct. These facts indicate that the photoproduct at alkaline pH interacts with the protein, analogously to those in L(550) under these conditions, while keeping its Schiff base protonated. pH dependence of the formation of this photoproduct that is stable at alkaline pH shows that the dissociation of a protein residue with a pk$_a$ value of 9 is responsible for. Corresponding change in the protein in terms of the UV absorption spectrum indicates that the residue involved is not tyrosine; a signal related to the dissociation of tyrosine is observed only in the region above pH 11. Resonance Raman studies with acetylated BR, in which all the free lysine residues are acetylated [19], exclude a possibility for the involvement of lysine. Only a protonated aspartic acid that was shown to be present in BR by ENGELHARD et al. [16] could be conceivable as a candidate. When protonated, this pK$_a$=9 residue acts to release proton. However, the long-lived intermediate is produced under its unprotonated state. The UV absorbance change observed at alkaline pH is a reflection of perturbation on the indole ring of tryptophan due to the appearance of a negative charge of the carboxylate arisen from the pK$_a$=9 aspartic acid described here.

As discussed above, proton liberation is forced from the carboxylic acid by the cationic protonated Schiff base that approached closely by light-induced isomerization of the chromophore. This is a short-circuiting between the insulated carboxylic acid in hydrophobic environment and the cationic protonated Schiff base. The carboxylate thus dissociated in the hydrophobic environment soon accepts proton from the protonated Schiff base and regains the form of protonated carboxylic acid. M(412) having the unprotonated Schiff base is produced conversely [16] (Fig.2c).

In this context, stabilization of the L-like intermediate at alkaline pH is puzzling in view of the fact that the rate for the conversion of L(550) to M(412) is very rapid at alkaline pH [20-22]. Its pH dependence follows a titration-like curve with a midpoint for the conversion at around pH 9. Accumulated L-like intermediate at alkaline pH in the spinning cell is relatively long-lived intermediate different from L(550) and may have a similar absorption spectrum to that of all-trans BR [18]. Recently, DRACHEV et al. described [23] the presence of an intermediate called P that arises before the conversion of M(412) to BR(568) in the final step of the photocycle. This shows the maximum absorption at 570 nm and seems to become long-lived under alkaline conditions.

Rhodopsin in eyes also shows similar changes in the IR spectrum of the carbonyl band upon its conversion to metarhodopsin I [24]. It is interesting that metarhodopsin I is stable even after dehydration that causes the deprotonation of the Schiff base of the unphotolyzed rhodopsin [25]. This could be understood if the carboxylic acid interacting with the isomerized protonated Schiff base is present in hydrophobic environment as the pK$_a$=9 aspartic acid in BR. A process analogous to that in BR may be working also in rhodopsin, though not leading to proton pumping. Uptake of proton is shown to occur at the stage of metarhodopsin II having the unprotonated Schiff base [26].

3 Highly Stable Protonated Schiff Base of BR

A subsequent step for the rebinding of proton from the opposite side starts at M(412), which lacks its Schiff base proton. The reprotonation of M(412) proceeds in biphasic kinetics with a fast component and a slow one [27-29]. The uptake of proton from the outside of the membrane is slower than the decay rate of the fast component of M(412) but is very similar to the rate for the slow component of M(412). The slower component has been shown to relate intimately to the vectorial proton transport process [30,31]. In other words, the recovery of the system surrounding the protonated Schiff base must be necessary for the uptake of proton. It is therefore important to analyze the system surrounding the protonated Schiff base in unphotolyzed BR.

Alkaline titration of all-trans BR shows no significant spectral shift up to pH 12, above which some blue shifted absorption band appears in the region around 460 nm. DRUCKMANN et al. have shown [32] that the Schiff base of the alkaline form of BR, BR(460), is unprotonated in view of the resonance Raman spectrum at liquid nitrogen temperature. They estimated its pK_a value to 13.3.

Two additional interesting results on the stability of the protonated Schiff base have been obtained. One is the anion dependence of the Schiff base pK_a value of HR [33,34]. When monovalent anion is deleted from HR, its Schiff base cannot keep the proton above pH 8. On readdition of monovalent anion, its pK_a value is increased to around 9; nevertheless the protonated Schiff base is very unstable in comparison with that of BR. SCHOBERT and LANYI have shown [35] that anion exerts its effect through electrostatic coupling on the protonated Schiff base. These results imply that an increase by about two units of pK_a value of the protonated Schiff base can be attained by electrostatic interaction with anions like chloride. However, several other interacting systems or the anion fixed tightly in the protein are required for further increase of the pK_a value of the Schiff base to a higher value such as that attained with BR.

4 13-Cis Chromophore is More Stable

The other is that the deprotonation of the Schiff base occurs preferentially from the all-trans chromophores as shown by retinal analogs having the aromatic ring instead of the ionone ring [36]. Upon binding to apoprotein of BR, all-trans phenyl retinal first gives a blue shifted absorption band around 480 nm; it then converts slowly to the red shifted band around 520 nm, eventually leading to a structured absorption spectrum with two peaks at 480 and 520 nm. The latter is produced at the expense of the former. Upon irradiation of the 480-nm band by blue light, the spectral shape turns to be smooth with the maximum at 510 nm and the chromophore isomerizes to the 13-cis form. In contrast, the 13-cis form gives a smooth shape with a single band at 510 nm both in the dark and in the light. The alkaline titration of all-trans phenyl BR reveals that its 480 nm absorption band is the unprotonated form derived from the 520 nm absorption band. In accordance with this, an irreversible uptake of the proton occurs upon irradiation of all-trans phenyl BR, while not of its 13-cis form. Similar trends were observed with other aromatic BR analogs.

In the two absorption bands of all-trans aromatic bR, the red-shifted one is supposed to be the form with the protonated Schiff base and the blue-shifted one with the unprotonated Schiff base. Both exists together at neutral pH. A single absorption band due to the 13-cis protonated Schiff base emerges between these two. The protonated Schiff base is thus more stable for the 13-cis chromophore than for the all-trans chromophore in the case of aromatic BR.

As stated above, BR with the native chromophore dissociates its Schiff base proton only above pH 12. Even with the native chromophore, however, LEMKE and OESTERHELT [37] have shown that the blue shift in the absorption spectrum presumably due to the deprotonation of the protonated Schiff base occurs in the pH region around 9, when the protein part was modified with tetranitromethane. A larger shift was obtained for light-adapted form of nitrated BR than for its dark-adapted form [38], indicating that the deprotonation of the Schiff base in the relatively weak alkaline pH region occurs preferentially from the all-trans chromophore rather than from the 13-cis one. It should be noticed that the effect is probably not a direct consequence of nitration of tyrosine since the effect is observed even after reduction of nitrated BR by dithionite. SCHERRER and STOECKENIUS [39] have reported that selective nitration of tyrosine 26 reversibly causes the blue shift in the absorption spectrum with a midpoint for the conversion at pH 10.5. The mechanism for the pK_a suppression by nitration of tyrosine 26 may be different from a large pK_a decrease for heavily nitrated BR described above.

These results strongly suggest that the Schiff bases of the all-trans and 13-cis chromophores interact differently with the protein residues, respectively. Recently, ROTHSCHILD et al. [40] have proposed an idea that the counterion of the protonated Schiff base is tyrosinate. For the stabilization of tyrosinate, an H-bond network is required. HILDEBRANDT and STOCKBURGER [41] have suggested the importance of solvation for the stabilization of the protonated Schiff base of the all-trans chromophore. Dehydration of purple membrane enforces the Schiff base to unprotonated form along with the conversion to the 13-cis form. Several stabilizing protein residues along with bound water may be involved for the interaction with the all-trans chromophore. In contrast, the 13-cis chromophore can stay stably without circumvention of these residues.

5 Two Anionic Charges Around the Protonated Schiff Base

These results can well be explained in terms of the presence of different counteranions to both the all-trans and the 13,15-cis chromophore of BR, respectively. These were denoted as A_1^- and A_2^- in Fig.2. A_1^- may be tyrosinate but A_2^- must be carboxylate. Existence of such two anions close to the protonated Schiff base has previously been shown on the basis of theoretical calculations by YOSHIHARA et al. [42] which intended to account for the data of the absorption spectra of several species in different isomeric states that are produced in acid [43,44]. The model proposed [42] is also compatible with a recent proposal for the absence of the isolated negative charge close to the ionone ring [45].

BR is known to turn its color blue by lowering the pH value below 3, yielding a red-shifted species, BR(605) [1,44,47]. Since the rate for the isomerization is quite rapid at acidic pH [43, 48], it can only be treated as the mixture of both all-trans and 13-cis species. Irradiation of BR(605) at 0°C with red light causes an extensive blue shift in absorption spectrum. Isomerization of the chromophore to 11-cis and 9-cis forms occurs successively [43], having the maximum absorbance at 565 and 495 nm, respectively.

Photoreaction of BR(605) at -72°C yields its blue shifted products [44], which regenerate the original BR(605) upon warming. Its difference spectrum shows a shape somehow different from the difference spectrum in the formation of 9-cis BR at 0°C. The chromophore did not isomerize to neither 9-cis nor 11-cis form. The intermediate thus produced is quite similar to L(acid) reported by OHTANI et al. [49] in view of the spectral parameters and the rate for its formation. They further showed that L(acid) is produced only from BR(605) having the all-trans chromophore but not from BR(605) having the 13-cis chromophore.

Among the two negative charges in the model, the one which is present in an anionic form even at pH 2 can be located close to C_{12} of retinal [42]. This place is too far away to interact with the protonated Schiff base of the all-trans chromophore and seems to correspond to A_2^- shown in Fig.2, the counterion to the protonated Schiff base of the 13,15-cis chromophore. Another negative charge, which couples with its positive countercharge and is protonated at pH 2, could come to the position of the protonated Schiff base of the all-trans chromophore at neutral pH. This could be A_1^- in Fig.2.

The isomerized photoproduct at acidic pH with either 9-cis (the max. at 495 nm), 11-cis (the max. at 565 nm) or 13-cis (the max. at 570 nm) [49] chromophores shows relatively blue shifted absorption spectrum. It is compatible with an idea that the protonated Schiff base of these photoproducts becomes free from the interaction with either A_1^- or A_2^- and acquires another interaction with aspartic acid of $pK_a=9$, A_3^-, like the interacting system in L(550). All these experimental results can also be accounted well by a model based on the theoretical calculation by YOSHIHARA et al. [42].

6 The Presence of Ionic Interaction in the Membrane Matrix

Interaction of A_1^- with the charged residues was studied by means of spectroscopic titration of BR treated with acetic anhydride. Rapid acetylation occurs in four to five free lysine residues out of six but a longer duration is required for complete acetylation [19]. The midpoint for the conversion to the blue membrane is increased by 1.4 pH units for acetylated BR in which all six free lysine residues are modified almost completely. The increase is not observable with ethylated BR, with which the lysine residues are modified by keeping positive charges, indicating an importance of the electrostatic interaction with the anion involved in color regulation.

The pH value for the conversion to the blue membrane depends on the ionic strength of the medium; a higher pH value with decreasing the ionic strengths of the medium. Ionic strength-dependence largely arises from the change in surface potential on the membrane [50]. For example, simple dialysis to distilled water is enough for developing blue color with acetylated BR because of the accumulation of proton in close proximity to the membrane surface due to negative surface potential. The pH values for the conversion to the blue membrane with acetylated BR are consistently higher than with native BR, independently of the ionic strengths. It is implied that the acetylatable residue is interacting with the anion electrostatically and causes the decrease in the pK_a value of the residue involved for the conversion to the blue membrane. The paired charges are located in the places insulated from the outer milieu. These results are consistent with a notion provided above for the presence of the negative charge, A_1^-, that is interacting with cationic residue.

At extremely acidic pH, at pH below 1, BR(605) again reverts to original purple color, yielding BR(565), probably as a result of the protonation of both A_1^- and A_2^-. The formation of BR(556) is dependent on the presence of chloride; perchlorate is not effective [51]. Chloride seems to bind to the site containing the lysine that is less accessible for acetylation by acetic anhydride [17].

7 Reprotonation of the Schiff Base

Isomerization of the chromophore by light breaks the interaction of A_1^- with the protonated Schiff base. This may lead to the protonation of A_1^-. Difference spectrum when BR is photolyzed [52-55] shows a deep trough around 275 nm and the maxima at 230 nm and 300 nm. This could not be due to the deprotonation of tyrosine, since the difference spectrum of tyrosine dissociation does not give such a deep trough around 275 nm. It would reflect a change of an electrical environment around the protonated Schiff base but has not so far been analyzed well. Protonation of A_1^-, the protonatable counterion, is expected to be responsible.

SMITH et al. [56] have argued that L(550), the antecedent to M(412), has the chromophore with 13-cis retinal Schiff base (Fig.2b). The same configuration is therefore supposed to be kept in M(412). They have further provided an idea that the reprotonation of the Schiff base will occur in the form of 13,15-cis (Fig.2d), the same form as bR(548), which is different from the 13-cis form in L(550). It was assumed that BR uses two different counterions, A_1^- and A_2^- for stabilizing the protonated Schiff base of the all-trans and 13-cis chromophores, respectively. A_2^- can stabilize the protonated Schiff base in the 13-cis chromophore without other interacting residues, in contrast to A_1^-, which needs further supporting residues for maintaining its stabilizing ability (Fig.2). In M(412), A_1^- was supposed to be protonated while A_2^- is not.

M(412) at neutral pH is produced by donating the Schiff base proton to the $pK_a=9$ aspartic acid. For its reprotonation, therefore, an establishment of a new interaction with another anionic residue is required. At the stage of M(412), A_1^- was inferred to be protonated. Even if not, it could be weak as the counterion for protonating the Schiff base. A_2^- would work as a strong anion for protonating the Schiff base by interacting with it (Fig.2). In order to attain the interaction with A_2^-, the retinal chromophore has to take 13,15-cis configuration like the chromophore in BR(548). In other words, the presence of A_2^- affords motivity for the isomerization of the C=N bond or the inversion at the nitrogen center postulated by SMITH et al. [56].

There are two forms of the M intermediate which can be distinguished from each other by the rate for the reprotonation of the Schiff base. The fraction of a slowly decaying component is predominant at neutral pH, whereas that of a fast decaying one at alkaline pH [28,31]. In the alkaline pH region, the $pK_a=9$ aspartic acid cannot work as a stable acceptor of the proton from the Schiff base. The Schiff base in M(412) can readily be reprotonated in virtue of the interaction with the dissociated form of the $pK_a=9$ aspartic acid, by keeping the form shown in Fig. 2c. No isomerization or the inversion at nitrogen is required, making possible for the rapid decay of M(412) at alkaline pH.

One of the consequences of this mechanism in the alkaline pH region would be an inability of the one way translocation of the proton. In this model, the protonated state of the $pK_a=9$ residue is critical in order to assure the vectorial transport of proton. This is in accordance with a recent observation for no vectorial proton transport at alkaline pH during the photocycle [57,58].

As a whole, light causes relaxation of both perturbed residues of aspartic acid with $pK_a=9$ and the protonated Schiff base with $pK_a=13.3$. These have unusually high pK_a values in the resting state. A_1^- also seems to be a perturbed residue if this is taken to be tyrosinate as suggested by ROTHSCHILD et al. [40]. The perturbated system is regenerated at the final step in the photocycle and serves to the next photocycle. It has been pointed out that electrically perturbed amino acid residues in the active site of enzyme and their relaxation upon substrate binding are important in considering the mechanism of enzyme action [59].

8. Structure of Halorhodopsin as Compared with Bacteriorhodopsin

As discussed above, the protonated Schiff base of BR may have three kinds of interaction with the protein residues. In the unphotolyzed all-trans state, it has an interaction with the protein residue, A_1^-, which has another electrostatic interaction with other residues (Fig.2a). This is in contrast to the partner of the 13,15-cis chromophore, A_2^-, which seems not to require other interations. The photolyzed BR changes the partner to the dissociated $pK_a=9$ aspartic acid, A_3^- (Fig.2b). This residue is present in the protonated state, A_3H, before the photoreaction. It would be helpful for considering the mechanism of chloride transport in HR in order further to understand the mechanism underlying in BR.

In contrast to BR, HR is present in the cytoplasmic membranes of the cell. Extraction by detergents and several steps of column chromatography are required in order to obtain a purified sample, especially free from the cytochromes that disturb the spectral measurements. We have extracted HR in Triton X-100 and separated it from other proteins by several repeated processes of column chromatography on octylsepharose [60]. In contrast to BR, the presence of salt is indispensable for maintaining its structure for the purified sample.

Experiments for analyzing the effect of salt reveal the importance of anionic species that affect the wavelength for maximum absorption of HR [11,60]. The maximum is found at 567 nm when chloride was replaced with nitrate. The addition of chloride or bromide to it causes the red shift in the spectrum, shifting the maximum to 576 nm. Thus a red shift by about 10 nm is resulted from the binding of chloride. Recently Lanyi has shown [6] that the shift is only resulted from HR having the all-trans chromophore. Any chloride-dependent red shift has not been obtained with the 13-cis form of HR. Chloride also affects the photoreaction of HR. The blue shifted intermediate, HR(520), is produced only in the presence of chloride [11,12]. The unprotonated intermediate of HR can be produced but less efficiently [61,62] in comparison with M(412) for BR.

HR has therefore at least two sites for the binding of chloride. The one is the site for the interaction in stabilizing the protonated Schiff base as discussed earlier. This will be referred to as Site 1. The effect is supposed to be electrostatic but may be exerted through conformation change of the protein [35]. Either chloride (and bromide) or non-halide anions are effective. The other is Site 2 specific to chloride (and bromide) and affects the photoreaction by producing HR(520) and the absorption spectrum of HR.

The specific effect of chloride was examined by resonance Raman spectroscopic method [63]. These spectra of both unphotolyzed HR in the presence of chloride, HR(576), and in its absence but in the presence of nitrate, HR(567), are compared with each other. Only a significant difference observed was a decrease in the frequencies of the C=NH stretching vibration from 1642 to 1635 cm^{-1} without accompanying changes with respect to the C=ND streching frequency at 1622 cm^{-1}. This has been interpreted as that chloride bound to the specific site (Site 2) does not interact with the protonated Schiff base by H-bonding [63]. The frequency of the C=NH stretching vibration of HR(576) in the presence of chloride at 1635 cm^{-1} is also lower than that of BR at 1642 cm^{-1} [63,64]. It is interesting to note that the wavelength for the maximum absorption in the absence of chloride is almost the same as that of BR. HR complexed with chloride is very unique in affecting the chromophore at the place close to the Schiff base.

9 An Implication for the Pumping Mechanism

It seems that the chloride at Site 1 in HR replace the site for the counteranion in BR, A_1^-. It is supposed that, upon light-induced isomerization, the protonated Schiff base leaves from the chloride at Site 1 and get another interaction with chloride at Site 2. This step is analogous to the process of BR(568) to L(550). In support of this, the resonance Raman spectrum of HR(520) is very similar to that of the L(550) of BR; the absence or the shift or the 1352 cm^{-1} line for the N-H bending, the presence of 1191 cm^{-1} line in the C-C stretching region and a twin bands of the C=C stretching vibrations at 1543 and 1551 cm^{-1} [22]. HR(520) is thus expected to have very similar structure to L(550) in BR. The chloride in Site 2 would then correspond to the pKa=9 aspartic acid. Different from BR is that the chloride thus concerned has also some interaction with the unphotolyzed chromophore in virtue of its effect on the C=NH stretching as discussed above.

The specific chloride would be located relatively close to the C=N bond because of its effect on the C=NH streching frequency and the different effect on the absorption spectrum depending on the all-trans and 13-cis chromophore configurations. This chloride may then give some interaction with the protonated Schiff base in HR(520) and move with an accompanied isomerization of the chromophore to Site 1. This would complete the process for the chloride pumping in the direction opposite to the proton pumping in BR. HR next resets the interaction with the chloride in Site 2 by taking it from the other side.

ACKNOWLEDGEMENTS

A part of this work was supported by a grant-in-Aids for Special Project Research on Molecular Mechanism of Bioelectric Response (59123002) from the Japanese Ministry of Education, Science and Culture.

REFERENCES

1. D. Oesterhelt & W. Stoeckenius: Nature New Biol. 233, 149 (1971)
2. W. Stoeckenius, R. H. Lozier & R. A. Bogomolni: Biochim. Biophys. Acta 505, 215 (1979)

3. W. Stoeckenius & R. A. Bogomolni: Annu. Rev. Biochem. 52, 587 (1981)
4. S. O. Smith, J. Lugtenburg & R. A. Mathies: J. Membrane Biol. 85, 95 (1985)
5. Y. Shichida, S. Matuoka, Y. Hidaka & T. Yoshizawa: Bioch m. Biophys. Acta 723, 240 (1983)
6. J. K. Lanyi: J. Biol. Chem. 261, 14025 (1986)
7. T. Yoshizawa: Ad. Biophys. 17, 5 (1984)
8. A. Maeda, T. Iwasa & T. Yoshizawa: J. Biochem. 82, 1599 (1977)
9. S. Seltzer & R. Zuckermann: J. Am. Chem. Soc. 107, 5523 (1985)
10. S. O. Smith, A. B. Myers, J. A. Pardoen, C. Winkel, P. P. J. Mulder, J. Lugtenburg & R. A. Mathies: Proc. Natl. Acad. Sci. U. S. A. 81, 2055 (1984)
11. J. K. Lanyi.: Annu. Rev. Biophys. Chem. 15, 11 (1986)
12. T. Ogurusu, A. Maeda, N. Sasaki & T. Yoshizawa: Biochim. Biophys. Acta 682, 446 (1982)
13. M. Steiner, D. Oesterhelt, M. Ariki & J. K. Lanyi: J. Biol. Chem. 259, 2179 (1984)
14. T. Iwasa, F. Tokunaga & T. Yoshizawa: Photochem. Photobiol. 33, 539 (1981)
15. J. K. Lanyi & V. Vodyanoy: Biochemistry 25, 1465 (1986)
16. M. Engelhard, K. Gerwert, B. Hess, W. Kreutz & F. Siebert: Biochemistry 24, 400 (1985)
17. T. Alshuth & M. Stockburger: Photochem. Photobiol. 43, 55 (1986)
18. A. Maeda, T. Ogura & T. Kitagawa: Biochemistry 25, 2798 (1986)
19. A. Maeda, Y. Takeuchi & T. Yoshizawa: Biochemistry 21, 4479 (1982)
20. A. Lewis, M. A. Marcus, B. Ehrenberg & H. Crespi: Proc. Natl. Acad. Sci. U. S. A. 75, 4642 (1978)
21. V. Rosenbach, R. Goldberg, C. Gillon & M. Ottolenghi: Photochem. Photobiol. 36, 197 (1982)
22. T. T. Ogura, A. Maeda, M. Nakagawa & T. Kitagawa; in this proceeding
23. L. A. Drachev, A. D. Kaulen, V. P. Skulachev & V. V. Zorina: FEBS Lett. 209, 316 (1986)
24. W. J. de Grip, J. Gillespie & K. J. Rothschild: Biochim. Biophys. Acta 809, 97–106 (1985)
25. C. N. Rafferty & H. Shichi: Photochem. Photobiol. 33, 229 (1981)
26. U. B. Kaupp, P. P. M. Schnetkamp & W. Junge: Biochemistry 20, 5500 (1981)
27. R. H. Lozier, W. Niederberger, R. A. Bogomolni, S. -B. Hwang & W. Stoeckenius: Biochim. Biophys. Acta 440, 545 (1976)
28. D. R. Ort & W. W. Parson: J. Biol. Chem. 253, 6158 (1978)
29. K. Ohno, Y. Takeuchi & M. Yoshida: Photochem. Photobiol. 33, 573 (1981)
30. K. Ohno, R. Govindjee & T. G. Ebrey: Biophys. J.: 43, 251 (1984)
31. Q. -Q. Li, R. Govindjee & T. G. Ebrey: Proc. Natl. Acad. Sci. U. S. A. 81, 7079 (1984)
32. S. Druckmann, M. Ottolenghi, A. Pande, J. Pande & R. H. Callender: Biochemistry 21, 4953 (1982)
33. T. Ogurusu, A. Maeda & T. Yoshizawa: unpublished experiments
34. B. Schobert, J. K. Lanyi & D. Oesterhelt: J. Biol. Chem. 261, 2690 (1986)
35. B. Schobert & J. K. Lanyi: Biochemistry 25, 4163 (1986)
36. A. Maeda, A. Asato, R. S. H. Liu & T. Yoshizawa: Biochemistry 23, 2507 (1984)
37. H. -D. Lemke & D. Oesterhelt: Eur. J. Biochem. 115, 595 (1982)
38. A. Maeda: unpublished experiments
39. P. Scherrer & W. Stoeckenius: Biochemistry 23, 6195 (1984)
40. K. J. Rothschild, P. Roepe, P. L. Ahl, T. N. Earnest, R. A. Bogomolni, S. K. Das Gupta, C. M. Mulliken & J. Herzfeld: Proc. Natl. Acad. Sci. U. S. A. 83, 347 (1986)
41. P. Hildebrandt & M. Stockburger: Biochemistry 23, 5539 (1984)
42. T. Yoshihara, H. Suzuki & A. Maeda: Photochem. Photobiol 33, 501 (1981)
43. A. Maeda, T. Iwasa & T. Yoshizawa: Biochemistry 19, 3825 (1980)
44. A. Maeda, T. Iwasa & T. Yoshizawa: Photochem. Photobiol. 33, 559 (1981)
45. J. Lugtenburg, M. Muradin-Szweykowska, C. Heeremans, J. A. Pardoen, G. S. Harbison, J. Herzfeld, R. G. Griffin, S. O. Smith & R. A. Mathies: J. Am. Chem. Soc. 108, 3104 (1986)
46. T. A. Moore, M. E. Edgerton, G. Parr, C. Greenwood & R. N. Perham: Biochem. J. 171, 469 (1978)
47. P. C. Mowery, R. H. Lozier, Q. Chae, Y. -W. Tseng, M. Taylor & W. Stoeckenius: Biochemistry 18, 4100 (1979)

48. K. Ohno, Y. Takeuchi & M. Yoshida: Biochim. Biophys. Acta 462, 575 (1977)
49. H. Ohtani, T. Kobayashi, J. Iwai & A. Ikegami: Biochemistry 25, 3356 (1986)
50. C. -H. Chang, R. Jonas, S. Melchiore, R. Govindjee & T. G. Ebrey: Biophys. J. 49, 731 (1986)
51. U. Fischer & D. Oesterhelt: Biophys. J. 28, 211 (1979)
52. C. N. Rafferty: Photochem. Photobiol. 29, 109 (1979)
53. R. A. Bogomolni, L. Stubbs & J. K. Lanyi: Biochemistry 17, 1037 (1978)
54. B. Hess & D. Kuschmitz: FEBS Lett. 100, 334 (1979)
55. J. K. Lanyi: FEBS Lett. 175, 337 (1984)
56. S. O. Smith, I. Hornung, R. v. d. Steen, J. A. Pardoen, M. S. Braiman, J. Lugtenburg & R. A. Mathies: Proc. Natl. Acad. Sci. U. S. A. 83, 967 (1986)
57. Q. Li, R. Govindjee & T. G. Ebrey: Photochem. Photobiol. 44, 515 (1986)
58. T. Kouyama, A. N-. Kouyama & A. Ikegami: in this proceeding
59. D. M. Blow & T. A. Steitz: Annu. Rev. Biochem. 39, 63 (1970)
60. T. Ogurusu, A. Maeda & T. Yoshizawa: J. Biochem. 95, 1073 (1984)
61. P. Hegemann, D. Oesterhelt & M. Steiner: EMBO J. 4, 2347 (1985)
62. T. Ogurusu, A. Maeda, N. Sasaki & T. Yoshizawa: J. Biochem. 90, 1267 (1981)
63. A. Maeda, T. Ogurusu, T. Yoshizawa & T. Kitagawa: Biochemistry 24, 2517 (1985)
64. S. O. Smith, M. J. Marvin, R. A. Bogomolni & R. A. Mathies: J. Biol. Chem. 259, 12326 (1984)

Picosecond Intermediates
in the Bacteriorhodopsin Photocycle

G.H. Atkinson

Department of Chemistry and Optical Sciences Center,
University of Arizona, Tucson, AZ 85721, USA

The molecular mechanism(s) by which bacteriorhodopsin (BR) utilizes the radiative
energy in the visible region absorbed by the retinal chromophore is of major importance in characterizing the bacterial photocycle and the associated transport of
protons across the BR membrane. The changes in retinal structure, electronic properties, and conformation which occur within the first 100 ps following excitation
are examined here using time-resolved resonance Raman (TR^3) and fluorescence (TRF)
spectroscopies with 5 ps time resolution. TR^3 spectra identify at least one major
change in retinal structure (all-trans to 13-cis isomerization) occurring during the
first 40 ps of the reaction, while TRF spectra reveal previously unreported fluorescence from a BR intermediate. These TR^3 and TRF data are used to construct a model
for retinal dynamics over the initial 100 ps interval of the BR photocycle.

1. Introduction

Chromophoric groups with the linear polyene structure are found in an extremely wide
range of biologically important systems. The retinal chromophores found in rhodopsin and bacteriorhodopsin (BR) and the carotenoid chromophores present in chlorophyll and bacteriochlorophyll are the two most commonly recognized examples. The
ubiquity with which polyenes appear in biological systems that have very different
functional roles suggests that the photo-induced structural and conformational
mechanisms in polyenes are sufficiently complex to fulfill a diverse set of biophysical requirements. As a consequence, the photochemistries of polyene chromophores have been examined extensively in recent years by many different experimental
methods.

The contribution of retinal in the BR photocycle is the subject treated in this
paper. Specifically, the changes in the molecular structure, electronic properties,
and conformation of retinal during the first 100 ps of the BR photocycle are examined. It is during the initial time period that the retinal chromophore utilizes
the absorbed visible radiation to drive the BR photocycle. It has been established
in previous work that both proton and ion transport occurs in the Halobacterium
halobrium system during the photocycle[1-3] and therefore, that the light-driven
cycle must create a substantial chemical potential across the membrane of the bacterium. The molecular mechanism by which retinal initiates this chemical potential
can be observed by monitoring the structural, electronic, and conformational transformations occurring in BR during this initial period.

At least three types of time-resolved spectroscopies with picosecond resolution
have been used recently to study the BR photocycle. Picosecond time-resolved resonance Raman (PTR^3) and picosecond time-resolved fluorescence (PTRF) spectroscopies
are described here. It is important, however, to correlate these results with those
reported from picosecond transient absorption (PTA) spectroscopy reported by other
workers[4]. Indeed, it should be emphasized that the molecular mechanism underlying
the BR photocycle can be clearly viewed only when all such measurements are inte-
grated into one model.

PTR[3] experiments are only recently available and yet may provide the most detailed view of the molecular mechanism since these data can monitor small changes in conformation and structure not readily observed in either PTRF or PTA spectroscopies. It is through these changes in conformation and structure particularly that retinal is thought to utilize the absorbed radiation. New results described here from PTR[3] and PTRF experiments support this preception.

2. Experimental

2.1 PTR[3] Experiments

A variety of experimental techniques have been used to record the vibrational Raman spectrum of transient, reaction species. The experimental requirements associated with initiating the reaction are often the determining factors in the choice of techniques. For example, when the absorption spectra of the transient species and the starting compound overlap, then one pulsed laser source may be used for TR[3] spectroscopy. In this case, one laser is used to both photolytically initiate the reaction and to generate the resonance Raman scattering from the resultant mixture of species[5,6]. There are several limitations to this experimental approach, not the least of which is that all transient species present during irradiation may contribute to the resultant TR[3] spectrum. The advantages and limitations of TR[3] spectroscopy using a single source have been described elsewhere[6,7]. Within the limitations of this single source technique, reaction intermediates with picosecond lifetimes have been observed by TR[3][8-11].

A more versatile experimental approach involves a pump-probe configuration derived from two laser sources. One is optimized for optically initiating the reaction and the other is optimized to generate resonance Raman scattering from the transient. Pump-probe experiments are especially important in the study of photolytically initiated, biochemical systems because of the need to precisely control a wide range of experimental parameters simultaneously including intensity, wavelength, pulsewidth, focusing volume, and spatial alignment. It is this two-laser, pump-probe experimental approach that is used in the work described here.

Pump-probe configurations have been used to record TR[3] spectra since approximately 1976 as shown in Fig. 1. Many early workers used a chemical or physical stimulus to initiate the reaction and sought time resolutions longer than a microsecond. Photolytically-initiated reactions were examined extensively with nanosecond time resolution often utilizing a 10 ns pulsed laser operating with repetition rates below 30 Hz and pulse energies in the several millijoule range. The

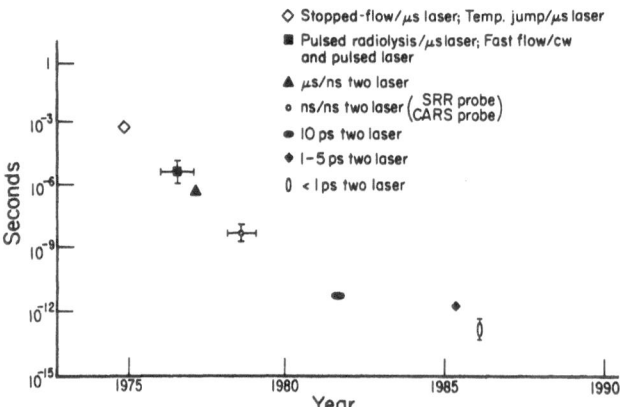

Figure 1. The increasing time resolution of several experimental techniques for recording time-resolved Raman spectra is plotted as a function of their historical development.

214

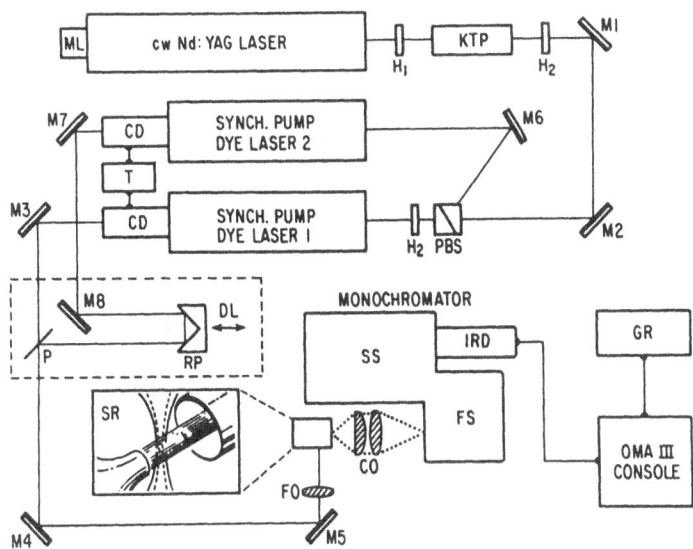

Figure 2. Two laser pump-probe PTR[3] instrumentation. ML: mode locker; H_1: halfwave plate (1064 nm); H_2: halfwave plate (532 nm); M1-M8: front surface dielectrically coated mirrors; PBS: polarizing beam splitter; CD: cavity dumper; T: timing synchronization; DL: delay line; RP: retroprism; P: dielectrically coated pellicle beam splitter; SR: sample region FO: focusing optics; FS: filter stage; SS: spectrometer stage; IRD: intensified reticon detector; GR: graphics. (Fig. taken from Ref. 12)

relative intensity used to initiate the reaction is a particularly important consideration in biochemical systems since it remains unclear whether the dynamics measured at higher excitation intensities have a direct relationship to the reactions occurring in the active biochemical system. Recent work on the BR photocycle illustrates this point explicitly[10]. Instrumentally, lower pulse energies can be used, but this usually must be accompanied by an increase in repetition rate in order to obtain TR[3] spectra with sufficiently high signal-to-noise ratios. Until recently, this type of laser technologically was not commonly available for the nanosecond time regime.

Interestingly, this genre of problem is addressed when the instrumental approach to recording TR[3] spectra with time resolution of <10 ps is considered. Mode-locked, cavity-dumped dye lasers typically operate with nanojoules per pulse energies and pulsewidths of a few picoseconds. As importantly, the repetition rates of these laser systems are variable from about 100 MHz to a few hertz. With this type of laser system, it is feasible to construct a pump-probe PTR[3] apparatus which emphasizes low intensity lasers operating at high repetition rates together with high sensitivity detection of Raman scattering. This is the instrumental approach shown in Fig. 2.

A mode-locked, cw Nd:YAG laser operating at 1.06 μ is used as a pumping source. The 100 ps pulses produced at a 76 MHz repetition rate are directed into a type II KTP crystal in order to obtain 532 nm radiation via frequency doubling. The resulting output typically is 1.2 W in the form of 70 ps pulses. The 532 nm radiation is split into equal parts (∿ 0.6 W) in order to synchronously pump two dye lasers. Each dye laser contains a three plate birefringent tuner and a secondary jet source designed to act as a saturable absorber. A cavity dumper selects a single pulse from the train of picosecond pulses in each dye laser. The output of the dye laser used for pumping (initiating the reaction) typically is 15-20 mW while the output of the probing laser used to generate resonance Raman scattering is <5 mW. Both dye laser outputs have approximately 10 ps autocorrelation pulsewidths and operate at

1 MHz repetition rates. The probe laser output is sent through an optical delay line in order to control time delays for probing the reaction dynamics.

The two beams are colinearly focused into the sample. The BR sample is driven through a capillary nozzle to form a liquid jet moving perpendicular to the direction of the two colinear laser beams. The flow rate is 20 m/s and the beam waist into which the lasers are focused is measured to be approximately 17 μm. The 1 MHz repetition rate of both the pump and probe lasers permits the liquid volume excited to be exchanged completely between successive pairs of pulses (pump-probe).

The RR scattering is collected by spherical optics located perpendicular to the plane formed by the two laser beams and the flowing liquid sample. The RR signal is focused onto the entrance slit of a triple monochromator. The first two stages are used as a wavelength selective filter while the concave grating in the third stage creates approximately 20 mm of spatially linear, dispersed radiation at the exit plane. The dispersed radiation is detected by an intensified reticon array and the resultant signal is processed by an optical multichannel analyzer. Spectral analysis is performed on a dedicated personal computer.

2.2 PTRF Experiments

The same picosecond apparatus designed for PTR^3 spectroscopy also is used to record PTRF spectra by changing the spectrometer/detection system. A 1 meter, single monochromator equipped with an 1200 groove/mm grating replaces the triple monochromator shown in Fig. 2. A high-gain photomultiplier (PMT) attached to the exit slit of the monochromator is utilized for detection. The PMT signal is processed by a lock-in amplifier and analyzed by a dedicated computer system. The only other change in the PTR^3 apparatus is the inclusion of a mechanical chopper in the probe laser beam. The chopping frequency is used as the reference for the lock-in amplifier and thereby, makes it feasible to monitor only the fluorescence signal produced by the probe laser at a given time delay after excitation. The spectrometer is scanned to record a fluorescence spectrum at a specific time delay after the reaction begins.

3. Results

3.1 PTR^3 Spectroscopy

It is essential that the probe laser by itself not significantly perturb the BR photocycle during the generation of a TR^3 spectrum. In the case of light-adapted BR, the fulfillment of these conditions can be best evaluated in the spectrum of BR itself. If the probe laser can be used to record a RR spectrum of BR only, then there is no significant amount of any reaction intermediate present during the pulsewidth of the probe laser. This assumes that the intermediates formed within 100 ps have absorption spectra which lie near that of BR and that as a consequence, the cross-section for RR scattering of each intermediate is similar to that of BR. Recent measurements of transient absorption spectra in the BR photocycle[4] show this assumption to be correct.

A RR spectrum of BR only was recorded using a laser operating at 590 nm with a 1 MHz repetition rate and pulses of 8 ps duration. This RR spectrum is compared in Fig. 3 with the analogous data obtained with a low power, cw dye laser operating at 514.5 nm. Except for small differences attributable to the changes in the excitation wavelengths, these spectra are the same and have been assigned to BR only. The experimental conditions under which this picosecond laser system can be used as a probe in PTR^3 experiments are defined by this result. An important aspect of these experiments involves the focusing parameters. Using a formalism which accounts for the laser power, excitation volume, and photochemical quantum yields, the focusing parameters associated with PTR^3 experiments can be quantified in terms of the value for $\ell_o t$[13,14]. The probe laser conditions used for the spectra in Fig. 3 give values for $\ell_o t$ near 0.2. PTR^3 experiments examining BR intermediates are designed on the basis of probe laser operating with <5 mW powers (i.e., $\ell_o t \sim 0.2$). Pump laser powers are selected between 15 mW and 20 mW.

The 1100-1240 cm^{-1} region of the RR spectra of BR is particularly useful for the study of picosecond intermediates since it has been assigned in a variety of previous studies to normal vibrational modes that are sensitive to retinal isomerization. The analysis of this so-called "fingerprint" region has been accomplished by the selective substitution of deuterium and ^{13}C isotopes into the retinal chromophore present in BR[15-18]. Although specific patterns (frequency positions and relative intensities) of RR bands in the fingerprint region have been assigned to the <u>all</u> <u>trans</u> and <u>13-cis</u> forms of retinal, there remains some uncertainty with respect to how quantitative these relationships are.

Nonetheless, this region of the RR spectrum reflects changes in retinal during the photocycle,as is demonstrated by the PTR^3 shown in Fig. 4. The probe-only spectrum is recorded with a low $\ell_0 t$ value and therefore can be assigned to BR (<u>all-trans</u> retinal). The 0 ps delay spectrum is recorded with the two 5 ps laser pulses overlapped in time, but with different intensities and wavelengths (see caption, Fig. 4). Small changes in the relative band intensities can be seen when the 0 ps delay and probe-only spectra are compared,especially in the 1184 cm^{-1} region. These differences are larger in the 40 ps delay spectrum. For example,the relative intensity of the 1184 cm^{-1} band increases by more than a factor of two relative to

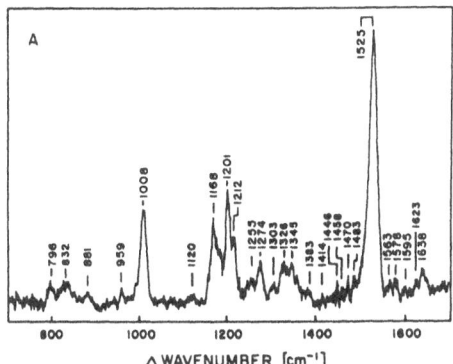

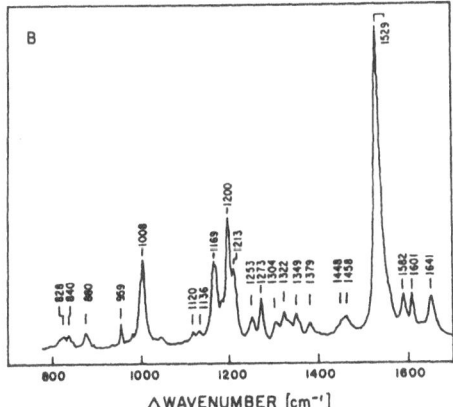

Figure 3. (A) PTR^3 spectrum of BR-570 in H$_2$O suspension (8ps pulsewidth, 590 nm, average power = 1.7 mW, repetition rate = 1 MHz, $\ell_0 t \sim 0.12$), (B) cw spectrum of BR-570 in H$_2$O suspension (514.5 nm, $\ell_0 t < 0.1$). (Fig. taken from Ref. 9)

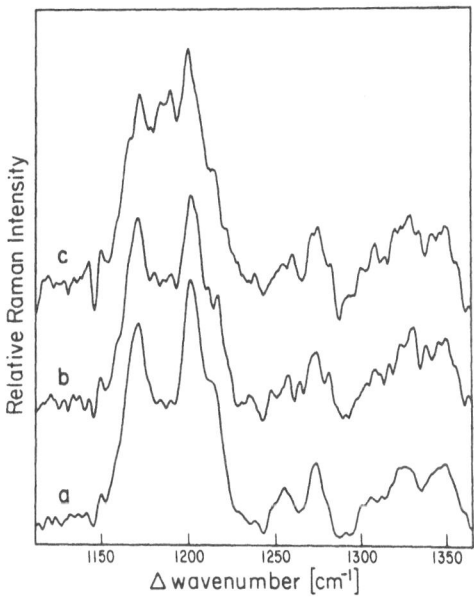

Figure 4. Two laser PTR^3 spectra of BR purple membrane in H$_2$O suspension for the fingerprint region: (a) 590 nm probe laser alone, (b) 575 nm excitation and 590 nm probe lasers (pulsewidth 5 ps) with 0 ps delay, (c) 575 nm excitation and 590 nm probe lasers (pulsewidth 5 ps) with 40 ps delay. Accumulation times for all three spectra are the same. The growth of the 1184 cm^{-1} 13-<u>cis</u> marker band should be noted. (Fig. taken from Ref. 12)

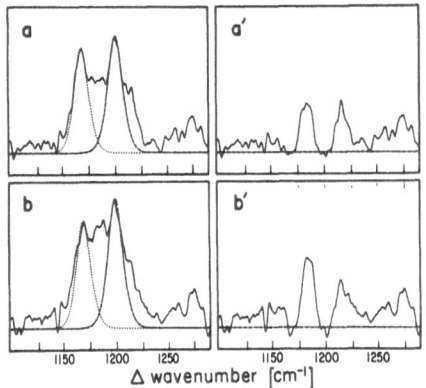

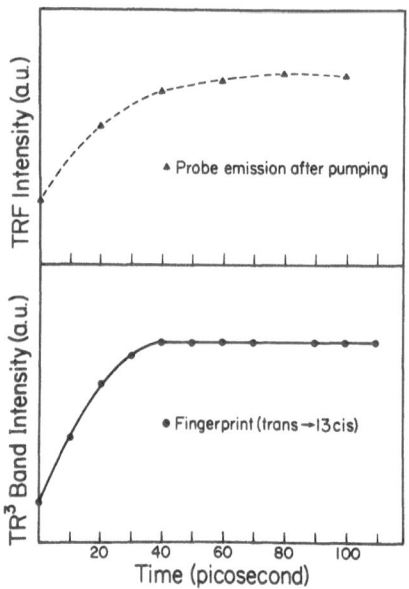

Figure 5. The fingerprint regions of the PTR³ spectra at (a) 0 ps and (b) 40 ps are displayed with Lorentzian-Gaussian band shapes (see text) fitted to the bands at 1170 cm⁻¹ and 1201 cm⁻¹. The fitting parameters were chosen to produce the best subjective fit to these bands in the 0 ps spectrum (a). The same band shapes were used on the 40 ps spectrum (b) with only a change in scaling. The result of subtracting the fitted bands from each spectrum is displayed in (a') and (b'), respectively. The growth of the band at 1184 cm⁻¹ between 0 ps delay and 40 ps delay should be noted. (Fig. taken from Ref. 12)

Figure 6. The PTR³ intensity of the 1184 cm⁻¹ (13-cis retinal marker band), is plotted as a function of the delay time between the pump and probe excitations (lower trace). The intensity of the fluorescence induced by the 590 nm TRF probe laser is plotted as a function of the delay between pump and probe excitations (upper trace). The PTR³ and TRF signals follow the same time development.

the probe-only spectrum. At delay times out to as long as 100 ps, however, no further changes in the fingerprint bands are found. These changes in relative band intensities can be observed even more clearly for the data in Fig. 5A where the bands at 1201 cm⁻¹ and 1170 cm⁻¹ are fitted with an analytic function (blended Gaussian-Lorentzian) for the 0 ps and 40 ps delay experiments. The PTR³ spectra obtained when these two analytic functions are subtracted are presented in Fig. 5B. A careful comparison of the residual band intensity at 1184 cm⁻¹ will confirm a substantial intensity change during the first 40 ps. PTR³ data recorded for time delays shorter than 40 ps demonstrate that the 1184 cm⁻¹ band intensity increases continuously, as shown in Fig. 6. Although other changes in RR band intensities occur in the fingerprint as well as other regions during this 40 ps interval, attention is focused on the 1184 cm⁻¹ band for this discussion, since it is the most evident and most readily assignable in terms of a normal coordinate analysis of the retinal chromophore.

3.2 PTRF Spectroscopy

The formation of several intermediates within a few picoseconds after the BR photocycle begins makes it experimentally difficult to distinguish emission from any specific species. Emission experiments previously reported concluded that only BR produced fluorescence at a detectable level while the intermediate K as well as those formed subsequently did not emit[18]. These experiments, however, were performed under excitation and/or temperature conditions which would not have necessarily resolved emission from BR intermediates formed within the initial 100 ps at room temperature, or would not have formed the same BR intermediates as thought generated at room temperature.

The fluorescence spectrum obtained with low power ($\ell_o t = 0.2$; 5 ps pulsewidth), probe-only excitation at 590 nm is presented in Fig. 7A. The general shape of this fluorescence spectrum corresponds with that reported previously for BR[18]. With the quantitative control over the excitation conditions described above, these data also are assigned to primarily BR. The 0 ps delay spectrum (Fig. 7B), however, exhibits a large (>30%) decrease in the fluorescence signal. The pulsewidths of the two lasers limit the time resolution of these results to about 5 ps interval. Nonetheless, a significant decrease in the total amount of fluorescence occurs during these initial 5 ps. Of equal significance is the result shown in Fig. 7C for the 40 ps delay. The fluorescence signal at 40 ps increases over the probe-only level by about 40% (i.e., approximately twice the probe-only signal). The fluorescence intensity remains at this higher level during at least the initial 100 ps period of the photocycle.

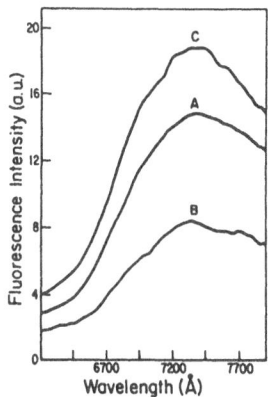

Figure 7. The time resolved fluorescence spectrum of BR produced by probe laser excitation when the sample is: (A) not excited by the pump laser (B) excited by the pump and probe beams simultaneously (0 ps delay; 5 ps pulsewidth) and (C) excited by a pump laser pulse 40 ps before probing.

It should be noted that there is no large change in the spectral shape or features in the three fluorescence spectra presented in Fig. 7, with the possible exception of the 40 ps spectrum over the 780-820 nm wavelength region. Unfortunately, these data were recorded with a PMT with limited sensitivity at wavelengths to the red of 820 nm and therefore, it is not feasible to examine the remainder of the fluorescence spectra for time-dependent differences. Such experiments are currently underway with a red-sensitive photomultiplier tube.

The time dependence of the fluorescence signals in the BR photocycle also can be measured by observing the fluorescence intensity over a selected spectral region as a function of delay time. Data of this type are shown in Fig. 8 for observations made at 732 nm. The decrease in fluorescence intensity at 0 ps time delay is difficult to analyze quantitatively because of the cross-correlation overlap in time between the pump and probe laser, but it can be quantitatively established that the signals decrease more when the pump laser intensity is larger. A similar relationship can be found for the degree of signal increase at 40 ps time delays (i.e., larger increases with larger pump laser intensities).

4. Discussion

The appearance of a prominent RR feature at 1184 cm^{-1} during the first 40 ps of the photocycle reflects the formation of a new isomeric intermediate. The vibrational assignments of the RR spectrum derived from studies of isotopically substituted retinals identify this intermediate as a 13-cis isomer[15-18]. The pattern of RR bands in the fingerprint region also has been used to assign the BR-548 form (i.e., dark-adapted) and 0-640 intermediate as both containing 13-cis isomer[15-18]. The

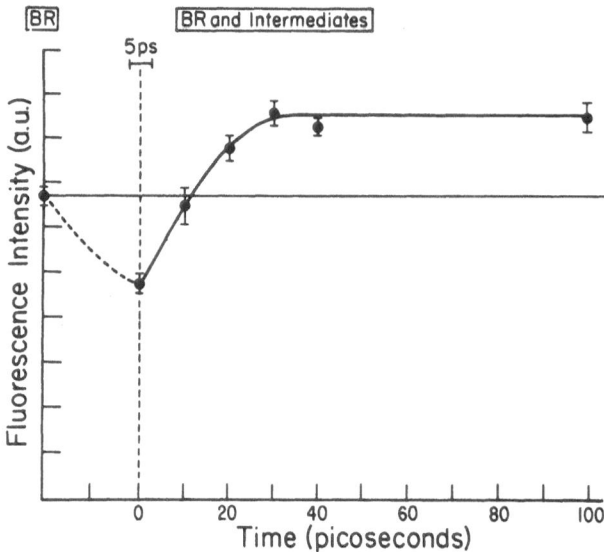

Figure 8. TRF intensity is plotted as a function of the delay between the pumping and probing excitation. The fluorescence level without prior pump excitation is indicated by the initial point.

increased intensity of the 1184 cm^{-1} band is only the most obvious feature in the fingerprint region of the 13-cis isomer and is thought to result from the repositioning of at least four RR bands assigned to the carbon-carbon, single bond stretching modes of retinal. Although these vibrational assignments need to be evaluated in terms of real-time measurements (i.e., PTR^3 data recorded on BR samples containing isotopically-substituted retinals), it appears clear that the major features of the RR spectrum recorded at a 40 ps time delay is characteristic of 13-cis retinal and that the time-dependent increase in the relative intensity of the 1184 cm^{-1} band accurately monitors the isomerization from all-trans to 13-cis retinal.

The formation of 13-cis retinal over a 40 ps period contradicts the conclusion reached from recent PTA measurements[4] in which the red-shifted absorption spectrum (maximum at 640 nm) observed within 0.5 ps of BR (40 fs excitation laser pulse at 570 nm) is attributed to a 13-cis retinal intermediate. The literature on transient absorption studies of BR adopts a notation which labels the 0.5 ps and 5 ps intermediates as J and K, respectively. Transient absorption data, however, contains no directly accessible information on either the isomeric or conformational form of the intermediates, but rather characterizes their respective electronic state properties. The conclusions concerning the isomeric form of the intermediate absorbing at a particular maximum must be derived from either comparisons with other spectroscopic studies (e.g., TR^3 spectroscopy) or with studies of structural related compounds in which chemical bonding is used to restrict molecular motions. Contrastingly, PTR^3 experiments directly monitor the vibrational degrees of freedom that are affected by structural (isomeric) and conformational changes.

The continuous increase in the 1184 cm^{-1} band intensity observed in PTR^3 data (Figures 4 and 5) over 40 ps is a more direct measure of the isomerization reaction than that afforded by PTA experiments. The absorption shifts appearing within 0.5 ps apparently arise from changes (e.g., reductions in the energy separations between electronic states or variations in vibrational and/or electronic state populations) other than isomerization at the C_{13}-C_{14} bond. The relationship between the absorption shifts reported to occur over 5 ps and the 40 ps RR data remains to be established.

220

The detection of increased fluorescence intensity which reaches a steady-state value at 40 ps has two interpretations which should be discussed here. First, it suggests that at least one intermediate in the photocycle emits and that the emission quantum yield is substantially larger than that measured for BR. This result alters the view of the emission spectroscopy previously held[19].

A comparison of the PTRF and PTR[3] results provides the opportunity to correlate data on electronic state properties with those measuring molecular structure and conformation. The time dependence of the emission signal at 732 nm region is plotted in Fig. 6 together with the 1184 cm^{-1} PTR[3] data. It is evident that both phenomena occur over the same time scale and may be related to the same molecular species. Since the PTR[3] data characterize this intermediate as containing 13-cis retinal it is tempting to assign the increased fluoresence to the formation of this specific isomer. Theoretically, it becomes important to explain why such a species has a larger emission yield than the all-trans isomeric form found in BR. It should be noted that other molecular or environmental factors may play a larger role in determining the emission rate than does the isomeric form of retinal.

The large (>30%) decrease in emission intensity observed within 5 ps (i.e., 0 ps delay data in Fig. 6) also provides a new characterization of the emission properties to be associated with the subpicosecond intermediates (e.g., J). If it is correct that the 5 ps intermediate is a ground-state species[4], then the reduced emission intensity is caused by the formation of an intermediate which has significantly lower emission yield (perhaps zero) than BR itself. The experimental approach used here to measure PTRF ensures that emission from only the species present at a given time delay is recorded. At a 40 ps time delay, the species contributing to the emission are primarily BR and the 13-cis retinal intermediate if the subpicosecond intermediates have an extremely small emission yield. In such a model, a semi-quantitative view of the photochemical quantum yield as well as the relative emission yields of different species can be obtained. The extent of emission decrease at 0 ps delay and increase at 40 ps and longer delays would be correlated to the net photochemical yield in the BR photocycle. The higher the percentage of BR molecules participating in the photocycle, the higher the concentrations of these intermediates formed and the larger the negative and positive changes in emission yields. Qualitatively, these relationships are followed when the emission intensities are measured as a function of excitation laser intensities, as is found in the data of the type presented in Fig. 8.

5. Concluding Remarks

The measurement of PTR[3], PTRF, and PTA data on the same sample of BR offers the opportunity to reach firm conclusions concerning the quantitative relationships between the electronic-state properties and molecular structures and conformations of reactive BR intermediates formed on the picosecond time scale. Results have been described here which illustrate this point for vibrational resonance Raman and fluorescence experiments. These comparisons have led to a clearer understanding of the isomerization at the C_{13}-C_{14} bond of retinal and the picosecond intermediates observed to emit. Many other characterizations of the molecular properties of retinal as well as its surrounding protein environment, together with data identifying changes in electronic state changes, are required in order to construct a more detailed mechanism for the BR photocycle.

6. Acknowledgements

I wish to gratefully acknowledge my coworkers in the research described here. These include Mr. T.L. Brack, Dr. H. Hayashi, Mr. D. Blanchard, Mr. D.A. Gilmore, Mr. H. LeMaire, and Mr. J. McConnel. This work was supported by a grant from the National Institutes of Health and the University of Arizona.

References

1. D. Osterhelt, W. Stoeckenius, Proc. Natl. Acad. Sci. USA, _70_, 289 (1973).
2. W. Stoeckenius, R.A. Bogomolni, Annu. Rev. Biochem., _51_, 587 (1982).
3. R.H. Lozier, R.A. Bogomolni, W. Stoeckenius, Biophys. J., _15_, 955 (1975).
4. H.J. Polland, M.A. Franz, W. Zinth, W. Kaiser, T. Kolleing, D. Oesterhelt, Biophys. J. _49_, 651 (1986).
5. G.H. Atkinson, In Advances in Infrared and Raman Spectroscopy, ed. by R.E. Hester and R.J.H. Clark, _9_, 1 (North Holland Publ. London, 1981).
6. I. Grieger, G.H. Atkinson, Biochemistry _24_, 5660 (1985).
7. G.H. Atkinson, In Chemical Dynamics: NATO Advanced Study Institute, ed. by P. Rentzepis and C. Capellos, (Academic Press, New York, N.Y. 1986).
8. S.O. Smith, M. Braiman, R.A. Mathies, In Time-Resolved Vibrational Spectroscopy, ed. by G.H. Atkinson, 219 (Academic Press, New York, N.Y. 1983).
9. G.H. Atkinson, I. Grieger, G. Rumbles, In Time-Resolved Vibrational Spectroscopy, ed. by M. Stockburger and A. Laubereau, 255 (Springer-Verlag, 1985).
10. I. Grieger, G.H. Atkinson, Biochemistry, _24_, 5660 (1985).
11. G.H. Atkinson, In Proceedings of the SPIE Conference: Laser Applications in Chemistry and Biophysics, ed. by M. El-Sayed, _620_, 82 (1986).
12. G.H. Atkinson, T.L. Brack, D. Blanchard, I. Greiger, G. Rumbles, L. Siemankowski, In Time-Resolved Vibrational Spectroscopy, ed. by G.H. Atkinson (Gordon and Breach, New York, N.Y., in press).
13. M. Stockburger, W. Klusmann, H. Gattermann, G. Massig, R. Peters, Biochemistry, _18_, 4886 (1979).
14. I. Grieger, Ph.D. Thesis, University of Gottingen (1980).
15. S.O. Smith, A.B. Myers, J.A. Pardoen, C. Winkel, P.P.J. Mulder, J. Lugtenburg, R. Mathies, Proc. Natl. Acad. Sci. USA, _81_, 2055 (1984).
16. R.A. Mathies, In Spectroscopy of Biological Molecules, ed. by C. Sandorfy and T. Theopanides, (D. Reidel Publishing Co., New York, N.Y. 1984).
17. M. Braiman, R. Mathies, Proc. Natl. Acad, Sci. USA, _79_, 403 (1982).
18. G. Eyring, B. Curry, A. Broek, J. Lugtenburg, R. Mathies, Biochemistry, _21_, 384 (1982).
19. A. Lewis, J.P. Spoonhower, G.J. Perreault, Nature (Lond.) _260_, 675 (1976).

The Role of Metal Ions in Bacteriorhodopsin Function

T.C. Corcoran, E.S. Awad, and M.A. El-Sayed

Department of Chemistry and Biochemistry, University of California,
Los Angeles, CA 90024, USA

The literature on the simultaneous dissociation of an acid of pK_a ~8.6-10 and the protonated Schiff base and a possible coupling mechanism is reviewed. The effect of metal cations on the color of retinal and the photocycle is also summarized in light of this coupling. Studies using Eu^{3+} to probe the ion binding sites in bacteriorhodopsin by following the spectroscopy and kinetics of its emission are reported. It is shown that the Eu^{3+} binding constants for the different sites appear quite similar, and that the total intensity of the emission increases linearly with the amount of added Eu^{3+}. A model is presented incorporating these results, explaining how the retinal color and extent of deprotonation in the photocycle can change sigmoidally with the amount of added cations, based on a requirement that more than one metal ion must be bound to allow these processes to occur.

1. INTRODUCTION

Bacteriorhodopsin (bR), the light-harvesting purple membrane protein of *Halobacterium halobium*, [1,2] is the only photosynthetic system in nature besides chloropyll. It utilizes retinal as its chromophore [1], covalently bound to the ε-amino group of a lysine residue via a protonated Schiff base (PSB) linkage [3]. Upon absorption of light, the protein-chromophore complex is transformed through a series of intermediates [4], on timescales ranging from sub-picosecond to milliseconds:

$$bR_{570} \rightarrow K_{610} \rightarrow L_{550} \rightarrow M_{412} \rightarrow O_{640} \rightarrow bR_{570} .$$

BR functions in the cell by using light to create a transmembrane electrochemical gradient by the release protons from the exterior of the cell and the uptake of protons from the interior. The mechanism of this proton pump is still not completely understood. This gradient is used by the cell for metabolic processes such as ATP synthesis [5]. Much experimental evidence has been found linking the M_{412} intermediate with the translocation of protons, such as: a) transmembrane electric potential measurements by DRACHEV et al. show transients on the same time scale as the formation and decay of the M_{412} intermediate [6] b) resonance Raman spectroscopy shows that the M_{412} intermediate is the only one in which the Schiff base is unprotonated [7] c) a good correlation has been found between the slow-decaying form of M_{412} and the number of protons pumped [8].

2. THE INVOLVEMENT OF AROMATIC AMINO ACID RESIDUES

2.1 Transient Changes in Fluorescence and Absorbance

As early as 1973, it was reported by OESTERHELT & HESS [9] that aromatic amino acid residues in bR were affected by the photocycle. They observed that both fluorescence and excitation spectra in the near ultraviolet decreased in amplitude when bR was also illuminated at 570 nm. The fluorescing species was identified as tryptophan, and they attributed the decrease to changes in the protein conformation [9]. Using a double-pulse technique, FUKUMOTO et al. showed the quantum yield of this fluorescence reaches a minimum on the timescale of the M_{412} formation [10]. It has since been suggested that a tryptophan-tyrosine interaction sensitive to the deprotonation of tyrosine may play a role in the fluorescence decrease [11] since transient UV absorption spectra taken on the timescale of the M_{412} formation showed evidence for deprotonation of a tyrosine residue and possibly the perturbation of a tryptophan [11-14]. Specifically, the spectra show a broad absorbance decrease centered around 275 nm, and somewhat sharper increase centered at about 296 nm and 240 nm [11-14]. BOGOMOLNI showed that the latter two bands are notably decreased at pH 10.8 compared to 6.6, and that these differences were consistent with the dissociation of one tyrosine residue, accessible to the solution [12]. Much more recently, selective tyrosine nitration has shown that the nitrated Tyr-64 residue deprotonates on the M_{412} timescale [15]. However, evidence has also emerged that aspartic acid residues of quite high pK_a undergo protonation changes during the photocycle [16]. It is possible that these changes may cause the above-mentioned tryptophan perturbation. However, the magnitude of its contribution to the transient UV absorption might be small at 296 nm [13, 14]. Interpretation of the transient events in the near UV is complicated by the *trans-cis* isomerization of retinal which also shows transient absorbance in this region but appears on a faster timescale, consistent with the K_{610} intermediate [14].

2.2 The Coupling of Protein and Chromophore Dynamics

The rate constants for the formation of the UV transients and M_{412} have been subject to some disagreement, possibly due to the multiplicity of conditions under which these measurements were made. At -1.5°C in thin purple membrane layers at 95% humidity, biexponential formation rates were observed at 275, 295 and 412 nm [12]. The slow component at 295 nm attributed to tyrosine deprotonation was found to be slower than the slow component of M_{412}. The fast components observed in the UV were all faster than those of M_{412}. However no mention was made of the substantial sub-microsecond component at the UV wavelengths due to retinal isomerization later reported by the same authors under somewhat different conditions [14]. Based upon the rate constants of the fast components mentioned above and their observation that low-temperature formation of M_{412} was markedly increased at high pH, KALISKY et al. hypothesized that tyrosine deprotonation precedes, and is a prerequisite for PSB deprotonation (M_{412} formation) [17]. ROSENBACH et al. [18] expanded this hypothesis, arguing that the rate of M_{412} formation is controlled by tyrosine deprotonation at neutral pH, and exhibits a titration-like curve having an apparent pK_a of 10.3.

Formation rate constants at 405 nm and 297 nm were measured by HANAMOTO et al. [19], DUPUIS & EL-SAYED [20], and CORCORAN et al. [21] under a variety of pH, ionic strength, and temperature conditions. Under all these conditions (which included the normal physiological conditions of the bacterium), the formation rate at 297 nm attributed to tyrosine deprotonation (i.e., not considering the unresolved fast component due to retinal isomerization) was always similar to, but somewhat less than the slow component of the biexponential formation kinetics observed at 405 nm, even compensating for the effect of the decay of the M_{412} intermediate [19]. Hence, these data argued strongly against the hypothesis that tyrosinate deprotonation is a prerequisite for PSB deprotonation [19]. The data also indicated that it was not the rate constants, but rather the relative amplitudes of the two components which changed as a function of pH [19]. This led to the hypothesis that the biexponential formation kinetics of M_{412} were due to two independent populations (i.e., two retinal sites) of bacteriorhodopsin: one in which an acid group (possibly tyrosine, lysine or aspartic acid) near the PSB [19,20] *or along its reaction coordinate* [21] was already deprotonated (i.e., due to high solution pH) before light activates the photocycle, giving rise to the fast component, and a second in which this acid is still protonated, giving rise to the slow component. The larger rate constant of the fast component must be due to differences in its entropy rather than energy of activation, since the latter is comparable for both components [19]. The equilibration rate of this acid with its conjugate base must be slower than the rate of M_{412} formation for the populations to be independent [18,19]. Using this model, an apparent pK_a for this acid can be derived from the relative amplitudes of the fast and slow components and the solution pH [19]. This pK_a varies with the ionic strength of the medium from about 8.6 in

4.3M basal salts to 10.0 in 0.2mM NaCl [19-21]. This model would predict that a linear correlation should exist between the amount of the undissociated acid and the amplitude of the slow component of M_{412} formation as a function of pH and the apparent pK_a of the acid. If the amplitude of the 297 nm transient, either tyrosinate or tryptophan perturbed by aspartate formed during the cycle, can be considered as a measure of the amount of this undissociated acid (since species which are already deprotonated before the actinic flash cannot give rise to this transient), then such a correlation has been observed [21].

Additional evidence suggests that observed UV transient is somehow coupled to (but not the cause of [19]) PSB deprotonation. The activation energies for both processes are quite similar, comparable to H-bond energies [19,20], suggesting that both processes are controlled by protein conformational changes [19]. A very simple electrostatic model (the cation model) was proposed to account for the apparent coupling of these two processes [19]. If a positive charge within the protein, such as an arginine or lysine residue, or possibly a bound metal cation (see below), becomes within interaction distance with the PSB as a result of protein conformational changes, the resulting repulsion could induce its deprotonation. If the tyrosine (or aspartic acid) was simultaneously approached by a positive charge (the same one, or possibly another, being moved by conformational changes in the same manner) the deprotonation of this neutral residue could also be induced by the stabilizing attraction between the positive charge and the conjugate base of the dissociated residue. Consequently, both a positive (the PSB) and a neutral species (the amino acid) could be induced to deprotonate by a common mechanism involving the movement of positive charge controlled by protein conformational changes [19].

Atomic absorption shows that under normal conditions, Ca^{2+} and Mg^{2+} are bound to bR [22]. However, when the purple membrane (λ_{max} = 570 nm) is acidified, [2,23] or deionized on a cation exchange column [24] or treated by several other methods [22], it turns blue (λ_{max} = 608 nm) and divalent cations are missing [22]. Of particular interest is that both PSB deprotonation [6, 16,23,25] and the above-mentioned UV transient [26] are both inhibited in the blue membrane preparations, lending further support to the notion that these processes are controlled by a common mechanism. Also, it was found that the rates of PSB deprotonation and the formation rate of the UV transient were largely unaffected by substituting Na^+ or La^{3+} for the native Ca^{2+} and Mg^{2+} [21]. One might expect that if the metal ion were the the interacting positive charge proposed in the cation model above, that the rate of these processes would increase with a higher-valence cation due to its greater repulsion. Yet, if the rate-controlling step in each of these processes is the conformational change of the protein [19] as their activation energies strongly suggest, then this argument would not hold. Alternatively, the metal ion may simply act as a catalyst for these conformational changes, and the interacting positive charge is some other group [26]. These data cannot make the distinction [21]. Consequently, much recent work within this group has focussed on attempting to elaborate the role of the bound metal ions in the events that lead to proton pumping in bR.

3. THE ROLE OF METAL IONS IN THE PHOTOCYCLE OF BACTERIORHODOPSIN

3.1 Is Retinal Isomerization Inhibited in Blue Membrane?

If the M_{412} formation step is blocked in blue membrane preparations, then one must ask if this inhibition also applies to the retinal isomerization steps (formation of the K_{610} and L_{550} intermediates). Transient absorption studies by MOWERY et al. [23], KOBAYASHI et al. and OHTANI et al. [27] indicated the presence of a red-shifted intermediate, suggesting that a species analogous to K_{610} is formed in blue membrane. To ascertain unequivocally whether retinal isomerization was occurring, time-resolved resonance Raman spectroscopy studies were undertaken by CHRONISTER et al. [28,29] because of its sensitivity to structural changes of the chromophore. Two regions in the Raman spectrum are of interest: the ethylenic stretch region from 1530 to 1570 cm^{-1}, which shifts with the electronic absorption maximum, and the "fingerprint" region near 1200 cm^{-1} which is sensitive to isomerization of the chromophore [30]. Samples of native bR (i.e. unmodified purple membrane), deionized blue membrane, acid blue membrane (pH ~2) and acid purple membrane (pH ~0) [23] were studied. The ethylenic stretch band indicated the formation of K_{610}, L_{550}, and M_{412} in native bR, but only species analogous to K_{610} and L_{550} in all of the latter three samples [28, 29]. The "fingerprint" region showed evidence for retinal isomerization to the 13-*cis* form after photoexcitation in all four of the samples [29]. From this, it was concluded that: a) only the $L_{550} \rightarrow M_{412}$ transformation is inhibited in acid purple and blue membrane samples b) retinal isomerization in itself was insufficient to cause PSB deprotonation, indicating that other mechanisms must contribute to lowering its pK_a during the photocycle c) metal cations are directly or indirectly involved in the final lowering of this pK_a during the $L_{550} \rightarrow M_{412}$ transformation d) it is the all-*trans* isomer of retinal that undergoes isomerization following photoexcitation in the blue membrane e) the active sites of the blue membrane samples (acidified and deionized) were similar, but not identical, as shown by the Raman spectra of their intermediates.

3.2 The Correlation of the Chromophore Color with M_{412} Formation

As has been mentioned earlier, blue membrane preparations do not form the M_{412} intermediate. CHRONISTER et al. showed that for both acidified and deionized membrane a strong negative correlation was observed between the amount of blue species present and the amount of M_{412} formed during the photocycle as a function of pH or Ca^{2+} concentration [28]. Furthermore, a protons-cations equilibrium was proposed to explain both of these processes as leading to the same blue species.

3.3 Probing the Metal Ion Sites using Europium Fluorescence

3.3.1 Spectroscopic Properties of Europium Emission

Recently, the metal cation environment and its changes during the cycle were studied [31]. Eu^{3+} is used as a probe. Eu^{3+} is a luminescent species which can isomorphously replace Ca^{2+} in proteins [32]. Changes in site symmetry and environment strongly affect its emission lifetime [33], so the number of components in it emission decay gives indication of the minimum number of Eu^{3+} sites in the protein [31]. Furthermore, the change in its observed lifetimes when ^2H_2O replaces H_2O can give a quantitative estimate of the number of water molecules coordinated to Eu^{3+} in each site [33,34]. The sensitivity of its emission kinetics to environmental changes has also been used to monitor what changes, if any, occur to the cation binding sites in bR during the photocycle as bR is transformed into M_{412} [31].

A comparison of the the emission spectra of aqueous $EuCl_3$ and Eu^{3+}-substituted bR showed that several Eu^{3+} transitions had become much more allowed in the protein compared to aqueous medium [31]. This strongly suggests that the crystal field symmetry around the protein-bound Eu^{3+} is greatly reduced from that found in aqueous Eu^{3+}, losing its center of inversion and causing the electric dipole transition to become allowed [35]. This is to be expected to be the case if Eu^{3+} is coordinated to one or more carboxylate groups in the protein [36] arranged in a low-symmetry site such as those found in thermolysin [32]. These groups induce the dominant portion of the crystal field, and strongly affect its anisotropy. Similar intensity changes in Eu^{3+} absorption spectra and radiative rates upon changes of chelates and solvents have been reported [35,37,38].

3.3.2 Eu^{3+} Emission Decays

Eu^{3+} in bR shows quite complex emission decay kinetics, which were fit to a sum of three exponentials [31]. This was felt to be a reasonable physical model for this system, since there is a great deal of evidence indicating that bR binds at least three cations [22,24,29,39,40] and each could have a different lifetime. Changes in radiative rate constants due to differences in site symmetry [35,38], and differences in nonradiative relaxation rates as well could account for this. The shorter-lived components had more intensity than the longer ones, favoring the argument that increased radiative rates are an important contribution to the short lifetimes. The lifetimes each differ by factors of 4 or more from the next component, making the triexponential fit possible. Consequently, the emission lifetime allows a microscopic probe of the different Eu^{3+} sites [31]. This interpretation is strengthened by the observation that the each component of the decay was quenched differently when H_2O was substituted for ^2H_2O [31], suggesting that the Eu^{3+} in each site has different numbers of coordinated H_2O molecules [33,34,41]. As might be expected, the sites with the faster emission rates, and by inference, the lowest symmetry (probably surface sites) showed the largest number of coordinated H_2O. This argues against attributing the form of the decay to some instrumental artifact.

3.3.3 Concentration dependence

Emission decays were measured as a function of Eu^{3+}/bR ranging from 1 to 3 [31]. Neither the relative amplitudes nor the lifetimes of the emission from the different sites varied significantly as the Eu^{3+}/bR ratio increased. Indeed, the decay curves appear to be nearly indistinguishable through this concentration range (Fig. 1). This implies that no single site is being filled first, but rather the fraction of the total Eu^{3+} bound to each site does not change significantly with the amount of ions added, suggesting that they all have similar binding constants. One cannot conclude from the amplitude of each component of the emission decay what the *absolute* fraction of ions bound to the corresponding site is, (since each site may have different molar extinction coefficients due to different symmetries, and different quantum yields) but we can see if *changes* are occurring. No saturation of an individual site was observed in this concentration range, strengthening the conclusion that there are at least three binding sites. Additionally, the total emission intensities immediately after the laser flash were found to increase linearly with the Eu^{3+}/bR ratio, lending further support (Fig. 2) to the conclusion that no saturation is occurring in this range.

While the data above indicate that Eu^{3+} binding increases in a linear fashion, it has been shown that the extent of PSB deprotonation [28] and the color change of the chromophore [22, 24,28,39] in deionized bR changes sigmoidally when titrated with ions. ARIKI & LANYI [39] have interpreted this data using a model for cation-chromophore interaction in which there are two binding sites with different binding constants, and an ion binding to the weaker of these two causes the color to change.

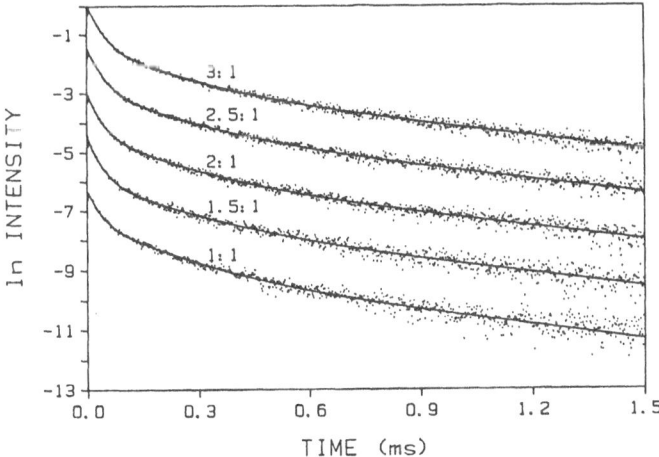

Fig. 1. Emission decays of Eu^{3+} in bacteriorhodopsin at Eu/bacteriorhodopsin ratios ranging from 1.0 to 3.0. There is no significant difference in their kinetics throughout this range.

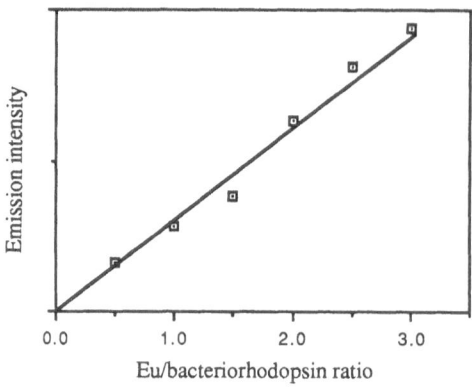

Fig. 2 The linear increase of Eu^{3+} emission as a function of Eu/bacteriorhodopsin ratio

This model has been demonstrated to fit the sigmoidal behavior mentioned above. Our work has indicated that the binding constants appear to be quite similar, and so we demonstrate here another very simple model (see APPENDIX), in which two sites have the same binding constants but binding to *both* *sites* is required for the chromophore color change. This model also gives sigmoidal titration curves (Fig. 3). It was suggested [31,28] that binding a minimum of two metal ions is probably needed to cause the protein to assume the conformation that allows PSB deprotonation and turns the chromophore purple. It is in this conformation that the interacting positive charge mentioned in the cation model is able to induce deprotonation during the photocycle [31].

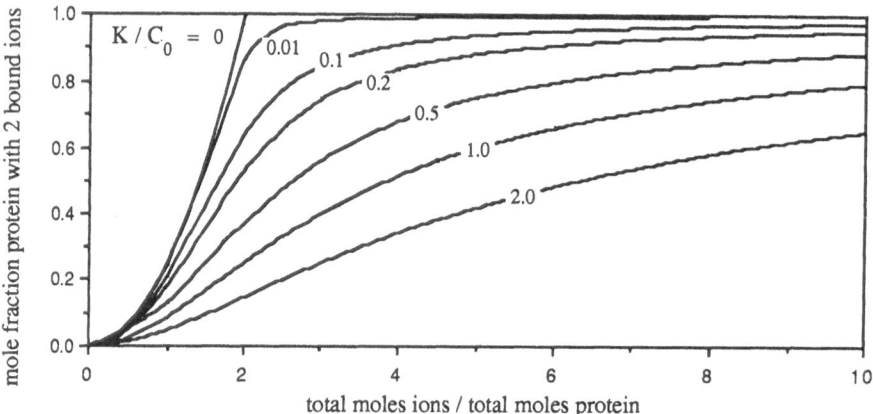

Fig. 3 The mole fraction of protein with two bound ions as a function of the total moles of ions added per mole protein. Calculations are described in the APPENDIX. The curves are given for several ratios of the dissociation constant, K to total concentration protein, C_0.

3.3.4 Effect of M_{412} formation on the Eu^{3+} environment

As previously mentioned, we have thus far been unable to distinguish whether a bound metal cation was the interacting partner in the PSB deprotonation. Here again the luminescent properties of Eu^{3+} afford us a microscopic probe. If this species were the metal cation, its environment would probably be perturbed by the approach of the Schiff base and the formation of the negatively charged acid residue that gives rise to the UV transient. This perturbation could affect its radiative properties or possibly the degree of its hydration. The emission lifetime of Eu^{3+}-substituted bR was measured suspended in glycerol-water at pH~6 and -40°C exposed to strong excitation at 570-nm [31]. Under

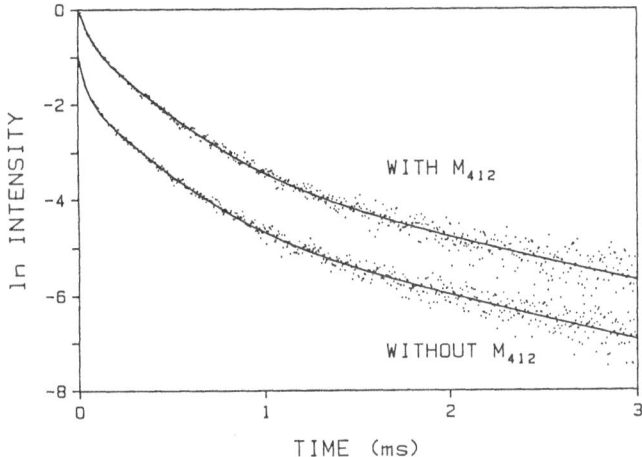

Fig. 4. Emission decays of Eu^{3+} in bacteriorhodopsin at -40°C, with >50% of the sample converted to the M_{412} intermediate, and without

these conditions, it was found that >50% of the sample was in the M_{412} intermediate. This is larger than one would expect at this pH [17,42], possibly because lanthanides tend to slow the decay of the M_{412} intermediate [6,22,39] The decay curves for this sample are shown compared with those for the same sample under the same conditions but without illumination at 570 nm (no M_{412} trapping) in Fig. 4. Within experimental errors, it seems that driving the cycle to form M_{412} does not perturb the luminescent sites of Eu^{3+}, casting doubt on the idea that a metal cation is the interacting partner. This again suggests that the involvement of metal cations is indirect, e.g., in inducing conformation changes [31].

Acknowledgement This work was supported by a grant from the U.S. Department of Energy (Office of Basic Energy Sciences).

4. REFERENCES

1. D. Oesterhelt and W. Stoeckenius: *Nature (London) New Biol.* **223,** 149 (1971)
2. For review, see W. Stoeckenius and R. A. Bogomolni: *Ann. Rev Biochem.* **52,** 587 (1982)
3. J. Bridgen and I. D. Walker: *Biochemistry,* **15,** 792 (1976)
4. R. Lozier, R. A. Bogomolni, and W. Stoeckenius: *Biophys. J.* **15,** 995 (1975)
5. For review, see R. A. Bogomolni, R. A. Baker, R. H. and W. Stoeckenius: *Biophys. Acta* **440,** 68 (1976)
6. L. A. Drachev, A. D. Kaulen, and V. P. Skulachev: *FEBS Lett.* **87,** 161 (1978)
7. A. Lewis, J. Spoonhower, R. A. Bogomolni, R. H. Lozier, and W. Stoeckenius; *Proc. Natl. Acad. Sci. USA* **71,** 4462 (1974); B. Aton, A. Doukas, R. H. Callender, B.Becher, and T. G. Ebrey: *Biochemistry* **16,** 2995 (1977)
8. Q.-Q. Li, R. Govindjee, and T. G. Ebrey: *Proc. Natl. Acad. Sci. USA* **81,** 7079 (1984)
9. D. Oesterhelt and B. Hess *Eur. J. Biochem.* **37,** 316 (1973)
10. J. M. Fukumoto, W. D. Hopewell, B. Karvaly, and M. A. El-Sayed: *Proc. Natl. Acad. Sci. USA* **78,** 252 (1981)
11. R. A. Bogomolni, L. Stubbs, and J. K. Lanyi: *Biochemistry* **17,** 1037 (1978)
12. B. Hess and D. Kuschmitz: *FEBS Lett.* **100,** 334 (1979)
13. R. A. Bogomolni: In *Bioelectrochemistry,* ed. by H. Keyzer and F. Gutmann, (Plenum, New York 1980) p.83
14. D. Kuschmitz and B. Hess: *FEBS Lett.* **138,** 137 (1982)
15. P. Scherrer and W. Stoeckenius: *Biochemistry* **24,** 7733 (1985)
16. M. Engelhard, K. Gewert, B. Hess, W. Kreutz and F. Siebert: *Biochemistry* **24,** 400 (1985)

17. O. Kalisky, M. Ottolenghi, B. Honig, and R. Korenstein: *Biochemistry* **20**, 649 (1981)
18. V. Rosenbach, R. Goldberg, C. Gilon, and M. Ottolenghi: *Photochem. Photobiol.* **36**, 197 (1982)
19. J. H. Hanamoto, P. Dupuis, and M. A. El-Sayed: *Proc. Natl. Acad. Sci. USA* **81**, 7083 (1984)
20. P. Dupuis and M. A. El-Sayed: *Can. J. Chem.* **63**, 1699 (1985)
21. T. C. Corcoran, P. Dupuis, and M. A. El-Sayed: *Photochem. Photobiol.* **43**, 655 (1986)
22. C.-H. Chang, J.-G. Chen, R. Govindjee, and T. Ebrey: *Proc. Natl. Acad. Sci. USA* **82**, 396 (1985)
23. P. C. Mowery, R. H. Lozier, Q. Chae, Y.-W. Tseng, M. Taylor, and W. Stoeckenius: *Biochemistry* **18**, 4100 (1979)
24. Y. Kimura, A. Ikegami, and W. Stoeckenius: *Photochem. Photobiol.* **40**, 641 (1984)
25. C.-H. Chang, C.-K. Suh, R. Govindjee, and T. Ebrey: *Biophys. J.* **45**, 210a (abstr.) (1984)
26. P. Dupuis, T. C. Corcoran, and M. A. El-Sayed: *Proc. Natl. Acad. Sci. USA* **82**, 3662 (1985)
27. T. Kobayashi, H. Ohtani, J. Iwai, and A. Ikegami: In *Ultrafast Phenomena IV*, eds. Auston, D. H. and Eisenthal, K.B. (Springer, Berlin, Heidelberg 1984) p. 481; T. Kobayashi, H. Ohtani, J. Iwai, and A. Ikegami: In *Proceedings of the International Symposium on Fast Reactions in Biological Systems* (Kyoto 1984) p. 90; H. Ohtani, T. Kobayashi, J. Iwai, and A. Ikegami: *op. cit.* p. 100; H. Ohtani, T. Kobayashi, J. Iwai, and A. Ikegami: *Biochemistry* **25**, 3356 (1986)
28. E. L. Chronister, T. C. Corcoran, L. Song, and M. A. El-Sayed: *Proc. Natl. Acad. Sci. USA* **83**, 8580 (1986)
29. E. L. Chronister and M. A. El-Sayed: *Photochem. Photobiol.* in press
30. J. Terner, C.-L. Hsieh, A. R. Burns, and M. A. El-Sayed: *Proc. Natl. Acad. Sci. USA* **76**, 3046 (1979); B. Aton, A. Doukas, R. H. Callender, and T. G. Ebrey: *Biochimica et Biophysica Acta* **576**, 424 (1979); D. L. Narva, R. H. Callender, and T. G. Ebrey: *Photochem. Photobiol.* **33**, 567 (1981)
31. T. C. Corcoran, K. Z. Ismail and M. A. El-Sayed: *Proc. Natl. Acad. Sci. USA* in press
32. B. W. Matthews and L. H. Weaver: *Biochemistry* **13**, 1719 (1974)
33. J. L. Kropp and M. W. Windsor: *J. Chem. Phys.* **42**, 1599 (1965)
34. W. DeW. Horrocks,Jr., G. F. Schmidt, D. R. Sudnick, C. Kitrell, and R. A. Bernheim: *J. Am. Chem. Soc.* **99**, 2378 (1977)
35. E. V. Sayre, D. G. Miller, and S. Freed: *J. Chem. Phys.* **26**, 109 (1957)
36. C.-H. Chang, R. Jonas, S. Melchiore, R. Govindjee, and T. Ebrey: *Biophys. J.* **49**, 731 (1986)
37. D. G. Miller, E. V. Sayre, and S. Freed: *J. Chem. Phys.* **29**, 454 (1958); W. R. Dawson, J. L. Kropp, and M. L. Windsor: *J. Chem. Phys.* **45**, 2410 (1966)
38. J. L. Kropp and M. W. Windsor: *J. Phys. Chem.* **71**, 477 (1967)
39. M. Ariki, and J. K. Lanyi: *J. Biol. Chem.* **261**, 8167 (1986)
40. N. V. Katre, Y. Kimura, and R. M. Stroud: *Biophys. J.* **50**, 277 (1986)
41. W. DeW. Horrocks,Jr., and D. R. Sudnick: *Acc. Chem. Res.* **14**, 384 (1981)
42. B. Chance, M. Porte, B. Hess, and D. Oesterhelt: *Biophys. J.* **15**, 913 (1975)

5. APPENDIX

The following two-site model gives rise to a sigmoidal binding curve. Consider a protein molecule with sites A and B which bind a ligand L. The system is governed by the equilibria:

$$AB + L \rightleftharpoons ALB \tag{1a}$$
$$AB + L \rightleftharpoons ABL \tag{1b}$$
$$ALB + L \rightleftharpoons ALBL \tag{1c}$$
$$ABL + L \rightleftharpoons ALBL . \tag{1d}$$

The corresponding dissociation constants are:

$$K_1 = [AB][L] / [ALB] \tag{2a}$$
$$K_2 = [AB][L] / [ALB] \tag{2b}$$
$$K_3 = [ALB][L] / [ALBL] \tag{2c}$$
$$K_4 = [ABL][L] / [ALBL] . \tag{2d}$$

AB represent the protein molecule with no ligand. In ALB, ligand is bound to site A. In ABL, ligand is bound to site B. In ALBL, both sites are occupied. The stoichiometric concentration of protein is

$$C_0 \;=\; [AB] \;+\; [ALB] \;+\; [ABL] \;+\; [ALBL] \;. \tag{3}$$

Let the relative concentrations, x and y, be defined by the equations:

$$0 \;\leq\; x \;=\; [AB]/C_0 \;\leq\; 1 \;, \tag{4a}$$
$$0 \;\leq\; y \;=\; [ALBL]/C_0 \;\leq\; 1 \;. \tag{4b}$$

From (2a, 2c) and (2b, 2d) it follows that

$$[AB]\,[L]^2/[ALBL] \;=\; K_1 K_3 \;=\; K_2 K_4 \;. \tag{5}$$

Let $\quad K \;\equiv\; (K_1 K_3)^{1/2} \;=\; (K_2 K_4)^{1/2} \;. \tag{6}$

Then $\quad [L] \;=\; K\,[ALBL]^{1/2}/[AB]^{1/2} \;=\; K\,(x/y) \;. \tag{7}$

From (2a, 2b) and using (6)

$$[ALB] \;=\; (K_3/K_1)^{1/2}\,[ALBL]^{1/2}\,[AB]^{1/2} \;, \tag{8a}$$
$$[ABL] \;=\; (K_4/K_2)^{1/2}\,[ALBL]^{1/2}\,[AB]^{1/2} \;. \tag{8b}$$

Hence

$$C_0 \;=\; [AB] \;+\; [ALBL] \;+\; R\,[ALBL]^{1/2}\,[AB]^{1/2} \;, \tag{9}$$

where $\quad R \;\equiv\; (K_3/K_1)^{1/2} \;+\; (K_4/K_2)^{1/2}. \tag{10}$

Dividing (9) by C_0 and using (4a, 4b) and (10), we have

$$1 \;=\; x \;+\; y \;+\; R\,(xy)^{1/2}. \tag{11}$$

Writing (11) as a quadratic in $x^{1/2}$ and solving gives

$$x^{1/2} \;=\; (R/2)\,y^{1/2}\,(Y-1) \;, \tag{12}$$

where $\quad Y \;\equiv\; \{\,1 \;+\; 4(1-y)/R^2 y\,\}^{1/2}. \tag{13}$

The stoichiometric concentration of ligand is

$$l_0 \;=\; [L] \;+\; [ALB] \;+\; [ABL] \;+\; 2\,[ALBL] \;. \tag{14}$$

Dividing (14) by C_0 and using (7), (8a, 8b) and (10), we obtain the ratio of stoichiometric concentrations

$$n \;\equiv\; l_0/C_0 \;=\; y^{1/2}\,\{\,[\,K/(C_0\,x^{1/2})\,] \;+\; Rx^{1/2} \;+\; 2y^{1/2}\,\} \;. \tag{15}$$

Substituting for $x^{1/2}$ we obtain

$$n \;=\; (2K/C_0 R)/(Y-1) \;+\; y\,\{\,(R^2/2)(Y-1) \;+\; 2\,\}. \tag{16}$$

We examine the case where we set $K_1 = K_2 = K_3 = K_4$. Consequently, $R = 2$, and (11) becomes

$$1 = x^{1/2} + y^{1/2}. \tag{17}$$

Hence (15) becomes

$$n = y^{1/2} \{ K/C_0 (1 - y^{1/2}) + 2 \}. \tag{18}$$

Equation (18) may be rearranged to

$$y^{1/2} = (m/4) \{ 1 - (1 - 8n/m^2)^{1/2} \}, \tag{19}$$

where $m \equiv K/C_0 + 2 + n$. $\tag{20}$

A plot of y, the mole fraction of protein molecules with both sites bound vs. n, the total moles of ligands added (i.e., metal cations in the case of bacteriorhodopsin) per mole of total protein is shown in Fig. 4 for several values of K/C_0.

Transient Resonance Raman Spectra
of Bacteriorhodopsin and Halorhodopsin

T. Ogura[1], *A. Maeda*[2], *M. Nakagawa*[1], *and T. Kitagawa*[1]

[1]Institute for Molecular Science, Okazaki National Research Institutes,
Myodaiji, Okazaki 444, Japan
[2]Department of Biophysics, Faculty of Science, Kyoto University,
Sakyo-ku, Kyoto 606, Japan

1. Introduction

Bacteriorhodopsin (bR) and halorhodopsin (hR) are light-dependent
electrogenic proteins in the membrane of Halobacterium halobium (see
Ref. 1 for a review). For both proteins, the species which have
all-trans retinal chromophore linked to ε-nitrogen of lysine through
a protonated Schiff base undergo the cis-trans isomerization to the
13-cis form upon illumination with light. The photo-isomerization
of the chromophore is followed by a structural change of a protein
moiety to compensate for the energy increase due to the charge
separation between the protonated Schiff base and its counter anion.
In the mechanism of the structural change a difference appears
between the two proteins: unidirectional proton and chloride
translocation through the membrane for bR and hR, respectively. In
the case of bR, the Schiff base proton is transferred to the protein
moiety upon the change from the L to M intermediates, which occurs
on a μs time scale, but the photoreaction of hR is not accompanied
by deprotonation of the Schiff base. It is interesting to study the
structural details of the transient species of these two proteins on
a μs time scale.

Resonance Raman (RR) scattering from retinoid proteins reveals the
vibrational spectra of the retinal Schiff base [2]. This technique
becomes powerful when the vibrational assignments of Raman lines are
established. Lewis et al. [3] assigned the Raman line of bR at 1646
cm^{-1} to the C=N stretching vibration ($\nu_{C=N}$) of the protonated Schiff
base. Since this band shifts to ~ 1620 cm^{-1} upon deprotonation, it
can be used as a diagnostic for protonation of the Schiff base.
Mathies and coworkers have analyzed the RR spectra of various
isotope-substituted retinals [4] and a few intermediates of bR
reconstituted with them [5]. It has been also demonstrated that the
in-phase C=C stretching ($\nu_{C=C}$) frequencies of retinoid proteins and
retinal Schiff base derivatives in organic solutions exhibit an
identical linear correlation with regard to the wavelengths of their
visible absorption maxima [6-8]. On the basis of these results the
RR data can be translated into structural terms and the mechanism of
the vectorial proton translocation has been discussed (see Ref. 9
for a review). Previously, we investigated RR spectra of bR [10]
and hR [11] in the unphotolyzed state. Here we report the RR
spectra of transient species of bR and hR in a μs time interval.

2. Experimental Procedures

Purple membrane was obtained from H. halobium R_1M_1 by the standard
method [12] and light-adapted bR in the membrane was exposed to a
0.2 M KCl solution of either 10 mM phosphate (pH 7.0) or carbonate
buffer (pH 10.5) just before Raman experiments. hR was purified
from a bR-deficient strain of H. halobium L-33 [13] according to
Ogurusu et al. [14]. The purified hR was dialyzed for 16 hr against
100 volumes of 10 mM phosphate buffer (pH 6.8) containing either 1 M

Dual-Beam Apparatus

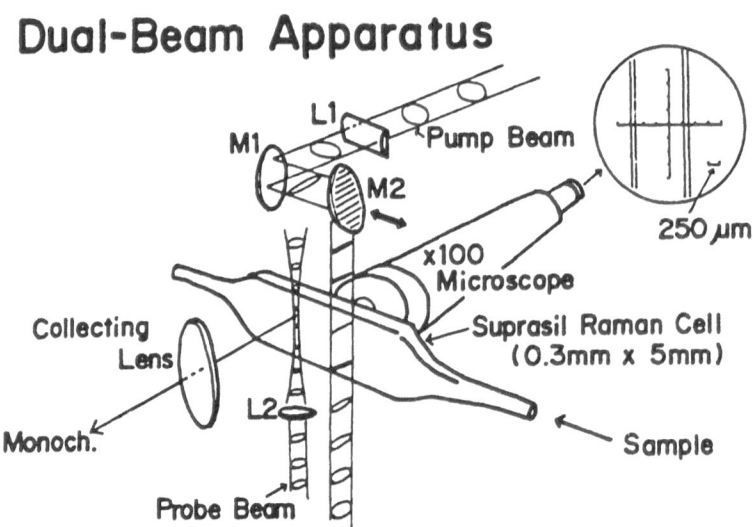

Fig. 1. Dual beam transient Raman apparatus. Pump beam is made
flat by a cylindrical lens (L1) and reflected into the upstream side
of the probe beam at the flow cell by mirrors, M1 and M2. The light
plane is perpendicular to the flow direction. The inset figure
illustrates a microscope image observed from the back side of the
sample-illuminating part. Distance between the two laser beams can
be changed by the moving mirror (M2) held on a X stage. The flow
cell has a thickness of 0.3 mm and height of 5 mm. The design is
very close to that reported by Smith et al. [15].

NaCl or $NaNO_3$. The preparations with 1 M NaCl and 1 M $NaNO_3$ will be
hereafter designated as hR(Cl) and hR(NO_3), respectively.

Conventional Raman spectra were observed with the 514.5 nm line
(30 mW) of an Ar^+ ion laser (NEC GLG3200) and a JEOL 400D Raman
spectrometer equipped with a cooled RCA 31034a photomultiplier.
Samples were contained in a spinning cell (1800 rpm, diameter = 2
cm) and kept at 5 °C by flushing with cold N_2 gas. Transient Raman
spectra were observed for a flowing sample by using the dual beam
apparatus illustrated in Fig. 1, which is close to the design
originally proposed by Smith et al. [15]. The shape of the pump
beam is made flat by a cylindrical lens (L1) and introduced into the
the flow cell at the upstream side of the probe beam by the fixed
(M1) and moving (M2) mirrors. This light illuminates all the sample
which flows through the cell with thickness of 0.3 mm. The probe
beam is focused by a lens (L2) to the downstream side. The center-
to-center distance between the pump and probe beams (d_0) is
determined by observing the illuminating part from the back side
with a microscope as illustrated in the inset. With a flow speed
(v) calculated from the flow rate and the cross section of the cell,
the delay time is determined as d_0/v. The time resolution depends
on the width of the laser beam ($2r_0$) and the flow speed, that is,
$2r_0/v$. About 14 mL of the bR solution with A_{568} = 1.4 (absorbance
for thickness of 1 cm) was pumped with a Micro Pump (Model 185-415)
at the rate of 140 mL/min (v = 155 cm/s) and the laser beam was
focused to $2r_0$ = 0.05 mm. Accordingly, the time resolution is 32
μs. The sample reservoir was kept at 0 °C. Raman spectra for the
flowing sample were obtained with an OMA-2 system (PAR 1215) and a
diode array detector (PAR 1420) attached to a double monochromator
(Spex 1404). Raman frequencies were calibrated with indene for both
kinds of measurements.

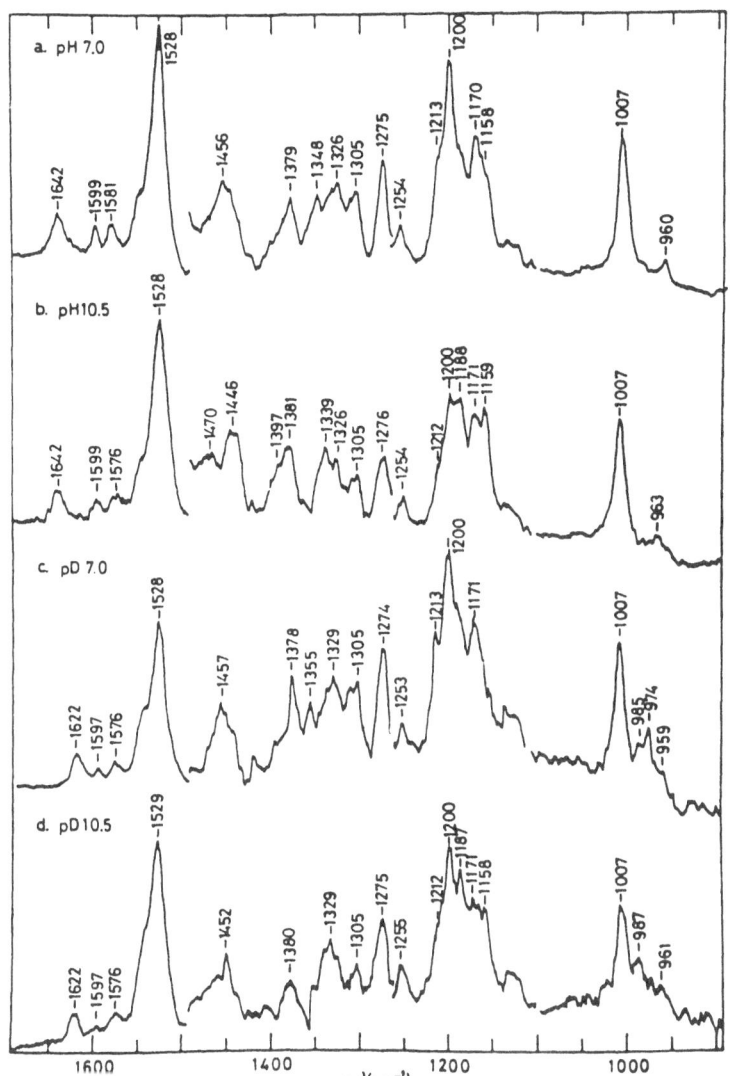

Fig. 2. Resonance Raman spectra of bR (a) at pH 7.0 in H_2O, (b) at pH 10.5 in H_2O, (c) at pD 7.0 in D_2O, and (d) at pD 10.5 in D_2O. The spectra were observed by using the spinning cell (1800 rpm) and 514.5 nm excitation line (transferred from Ref. 10).

3. Results and Discussion
3.1. Bacteriorhodopsin
The RR spectra of bR observed with the spinning cell are shown in Fig. 2, where the spectra for the H_2O (a) and D_2O(c) solutions at pH 7.0 are compared with those for the H_2O (b) and D_2O(d) solutions at pH 10.5. Spectra (a) and (c) are in agreement with those of bR_{568} reported by other groups [9,16-18], indicating that bR is little photolyzed under the present experimental conditions. However, when the alkaline solution was measured under identical instrumental conditions, the apparent spectra were distinct from those of the

neutral solutions; the 1456 cm^{-1} band of (a) is shifted to a lower frequency in (b), a shoulder at 1397 cm^{-1} is intensified in (b), the 1348 cm^{-1} band of (a) disappears in (b) and a new band appears at 1188 cm^{-1} in (b). These features are noticeably close to the RR spectral characteristics of the L intermediate reported [16,18,19]. Nevertheless, the $\nu_{C=C}$ band of spectrum (b) exhibited no shift despite the fact that the L intermediate was supposed to give rise to the split bands at ∿1540 and ∿1550 cm^{-1}. Since the acceptor residue of the Schiff base proton is considered to have its pK$_a$ value around 9 [20], we thought it important to elucidate the difference between spectra (a) and (b).

The first problem to be solved is whether spectrum (b) reflects the unphotolyzed species or not. To answer this question RR spectra of the flowing solution were measured. The left (A) and right (B) sides in Fig. 3 shows the results at pH 7 and 10.5, respectively. Spectra A1 and B1 were obtained by using a single beam (without pump beam). Although the spectral resolution is not as good as that of Fig. 2, the essential features of spectrum A1 agree with those of spectrum (a) in Fig. 2. On the other hand, spectrum B1 for the solution at pH 10.5 is distinct from spectrum (b) in Fig. 2 and is close to spectrum A1. This strongly suggests that unphotolyzed bR at pH 10.5 resembles unphotolyzed bR at pH 7.

The spectral difference between Figs. 2(b) and 3(B1) should arise from the difference in experimental conditions; a given molecule is illuminated by laser light for 55 µs every 33 ms with the spinning cell and for 32 µs every 6 s with the flow cell

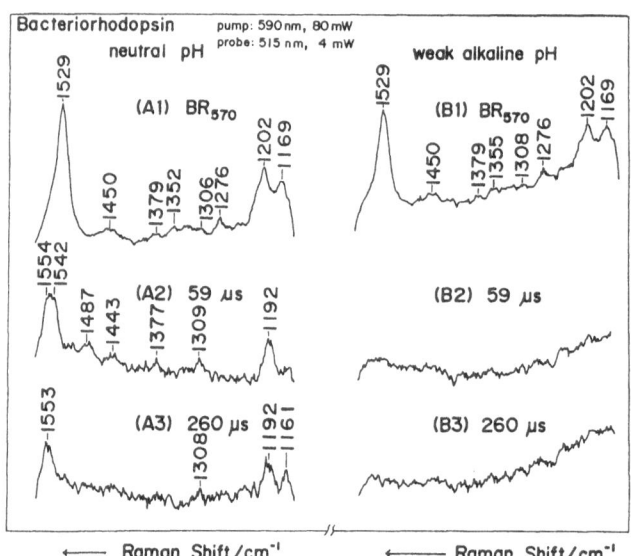

Fig. 3. Resonance Raman spectra of bR at neutral (left) and alkaline pH (right) obtained by using the transient Raman apparatus shown in Fig. 1. Traces A1 and B1 are raw spectra observed only with the probe beam (514.5 nm, 4 mW) and others are difference between the spectra observed in the presence of the pump beam (590 nm, 80 mW) and spectra A1 (neutral) or B1 (alkaline). Delay times between the pump and probe beams are specified in the figure. All spectra were observed under an identical optical alignment in order A1, A2, A3, B3, B2, and B1; flow speed, 155 cm/s; beam width, 50 µm.

technique. To examine a possibility that spectrum (b) in Fig. 2
contains a significant contribution from some intermediates, the
pump-probe experiments were carried out for the flowing sample and
the results for the time delays of 59 μs and 260 μs are shown in the
lower part of Fig. 3, where the differences between the observed
spectra and the spectrum A1 or B1 for the neutral or alkaline
solution, respectively, are displayed. The spectrum A2 exhibits
the characteristic features of the L intermediate [16,18,19]. It is
interesting to note that the 1542 cm^{-1} band of spectrum A2 is
missing in spectrum A3, suggesting that the two bands around 1555-
1540 cm^{-1} arise from different molecular species. On the other
hand, the corresponding difference spectra at pH 10.5 (B2 and B3)
are very weak for both at 59 and 260 μs delay. This resulted from
the fact that the spectra observed with some time delay are
essentially the same as spectrum B1 except for the absolute
intensity. Therefore, it is highly likely that spectrum (b) in Fig.
2 indicates a RR spectrum of another intermediate which has a longer
lifetime and is therefore accumulated in the spinning cell. Since
the Raman spectrum of the M intermediate is not resonance enhanced
with this excitation wavelength, this observation suggests either
that the lifetime of the L intermediate at pH 10.5 is very short or
that the L intermediate is not generated at pH 10.5.

To determine which alternative is correct, laser-power difference
spectra were examined. Figure 4 shows the spectra observed with a
single beam of 3 mW (A) and 100 mW (B) for the neutral solution in
the spinning cell and their difference (C). In this experiment all
intermediates generated within 55 μs of photolysis contribute to
the difference but the dominant contribution comes from the L
intermediate which gives bands around 1191 and 1550 cm^{-1}. Since

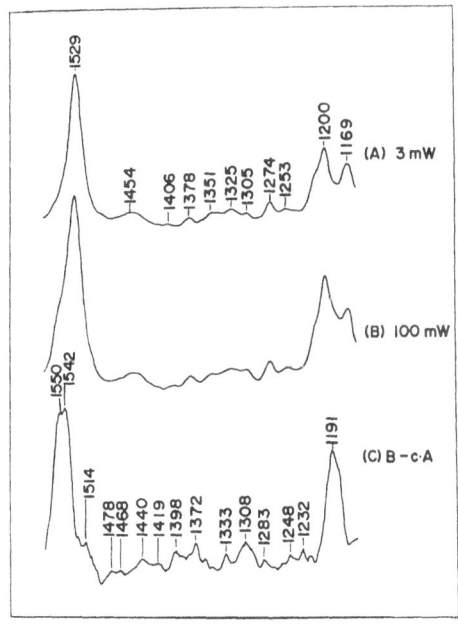

← RAMAN SHIFT/cm^{-1}

Fig. 4. Laser power dependence of resonance Raman spectra of bR at
neutral pH obtained with the spinning cell. Laser power is 3 mW for
(A) and 100 mW for (B) at 514.5 nm. Trace (C) denotes the
difference, (B) - 4.7*(A).

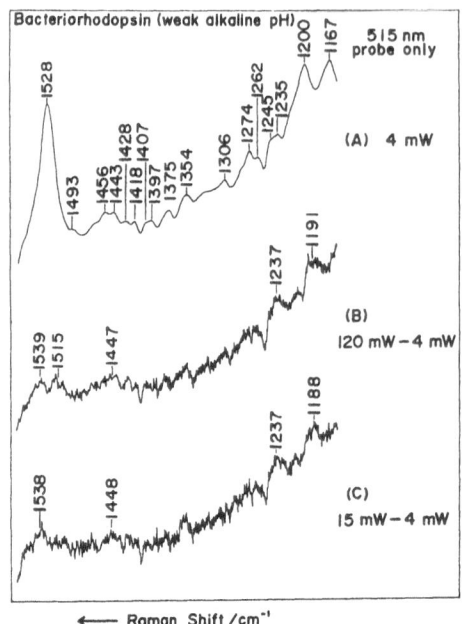

Fig. 5. Laser power dependence of resonance Raman spectra of bR at alkaline pH obtained with the flow technique. Trace (A) is the raw spectrum observed with the laser power of 4 mW. Trace (B) or (C) is difference between the spectra observed with the laser power of 120 or 15 mW, respectively, and spectrum (A); flow speed, 155 cm/s; beam width, 50 μm.

this method was confirmed to be satisfactory, similar power difference spectra were measured for the flowing alkaline bR and the results are shown in Fig. 5. When the laser power was made higher, the K intermediate appreciably contributes to the difference spectrum (B) but its contribution becomes negligible with a low laser power as judged from the absence of the 1515 cm^{-1} line in spectrum (C). It is emphasized that spectrum (C) in Fig. 5 gives rise to the marker band of the L intermediate at 1191 cm^{-1} and the $\nu_{C=C}$ band at 1538 cm^{-1} which corresponds to the fast-decay component of the doublet of spectrum A2 in Fig.3. This observation indicates that the L intermediate is generated even at pH 10.5 similar to that at pH 7 but its lifetime is markedly short at the alkaline pH, presumably due to deprotonation of the residue with pK$_a$ = 9 [20]. This is consistent with the reported increase of the rate constant for the formation of the M intermediate at alkaline pH [21].

The transient spectra displayed in Figs. 3 and 5 suggest that the L intermediate has a single $\nu_{C=C}$ band at 1539-1542 cm^{-1} and that the 1554 cm^{-1} component arises from an intermediate different from the L intermediate. This was first pointed out by Marcus and Lewis [22] who called it the X intermediate and assigned it to the deprotonated 13-cis form, although the existence of the X intermediate has not been established yet [16,18,19]. Further detailed analysis of the μs intermediates of bR is necessary and now under investigation in this laboratory.

3.2. Halorhodopsin
The RR spectra of hR in the presence of chloride ions were reported previously [11,24,25]. The absorption maximum of hR shifts from 578

238

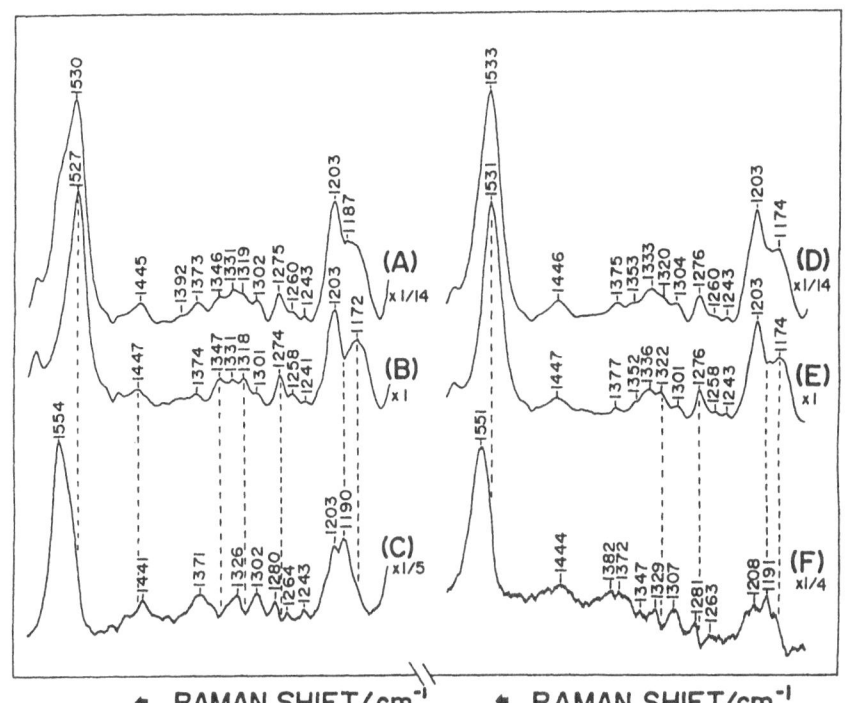

Fig. 6. Transient resonance Raman spectra of hR in the presence (A - C) and absence (D - F) of chloride ions obtained with a spinning cell. Laser power (514.5 nm) is 120 mW for (A) and (D), and 2 mW for (B) and (E). Traces (C) and (F) denote the difference spectra; (C) = [(A) - 13*(B)]/5, (F) = [(D) - 13.5*(E)]/4.

to 565 nm when chloride ions are removed from the solution [23]. Accordingly, RR spectra of hR exhibited small changes upon removal of chloride ions [11]. The photocycle of hR(Cl), which requires 10 ms, involves the trans to cis isomerization and subsequent relaxation [26]: $hR_{578} \rightarrow hR_{600} \rightarrow hR_{520} \rightarrow hR_{640} \rightarrow hR_{565} \rightarrow hR_{578}$. The hR_{565} intermediate is considered to be the same species as the unphotolyzed $hR(NO_3)$. The photocycle of $hR(NO_3)$ takes 1.5 ms and involves all-trans to 13-cis isomerization but only two intermediates [27]: $hR_{565} \rightarrow hR_X \rightarrow hR_{640} \rightarrow hR_{565}$. It is interesting to compare the transient RR spectra of hR in the presence and absence of Cl^-.

Figure 6 shows the transient RR spectra of hR(Cl) (left) and $hR(NO_3)$ (right) obtained from the power difference method; spectra A and D were observed with the laser power of 100 mW and spectra B and E were observed with the laser power of 2 mW for samples in the spinning cell. Although the results are not displayed here, it is noted that Raman lines due to the Schiff base C=N stretching mode observed in this study under the same experimental conditions as those for spectra (B) and (E) could reproduce the chloride effect reported previously [11]. The difference spectra shown by spectra (C) and (F) resembled the spectrum of the L intermediate of bR [11,18,19], although the peak intensity of spectrum (F) is about 0.6 of that of spectrum (C). It is stressed that the $\nu_{C=C}$ frequency of the hR intermediate is close to that of the X intermediate of bR noted previously.

With regard to the photocycle of $hR(NO_3)$ the existence of the hR_{640} intermediate seems fairly certain, but it is considered that this species does not give rise to Raman lines in the present experiments due to poor resonance conditions. The intermediate between hR_{565} and hR_{640} has not been identified yet. This appears to be contradictory to the results shown in Fig. 6. One explanation is that spectrum F is caused by incomplete removal of Cl^- ions from the solution, and the other is to assume that spectrum (F) arises from an intermediate which may have its absorption maximum at 460-490 nm. Since the laser power difference spectrum with a single laser beam is likely to catch an intermediate with a shorter lifetime if its absorption maximum is close to the excitation wavelength, the latter possibility cannot be ruled out completely.

4. Conclusion

bR at pH 10.5 creates an intermediate with a longer lifetime (100 - 1000 ms) during the RR measurement using the spinning cell and this caused the apparent difference between the spectra obtained with a spinning cell and the flow technique. Unphotolyzed bR at pH 10.5 gives almost the same spectrum as that at pH 7. However, the lifetime of the L intermediate is significantly shorter with the alkaline bR. The two C=C stretching RR bands of the μs intermediate of neutral bR at 1551 and 1539 cm^{-1}, which have been so far considered to arise from the L intermediate, exhibited different time behavior, suggesting that these two bands arise from different species. The RR spectra of intermediates of $hR(Cl)$ and $hR(NO_3)$ bore close similarity to those of bR, but their $\nu_{C=C}$ frequencies were noticeably close to that of the X intermediate of bR.

Acknowledgment

The authors are grateful to Dr. T. Ogurusu for his help in preparation of halorhodopsin and also to Prof. T. Yoshizawa for his encouragement.

References

1. W. Stoeckenius, R. A. Bogomolni: Annu. Rev. Biochem. 52 587 (1982).
2. L. Rimai, R. G. Kilponen, D. Gill: Biochem. Biophys. Res. Commun. 41 492 (1970).
3. A. Lewis, J. Spoonhower, R. A. Bogomolni, R. H. Lozier, W. Stoeckenius: Proc. Natl. Acad. Sci. U.S.A. 71, 4462 (1974).
4. S. O. Smith, A. B. Meyers, R. A. Mathies, J. A. Pardoen, C. Winkel, E. M. M. van der Berg, J. Lugtenburg: Biophys. J. 47, 653 (1985).
5. R. A. Mathies: In Spectroscopy of Biological Molecules (C. Sandorfy, T. Theophanides, eds.) p.303 (1984).
6. M. E. Heyde, D. Gill, R. G. Kilponen, L. Rimai: J. Am. Chem. Soc. 93 6776 (1971).
7. B. Aton, A. Doukas, R. H. Callender, B. Becher, T. G. Ebrey: Biochemistry 16 2955 (1977).
8. T. Sugihara, T. Kitagawa: Bull. Chem. Soc. Jpn.: 59 2929 (1986).
9. M. Stockburger, T. Alshuth, D. Oesterhelt, W. Gartner: in Spectroscopy of Biological Systems (R. J. H. Clark, R. E. Hester, eds.) Wiley & Sons, p.483 (1986).
10. A. Maeda, T. Ogura, T. Kitagawa: Biochemistry 25 2798 (1986).
11. A. Maeda, T. Ogurusu, T. Yoshizawa, T. Kitagawa: Biochemistry 24, 2517 (1985).
12. D. Oesterhelt, W. Stoeckenius: Methods Enzymol 31A 667 (1974).
13. G. Wagner, D. Oesterhelt, G. Krippahl, J. K. Layni: FEBS Lett. 131 341 (1981).
14. T. Ogurusu, A. Maeda, T. Yoshizawa: J. Biochem. 95 1073 (1984).
15. S.O. Smith, J. A. Pardoen, P. P. J. Mulder, B., Curry, J. Lugtenburg, R. A. Mathies: Biochemistry 22, 6141 (1983).

16. J. Terner, C. -L. Hsieh, M. A. El-Sayed: Biophys. J. 26, 527 (1979).
17. S. O. Smith, A. B. Myers, J. A. Pardoen, C. Winkel, P. P. J. Mulder, J. Lugtenburg, R. Mathies: Proc. Natl. Acad. Sci. U.S.A. 81 2055 (1984).
18. T. Alshuth, M. Stockburger: Photochem. Photobiol. 43 55 (1986).
19. P. V. Argade, K. J. Rothschild: Biochemistry 22, 3460 (1983).
20. A. Maeda: this book.
21. A. Lewis, M. A. Marcus, B. Ehrenberg, H. Crespi: Proc. Natl. Acad. Sci. U.S.A. 25 4642 (1978).
22. M. A. Marcus, A. Lewis: Biochemistry 17, 4722 (1978).
23. T. Ogurusu, A. Maeda, J. Sasaki, T. Yoshizawa: Biochim. Biophys. Acta 682 446 (1982).
24. S. O. Smith, M. J. Marvin, R. A. Bogomolni, R. A. Mathies: J. Biol. Chem. 259 12326 (1984).
25. T. Alshuth, M. Stockburger, P. Hagemann, D. Oesterhelt: FEBS Lett. 79 55 (1985).
26. J. K. Lanyi, V. Vodyanoy: Biochemistry, 25, 1465 (1986).
27. B. Schobert, J. K. Lanyi, E. J. Cragoe, Jr.: J. Biol. Chem. 258 15158 (1983).

Index of Contributors